Ultrasound in Critical Care

Ultrasound in Critical Care

Editor
Puneet Khanna MBBS MD (Anesthesiology) AIIMS FAGE
Assistant Professor
Department of Anesthesiology
Pain Medicine and Critical Care
Former Senior Research Associate (CSIR)
All India Institute of Medical Sciences
New Delhi, India

Foreword
Colin Royse MBBS MD FANZCA

The Health Sciences Publisher
New Delhi | London | Panama

 **Jaypee Brothers Medical Publishers (P) Ltd**

Headquarters

Jaypee Brothers Medical Publishers (P) Ltd
4838/24, Ansari Road, Daryaganj
New Delhi 110 002, India
Phone: +91-11-43574357
Fax: +91-11-43574314
Email: jaypee@jaypeebrothers.com

Overseas Offices

J.P. Medical Ltd
83 Victoria Street, London
SW1H 0HW (UK)
Phone: +44 20 3170 8910
Fax: +44 (0)20 3008 6180
Email: info@jpmedpub.com

Jaypee-Highlights Medical Publishers Inc
City of Knowledge, Bld. 235, 2nd Floor, Clayton
Panama City, Panama
Phone: +1 507-301-0496
Fax: +1 507-301-0499
Email: cservice@jphmedical.com

Jaypee Brothers Medical Publishers (P) Ltd
17/1-B Babar Road, Block-B, Shaymali
Mohammadpur, Dhaka-1207
Bangladesh
Mobile: +08801912003485
Email: jaypeedhaka@gmail.com

Jaypee Brothers Medical Publishers (P) Ltd
Bhotahity, Kathmandu, Nepal
Phone +977-9741283608
Email: kathmandu@jaypeebrothers.com

Website: www.jaypeebrothers.com
Website: www.jaypeedigital.com

© 2019, Jaypee Brothers Medical Publishers

The views and opinions expressed in this book are solely those of the original contributor(s)/author(s) and do not necessarily represent those of editor(s) of the book.

All rights reserved. No part of this publication may be reproduced, stored or transmitted in any form or by any means, electronic, mechanical, photocopying, recording or otherwise, without the prior permission in writing of the publishers.

All brand names and product names used in this book are trade names, service marks, trademarks or registered trademarks of their respective owners. The publisher is not associated with any product or vendor mentioned in this book.

Medical knowledge and practice change constantly. This book is designed to provide accurate, authoritative information about the subject matter in question. However, readers are advised to check the most current information available on procedures included and check information from the manufacturer of each product to be administered, to verify the recommended dose, formula, method and duration of administration, adverse effects and contraindications. It is the responsibility of the practitioner to take all appropriate safety precautions. Neither the publisher nor the author(s)/editor(s) assume any liability for any injury and/or damage to persons or property arising from or related to use of material in this book.

This book is sold on the understanding that the publisher is not engaged in providing professional medical services. If such advice or services are required, the services of a competent medical professional should be sought.

Every effort has been made where necessary to contact holders of copyright to obtain permission to reproduce copyright material. If any have been inadvertently overlooked, the publisher will be pleased to make the necessary arrangements at the first opportunity.
The **CD/DVD-ROM** (if any) provided in the sealed envelope with this book is complimentary and free of cost. **Not meant for sale.**

Inquiries for bulk sales may be solicited at: jaypee@jaypeebrothers.com

Ultrasound in Critical Care

First Edition: **2019**

ISBN: 978-93-5270-168-1

Dedicated to

*The Teachers and
Residents in Anesthesiology and Critical Care for their
endeavor to dissipate and acquire knowledge
My parents for their blessings.
My daughter Ziva
for her patience and tolerance shown to
loss of many precious moments.
And finally to my wife Ishita
for her understanding and encouragement.*

Contributors

Rahul K Anand
MD (Anesthesiology) EDIC
Assistant Professor
Department of Anesthesiology
Pain Medicine and Critical Care
All India Institute of Medical Sciences
Member
Europian Society of Intensive Care Medicine (ESICM)
Indian Society of Critical Care Medicine (ISCCM)
New Delhi, India

Dalim Kumar Baidya
MD EDIC
Associate Professor
Department of Anesthesiology
Pain Medicine and Critical Care
All India Institute of Medical Sciences
New Delhi, India

Sonali Bansal
MD (Anesthesiology)
Senior Resident
Department of Anesthesiology
Jawaharlal Nehru Medical College
Aligarh, Uttar Pradesh, India

Shailaja Behera
MD (Anesthesiology) PDCC Critical Care Medicine (BHU)
PDF Neuro Critical Care (NIMHANS)
DM Critical Care Medicine
Senior Resident
Department of Anesthesiology
Pain Medicine and Critical Care
All India Institute of Medical Sciences
New Delhi, India

Rashmi Bhatt
MD Anesthesia
Consultant Anesthesia
Madhukar Rainbow Children's Hospital
New Delhi, India

Sulagna Bhattacharjee
MBBS MD DNB (Anesthesiology)
DM Critical Care Medicine
Senior Resident, Department of Anesthesiology
Pain Medicine and Critical Care
All India Institute of Medical Sciences
New Delhi, India

Rakesh Garg
MD DNB FICA PGCCHM MNAMS CCEPC
FIMSA Fellowship in Palliative Medicine
Associate Professor
Department of Anesthesiology
Intensive Care
Pain and Palliative Medicine
Dr BR Ambedkar Institute of
Rotary Cancer Hospital
All India Institute of Medical Sciences
New Delhi, India

Sakshi Gera
MD DNB (Anesthesia)
Attending Consultant Anesthesia
Venkateshwar Hospital
New Delhi, India

Vijay Hadda
MD
Assistant Professor
Department of Pulmonary Medicine and Sleep Disorders
All India Institute of Medical Sciences
New Delhi, India

Karthik Iyer
MBBS MD (Anesthesiology)
Senior Resident
Department of Anesthesiology
Pain Medicine and Critical Care
All India Institute of Medical Sciences
New Delhi, India

Manisha Jana
MD DNB FRCR
Associate Professor
Department of Radiodiagnosis
All India Institute of Medical Sciences
New Delhi, India

Devasenathipathy Kandasamy
MD DNB FRCR
Associate Professor
Department of Radiodiagnosis
All India Institute of Medical Sciences
New Delhi, India

Puneet Khanna
MBBS MD (Anesthesiology) AIIMS FAGE
Assistant Professor
Department of Anesthesiology
Pain Medicine and Critical Care
Former Senior Research Associate (CSIR)
All India Institute of Medical Sciences
New Delhi, India

Arvind Kumar
MBBS MD (Medicine)
Dip Echocardiography
Assistant Professor
Department of Medicine
All India Institute of Medical Sciences
New Delhi, India

Niraj Kumar
MBBS MD (Anesthesiology)
Associate Professor
Department of Neuroanesthesiology and Critical Care
All India Institute of Medical Sciences
New Delhi, India

Rohit Kumar
MD DNB MRCP DM
Specialist
Department of Pulmonary and Critical Care Medicine
Vardhman Mahavir Medical College and Safdarjung Hospital
New Delhi, India

Riddhi Kundu
MD (Anesthesiology)
DM Critical Care Medicine
Senior Resident
Department of Anesthesiology
Pain Medicine and Critical Care
All India Institute of Medical Sciences
New Delhi, India

Shyam Madabhushi
MD DNB (Anesthesiology) Classified Specialist
(Anesthesiology) DM Critical Care Medicine
Senior Resident
Department of Anesthesiology
Pain Medicine and Critical Care
All India Institute of Medical Sciences
New Delhi, India

Souvik Maitra
MD DNB (Anesthesiology) EDIC
Assistant Professor
Department of Anesthesiology
Pain Medicine and Critical Care
All India Institute of Medical Sciences
New Delhi, India

Anil Malik
MBBS MD (Anesthesiology)
DM Critical Care Medicine
Senior Resident
Department of Anesthesiology
Pain Medicine and Critical Care
All India Institute of Medical Sciences
New Delhi, India

Ananya Panda
MD DNB FRCR
Fellow, MRI Division
Department of Radiology
Case Western Reserve University
Cleveland, Ohio, USA

Nitish Parmar
MD (Anesthesiology)
PDCC Critical Care Medicine (SGPGI)
DM Critical Care Medicine
Senior Resident
Department of Anesthesiology
Pain Medicine and Critical Care
All India Institute of Medical Sciences
New Delhi, India

Bikash Ranjan Ray
MBBS MD (Anesthesiology)
Assistant Professor
Department of Anesthesiology
Pain Medicine and Critical Care
All India Institute of Medical Sciences
New Delhi, India

Soumya Ray
MD DNB FCICM Cert Clin Edu DDU (Crit Care)
ICU Consultant
Royal Perth Hospital
Wellington Street
Perth, Australia

Sandeep Sahu
MD PDCC MAMS FACEE FICCM
Additional Professor
Department of Anesthesiology and
Emergency Medicine
Sanjay Gandhi Postgraduate
Institute of Medical Sciences
Lucknow, Uttar Pradesh, India

Tanvir Samra
MBBS MD (Anesthesia)
Assistant Professor
Department of Anesthesia and Intensive Care
Postgraduate Institute of
Medical Education and Research
Chandigarh, India

Rajkumar Subramanian
MD DNB (Anesthesia)
Attending Consultant Anesthesia
Max Superspecialty Hospital
Saket, New Delhi, India

Sarvesh Pal Singh
MD (Anesthesiology) DM (Cardiac Anesthesiology)
FIACTA FTEE
Assistant Professor
Department of Intensive Care for
Cardiothoracic and Vascular Surgery
All India Institute of Medical Sciences
New Delhi, India

Akhil Kant Singh
MBBS MD (Anesthesiology)
Fellowship in Liver Transplantation (University of Toronto)
Assistant Professor
Department of Anesthesiology
Pain medicine and Critical Care
All India Institute of Medical Sciences
New Delhi, India
Outcomes Research Consortium
Cleveland, Ohio, USA

Gyaninder Pal Singh
MBBS MD DM
Associate Professor
Department of Neuroanesthesiology and
Critical Care
Neurosciences Center
All India Institute of Medical Sciences
New Delhi, India

Ajay Singh
MBBS MD (Anesthesiology)
Senior Resident
Department of Anesthesiology
Pain Medicine and Critical Care
All India Institute of Medical Sciences
New Delhi, India

Manish Soneja
MBBS MD (Medicine)
Associate Professor
Department of Medicine
All India Institute of Medical Sciences
New Delhi, India

Kapil Dev Soni
MD (Anesthesia and Critical Care)
Associate Professor
Critical and Intensive Care
JPN Apex Trauma Center
All India Institute of Medical Sciences
New Delhi, India

Rashmi Soori
MBBS MD (Anesthesiology)
Senior Resident
Department of Anesthesiology and
Emergency Medicine
Sanjay Gandhi Postgraduate
Institute of Medical Sciences
Lucknow, Uttar Pradesh, India

Alka Verma
MBBS MD (Anesthesiology)
Lecturer, Department of Anesthesiology
Government Medical College
Kannauj, Uttar Pradesh, India

Foreword

Over the past 30 years, ultrasound has moved from the exclusive domain of the cardiologist and radiologist, to many other specialist disciplines. Indeed, basic, clinical ultrasound is being taught in many medical schools and it will not be long before it is simply a core skill for all graduating doctors. The critical care specialties such as anesthesiology, intensive care and emergency medicine have been the leaders in the adoption of clinical ultrasound outside of cardiology and radiology laboratories. The reason is very simple—the additional information provided by bedside ultrasound allows the clinician to make better decisions in real-time. We know that clinical assessment of cardiovascular disease is at least 50% inaccurate, and the medical student with a handheld ultrasound is more accurate than a cardiologist with a stethoscope! This is similar for other body parts such as the lungs, deep veins, and abdominal examination. We are transitioning from thinking of isolated ultrasound examinations (echocardiography, lung ultrasound, abdominal ultrasound) to being far more generic and simply considering it as "ultrasound-assisted examination". This does not mean that ultrasound replaces good clinical examination and clinical judgment, but rather improves the diagnostic accuracy, which in turn allows the clinician to make better-informed management decisions.

The same concept applies to procedural accuracy. There is no doubt that directing needles with ultrasound guidance is more accurate than attempting blind cannulation. This skill set required for vascular access is very similar to regional anesthesia, which is not that different from pleural drain insertion or pericardial drain insertion. We can consider the use of ultrasound in a more generic manner as simply "ultrasound-guided procedures".

What is truly disruptive is the introduction of handheld devices at low cost. This will only increase as technology improves. When I first learned ultrasound in 1995, and echocardiography machine was the price of a house. 10 years later, a more portable machine was the price of a European car. Now a handheld device is the price of a motorbike or a cheap car and will not be long before it is only a little more expensive than the smart phone that everyone carries. Personal ultrasound is already here, and likely to become much more widespread in the future. This means that physicians everywhere will be able to perform an ultrasound-assisted examination as part of their routine clinical evaluation.

Like any new technology, it is important that people are trained in its use. Whilst I envisage that in the future, it will be a core skill taught at all medical students, there is currently a large gap in training for existing doctors. How do we overcome this? Firstly, it is important to accept that not everyone has to be an expert. Everyone needs to be competent at a basic ultrasound of multiple body parts, but only a few need to progress to more expert levels. Secondly, each medical discipline must determine the scope of practice using ultrasound and train accordingly. Thirdly, everyone has to start somewhere—even experts start at the basic level! Small packages of learning to teach basic ultrasound examination, interpretation, which are low-cost will help train a large number of practitioners. Books such as *Ultrasound in Critical Care*, help to perform this function and additionally provide guidance to critical care physicians about the multiple uses of ultrasound in their specialty discipline. Other resources such as e-learning programs allow doctors to acquire knowledge in their own time and place, enabling learning and practice. Fourthly, it is important to remove barriers to learning. Institutions, societies and colleges must work out ways to make it easy for people to learn, rather than making it difficult to learn—thereby reducing the number of people who will adopt the technology.

I congratulate Dr Puneet Khanna and his co-authors on the production of this excellent text. It will help guide critical care physicians and their trainees and exploring the incredible utility of point-of-care ultrasound, and hopefully increase adoption of this important technology into critical care practice.

Colin Royse MBBS MD FANZCA
Professor, Co-Director of the Ultrasound Education Group
The University of Melbourne
Melbourne, Australia

Preface

Knowledge and practice of ultrasound in critical care is expanding at very rapid pace. This text is invaluable for emergency physicians, general physicians, intensivists, pulmonologists and anesthetists those practicing cardiac critical care, neurocritical care, trauma critical care, trauma emergency departments at all levels. The real strength of this textbook is its clinical focus. The book is meant to be read in entirety in several hours, and hopefully impart you to a foundation on which to grow and enjoy this beautiful and ever-changing specialty.

Ultrasound in Critical Care forms a sizeable portion of day-to-day critical care practice. The book will be of help to MD, DNB (Anesthesia, Medicine, Emergency Medicine); FNB, DM, PDCC, Fellowship (Critical care, Trauma, Emergency, Cardiac Critical Care and Neurocritical Care). However, to keep this book practically oriented, detailed theory has been purposely kept aside. While the book is aimed at providing information on basic principles and uses of ultrasound in critical care, it is hoped that it will also serve as handbook for those already in practice and for those engaged in teaching of the subject.

In spite of best efforts, a venture like this is unlikely to be error free. Constructive criticism and suggestions from the reader are invited for further improvement of the book. I would appreciate any recommendations and images that would improve the subsequent edition. You may email me at *k.punit@yahoo.com*.

Puneet Khanna

Acknowledgments

I would like to express my special thanks of gratitude to Jaypee Publications who gave me the golden opportunity to write the book on *Ultrasound in Critical Care*, which also helped me in doing a lot of research and I came to know about so many new things. I am really thankful to them.

I especially appreciate the constant support and encouragement of Mr Jitendar P Vij (Group Chairman) and Mr Ankit Vij (Managing Director) of Jaypee Brothers Medical Publishers (P) Ltd, New Delhi, India in publishing this textbook and also their associates particularly Ms Chetna Malhotra Vohra (Associate Director—Content Strategy) and Ms Nikita Chauhan (Senior Development Editor) who have been prompt, efficient and most helpful.

Last but not least, I would like to thank Dr Rashmi Bhatt who helped me in proofreading the book.

Contents

1A. Point of Care Ultrasound 1
Puneet Khanna, Shakshi Gera, Rashmi Bhatt
- Characteristics of POCUS *1*
- History of POCUS *1*
- Who uses POCUS? *1*
- How to Use POCUS *2*
- POCUS in Medical Education *6*
- Limitations *6*

1B. Applications in Critical Care Medicine 10
Arvind Kumar, Manish Soneja
- Head and Neck including Subclavian Vein *10*
- Ultrasonography of the Airway *13*
- Utility of Chest Sonography in Critical Care *13*
- Echocardiography in the Critical Care Ward *19*
- Basic Views in Critical Care *19*
- Assessment of Ventricular Function *20*
- Measurement of Cardiac Output *21*
- Measurement of Chamber Pressures *22*
- Assessment of Hypovolemia *22*
- The Hypotensive Patient *22*
- Deep Vein Thrombosis *23*
- Abdominal Ultrasound *25*

1C. Should Intensivist do the Ultrasonography: Guidelines, Qualification, Experience and Training? 31
Shailaja Behera, Puneet Khanna
- Training and Accreditation *31*

2A. Basic Physics of Ultrasound and the Doppler Phenomenon 34
Devasenathipathy Kandasamy, Ananya Panda
- Overview *34*
- Properties of Ultrasound *34*
- Basic Ultrasound Instrumentation *37*
- Safety Considerations *38*

2B. Modes of Ultrasonography **39**
Devasenathipathy Kandasamy, Ananya Panda
- Overview *39*
- A-Mode (Amplitude-Mode) Imaging *39*
- M-Mode (Motion-Mode) Imaging *40*
- B-Mode (Brightness-Mode) Imaging *40*
- Image Resolution *43*
- Image Optimization *44*
- Ultrasound Artifacts *47*

2C. Doppler Imaging **51**
Devasenathipathy Kandasamy, Ananya Panda
- Overview *51*
- The Doppler Principle *51*
- Modes of Doppler Imaging *51*
- Setting up and Optimizing Doppler Images *54*

3. Ultrasound for Vascular Access **57**
Akhil Kant Singh, Karthic Iyer
- History of Central Venous Access *57*
- The Evidence *58*
- General Considerations *58*
- The Procedure *60*
- After the Procedure *65*

4A. Airway Ultrasound **68**
Bikash Ranjan Ray
- Advantage of Ultrasound Imaging of Upper Airway *68*
- Challenges of Airway Imaging *68*
- Airway Anatomy: What Does Ultrasound Reveal? *68*
- Ultrasound Imaging of Airway *70*
- Clinical Application of Upper Airway Ultrasound *74*
- Future Aspects *79*

**4B. Ultrasound-guided Percutaneous Dilatational
Tracheostomy and Cricothyrotomy** **82**
Sulagna Bhattacharjee, Souvik Maitra
- Ultrasonographic Anatomy of Neck *83*
- Techniques *83*
- Confirmation of Placement *86*
- Complications *87*
- Ultrasound-guided Cricothyrotomy *87*

- Cricothyrotomy Technique 87
- Role of Ultrasound in Cricothyrotomy 88
- The Transverse TACA Technique (Thyroid Cartilage-Airline-Cricoid Cartilage-Airline) 88
- The Longitudinal "String of Pearl" Technique 89
- Which Technique to Follow? 89
- Complications 89

5. Lung Ultrasound 93
Soumya Ray
- Overview 93
- History 93
- Equipment for Lung Ultrasound 93
- How to Perform Lung Ultrasound? 94
- Lung Ultrasound in Intensive Care Unit 105
- Limitations of Lung Ultrasound 108

6. Chest Ultrasound in Acute Respiratory Distress Syndrome 110
Shyam Madabhushi, Puneet Khanna
- Lung Morphology in Acute Respiratory Distress Syndrome 110
- Differentiating Acute Respiratory Distress Syndrome from other Causes of Respiratory Failure 111
- Lung Ultrasound and Extravascular Lung Water 111
- Prognosis of Acute Respiratory Distress Syndrome Determined by Ultrasound 111

7. Ultrasonography in Cardiac Evaluation 115
Sarvesh Pal Singh, Sonali Bansal
- Indications for Performing Echocardiography in the Critically Ill Patient 115
- Advantages and Disadvantages of Echocardiography in the Intensive Care Unit 116
- General Points 116
- Common Transthoracic Echocardiography Views for Intensive Care Unit 117
- Assessment of Ventricular Function 122
- Assessment of Systolic Function of Left Ventricle 123
- Assessment of Diastolic Function of Left Ventricle 127
- Assessment of Right Ventricular Function 129

8. Volume Status Assessment and Predicting Volume Responsiveness 138
Tanvir Samra
- Static versus Dynamic Parameters 138

- Static Indices of Ventricular Preload *138*
- Dynamic Parameters: Predicting Volume Responsiveness *140*
- Passive Leg Raising: Prediction of Fluid Responsiveness in Spontaneously Breathing Adults *143*
- Current Role of Ultrasound for Volume Status Assessment and Prediction of Fluid Responsiveness *144*
- Ultrasound Protocols *144*

9. Ultrasound in Cardiopulmonary Resuscitation 150
Rakesh Garg
- Need of Ultrasound in Resuscitation *150*
- Ultrasound-guided Management *151*
- Limitations *162*

10. Ultrasound-guided Therapeutic Procedures in Intensive Care Unit 167
Nitish Parmar, Puneet Khanna, Devasenathipathy Kandasamy
- General Principles *167*
- Thoracocentesis *173*
- Pericardiocentesis *178*
- Paracentesis *183*
- Percutaneous Drainage of Intra-abdominal Collections *187*

11. Ultrasonography in Neurocritical Care 196
Gyaninder Pal Singh, Niraj Kumar
- Measurement of Optic Nerve Sheath Diameter *196*
- Measurement of Midline Shift of Brain *199*
- Assessing Adequacy of Cerebral Blood Flow with Transcranial Doppler *199*
- Assessing Pupils in Patients *203*

12. Focused Assessment Sonography of Trauma 208
Kapil Dev Soni
- Technique *209*
- Clinical Application *213*
- Limitations *215*

13. Abdominal Ultrasonography inThe Intensive Care Unit 216
Riddhi Kundu, Puneet Khanna, Manisha Jana
- Detection of Free-Fluid Accumulation (Focused Assessment with Sonography for Trauma and Extended Focused Assessment with Sonography for Trauma) *217*
- Ultrasound of the Kidneys and Urinary Bladder *221*

- Imaging the Major Vessels *223*
- Imaging the Gallbladder *225*
- Detection of Pneumoperitoneum *226*
- Guiding Therapeutic Procedures *226*

14. Blue, Falls, and Sesame Protocol 229
Dalim Kumar Baidya, Anil Malik
- The Blue Protocol (Bedside Lung Ultrasound in Emergency) *229*
- The Falls Protocol (Fluid Administration Limited by Lung Sonography) *234*
- The Sesame Protocol *235*

15A. Ultrasound of Diaphragm 238
Sandeep Sahu, Alka Verma, Rashmi Soori
- Applied Anatomy of Diaphragm *238*
- Variant Anatomy *239*
- Function *240*
- Dysfunction *240*
- Clinical Presentation of Diaphragm Dysfunction *240*
- Different Techniques for Evaluating the Diaphragm *241*
- Ultrasonography *241*
- Thickness of Diaphragm *243*
- Diaphragmatic Velocity *248*
- The Echogenicity of Diaphragm in Intensive Care Unit *248*
- Clinical Applications *248*
- Limitations of Ultrasonography *249*

15B. Ultrasonography for Assessment of Muscles Dysfunction in ICU 252
Vijay Hadda, Rohit Kumar
- Assessment of Muscle Function *252*
- Reliability of Ultrasonography for Assessment of Muscle Functions *257*
- Literature Supporting the Use of Ultrasonography for Assessment of Muscle Function in Intensive Care Unit *258*

15C. Ultrasonography in Deep Vein Thrombosis 260
Akhil Kant Singh
- Ultrasound *261*
- How I Scan the Leg? *264*

15D. Clinical Case Scenarios 268
Rahul K Anand, Kapil Dev Soni, Ajay Singh, Rashmi Bhatt
- Case 1 *268*

- Case 2 *269*
- Case 3 *270*
- Case 4 *271*
- Case 5 *273*
- Case 6 *274*

15E. Infection Control and Legal Issues in ICU Sonography 276
Puneet Khanna, Rajkumar Subramanian
- Organisms of Concern *276*
- Legal Issues in ICU Sonography *278*

Index *281*

Book Review

Over the past decade, ultrasound (USG) has established itself as an essential element in evaluating critically-ill patients in intensive care units, operating rooms, and emergency departments. Why has USG become such an important clinical tool in the evaluation of critically-ill patients?

Ultrasound is a unique imaging modality that provides both anatomical and functional information at the bedside. Critical care USG has progressively supplanted the use of invasive, yet blind, devices and procedures. Second, multiple scientific publications in peer-reviewed journals have validated USG for the evaluation of patients with circulatory and respiratory failure.

Improvements in technology and engineering have resulted in the recent emergence of full-feature, compact systems, which are highly portable for optimal use in crowded care units. Numerous intensivists and anesthesiologists have progressively become self-sufficient in performing and interpreting USG examinations as part of the acute management of critically-ill patients.

Point-of-care ultrasound (POCUS) refers to the use of portable ultrasonography at the patient's bedside for diagnostic and therapeutic purposes performed "real time" by the provider. The use of ultrasound has greatly expanded over the past years with its increasing usefulness being realized in different clinical areas.

Since a close collaboration among intensivists, anesthesiologists, and cardiologists is mandatory on clinical grounds in many countries, scientific societies and academic boards have organized national training programs dedicated to critical care USG.

Competence in critical care USG relies on a specific training and curriculum to achieve professional efficiency. This book is aimed at helping frontline physicians in developing competence in critical care USG, regardless of their personal background. It provides an in-depth, up-to-date, state-of-the-art review of the capabilities and limitations of USG in the specific settings of critical care medicine.

Coverage includes the latest applications of ultrasound for neurologic critical care; vascular problems; chest; hemodynamic monitoring; and abdominal and emergency uses, as well as assistance in a variety of specific procedures in critical care medicine. The book covers the full spectrum of conditions diagnosed using ultrasound and gives practical guidance on how to use ultrasound for common invasive procedures. Major applications are introduced using focused diagnostic questions and reviewing the image-acquisition skills needed to answer them. Images of positive and negative findings are presented, and scanning tips for improving image quality.

As such, the present book should be regarded as a valuable tool for intensivists and anesthesiologists, emergency physicians, general physicians, pulmonologists who wish to develop their proficiency in using critical care in their daily practice for the benefit of the severely ill. The international panel of authors and authors from AIIMS, PGIMER, Chandigarh,

SGPGI who contributed to this book reflects the close collaboration and large consensual approach of scientific and medical communities.

MK Arora MD (AIIMS)
Professor
Department of Anesthesia
Institute of Liver and Biliary Sciences
New Delhi

Formerly
Professor and Head
Department of Anesthesiology
Pain Medicine and Critical Care
All India Institute of Medical Science
New Delhi, India

Faculty Anesthesia
McGill University
Montreal, Canada

Chapter 1A

Point of Care Ultrasound

Puneet Khanna, Shakshi Gera, Rashmi Bhatt

Point-of-care ultrasound (POCUS) refers to the use of portable ultrasonography at the patient's bedside for diagnostic and therapeutic purposes performed "real time" by the provider. The use of ultrasound has greatly expanded over past years with its increasing usefulness being realized in different clinical areas. It has now become the point-of-care tool in intensive care units (ICU) with varied diagnostic and therapeutic applicability, rightfully earning the nomenclature of 'ultrasound stethoscope' (AIUM, 2004).[1]

CHARACTERISTICS OF POCUS

1. Aim is a well-defined purpose linked to improving patient outcomes.
2. Examination is focused and goal-directed.
3. Examination findings are easily recognizable.
4. Examination is easily learned.
5. Examination is quickly performed.
6. Examination is performed at the patient's bedside.

HISTORY OF POCUS

The history of POCUS dates back to the late 80s when it was first used in ICU and emergency department in 1989. A few years later, there was an increase in ultrasound-guided procedures, during the 1990s. Gradually, POCUS came to be incorporated in academic curriculum like emergency medicine in 1994 followed by US medical school curriculum for the first time in 2005. In 2010, it was integrated in the US residency curriculum. Since then, it has gradually been assimilated in working protocols as well as teaching programs in several countries.

WHO USES POCUS?

A generation of physicians will need to be trained to view this technology as an extension of their senses, just as many generations have viewed the stethoscope. That development will require the medical education community to embrace and incorporate the technology throughout the curriculum. Point-of-Care US in Medical Education, NEJM, 2014.[2]

HOW TO USE POCUS?

Point-of-care ultrasound (POCUS) is an integral element to clinical diagnosis and management. It is a vital link following a history and physical examination, which in turn can help review the initial assessment. It can help determine the further diagnostic work up required as well as carry out many therapeutic procedures as well. POCUS has now come to occupy an indispensable place in all the stages of patient management, i.e. diagnosis, treatment and monitoring.

POC ultrasound examinations differ from complete studies in that they are:
- Limited in scope, designed to achieve specific procedural aims (e.g. direct the needle to the correct location) or answer focused questions (e.g. does my patient have ascites?).
- Performed by the same care provider who will be using the information to direct immediate patient care management at the bedside.

The critically ill-patients present with complex clinical problems with the initial pathological process deteriorating into multiorgan involvement over days or even acutely. Whether as, the initial insult or secondary involvement, the cardiorespiratory failure in patients, is a challenging diagnostic dilemma encountered routinely. Similarly, diagnosing the cause of fever, assessing fluid responsiveness, detection of raised intracranial pressure in sedated and ventilated patient, achieving vascular access and insertion of intra-arterial catheter in edematous or injured patients pose challenge to an intensivist. Even an astute clinician will find this modality useful in overcoming these challenges. Its expanding clinical utility lies in the fact that it is a bedside technique with minimal recurring costs. This has allowed it to be incorporated in critical care set ups. It provides additional advantage for no technique related ionizing radiation exposure. The time is saved due to the obviation of the need to transport the patients, who are usually on multiple life support systems, to the radiology suites. There is less need for third party referrals. If it is done by the intensivist who is directly involved in the patient management, more clinically relevant and focused examination can be done which will positively influence the management of the patient.

IS POCUS A SUBSTITUTE FOR PHYSICAL EXAMINATION?
- It is just an additional tool, as an adjunct to physical examination.
- Usually, it is focused and used in a binary fashion, either to support or refute the hypothesis or clinical assumption generated by the history and physical examination.

Comprehensive Perioperative Ultrasound Examination

(Foresight)
(Adapted from Ref. 3)

{
- **F**ocused
- **P**eri**O**perative
- **R**isk
- **E**valuation
- **S**onography
- **I**nvolving
- **G**astro-Abdominal
- **H**emodynamic, and
- **T**rans-Thoracic Ultrasound
}

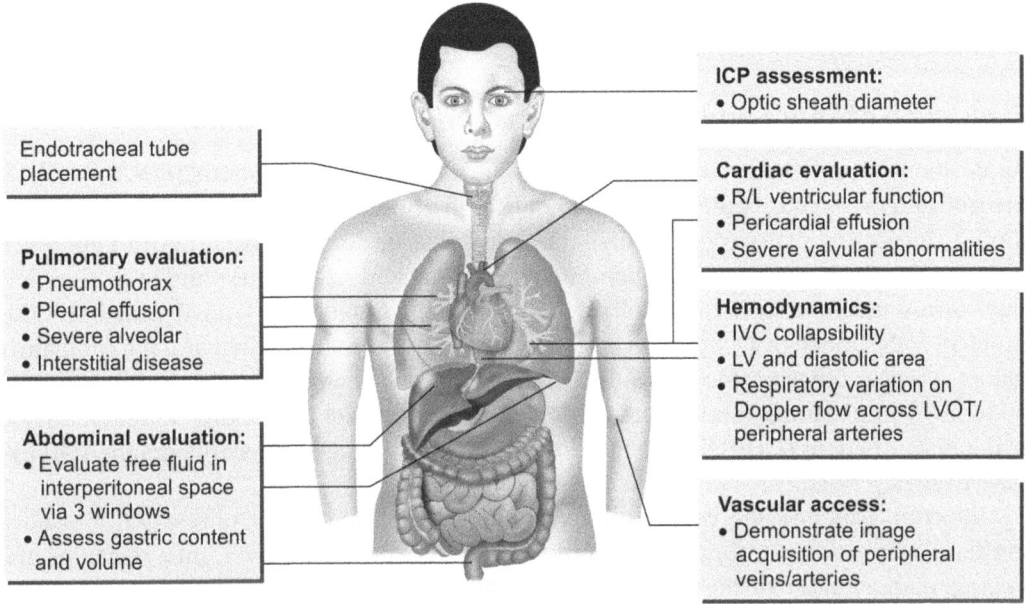

Fig. 1: Role of POCUS in ICU: scope of use.[3]
Abbreviations: ICP: Intracranial pressure; R/L: Right/Left; IVC: Inferior vena cava; LV: Left ventricle; LVOT: Left ventricular outflow tract.

Role of POCUS in ICU

The importance of ultrasound in ICU can be underscored by its utility in various scenarios (Fig. 1). These are listed below briefly; an elaborate description follows in the appropriate section:

Differential Diagnosis of Acute Respiratory Failure

Critically-ill patients usually suffer from wide spectrum and varied degree of respiratory involvement. The conventionally used roentgenogram has some pitfalls in ICU setting. The sensitivity and specificity of lung ultrasound for detection of pneumothorax has been variously reported but its diagnostic accuracy remains higher than chest X-ray.

The pooled sensitivity and specificity of chest ultrasound were 0.87 (95% CI: 0.81–0.92) and 0.99 (95% CI: 0.98–0.99), while those of chest X-ray were 0.46 (95% CI: 0.36–0.56) and 1.0 (95% CI: 0.99–1.0), respectively in a meta-analysis by Ebrahimi et al.[4]

A number of signs like the bat sign, vertical B lines, lung rockets corresponding to tissue fluid interface and interstitial syndrome, quad sign and sinusoid sign, presence of lung point, shred sign and tissue-like sign visualized on lung sonography bear significance.[5]

Lichtenstein et al. proposed a USG based BLUE protocol (bedside lung ultrasound in emergency) in 2004, describing the application of ultrasound in critically ill, in the differential diagnosis of acute respiratory failure.[6] An intensivist trained in USG can further carry out a more comprehensive examination and maneuvers to understand the pathology, e.g. assessment of lung overinflation, aeration loss and its distribution, assessment of lung recruitment following positive end-expiratory pressure (PEEP), etc.[6]

Evaluation of Shock and Application in Cardiac Arrest Situation

Application of cardiac ultrasound by other clinicians, such as intensivists (FOCUS), allows brisk diagnosis of acute cardiovascular pathologies in a noninvasive manner. The key elements during a focused cardiac ultrasound in shock patients can be evaluated using SIMPLE approach described by Mok et al.[7] Similarly, combination of lung and heart ultrasound in a schematic way, forms the fluid administration limited by lung sonography (FALLS) protocol.[8] This protocol as described by Lichenstein et al. serves as a tool for management of acute circulatory failure. Various other protocols like rapid ultrasound for shock and hypotension (RUSH),[9] abdominal and cardiac evaluation with sonography in shock (ACES),[10] undifferentiated hypotension patient (UHP),[11] or focused assessed transthoracic echocardiography (FATE)[12] have been described.

Ha et al. explored the combined application of lung and heart ultrasound for shock evaluation by using multi-organ point-of-care (MOPOCUS) strategy.[13] They classified the etiology of shock based on the diffuse interstitial and noninterstitial pattern on lung ultrasound. Various authors have described the application of USG in cardiac arrest settings.[14-17] Though controversial, when applied in a prompt and focused manner, it can yield valuable information (Fig. 2). The efficacy of goal directed or focused echocardiography is highlighted by the study of Manasia et al.[18] In their observational study of 90 patients, they found that intensivists formally trained in handheld echocardiographic system could successfully perform and correctly interpret a limited transthoracic echocardiography (TTE) in critically-ill patients, which could modify their management.

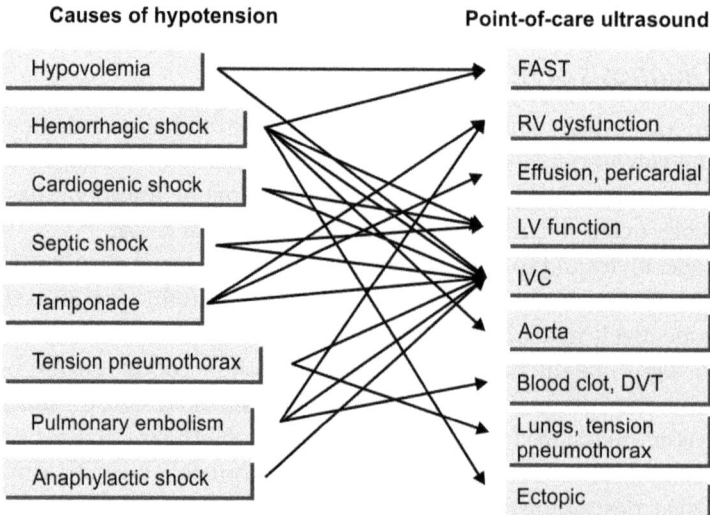

Fig. 2: The causes of hypotension and their ultrasound findings.
Abbreviations: FAST: Focused assessment with sonography in trauma; RV: Right ventricle; LV: Left ventricle; IVC: Inferior vena cava; DVT: Deep vein thrombosis.

Holistic approach to evaluation of heart: The training in point-of-care echocardiography for noncardiologists, such as intensivists is limited to basic elements. However, there is a new concept of automated speckle tracking echocardiography. Bagger et al. found a reasonable agreement between automated and conventional methods to determine left ventricular systolic function.[19] The recognition of left ventricular diastolic dysfunction (DD) in ICU patients is gaining importance as it might be linked to weaning failure in them. These patients are particularly likely to have DD, due to underlying disease or the applied therapies. However, its exact incidence and prognostic implications are unknown. Eisen et al. observed that heart failure with a normal LV ejection fraction (HFNEF) may account for more than 50% of heart failure cases.[20] The recognition of these conditions by a bedside ECHO may in fact become a factor affecting clinical outcome.

Fever in ICU

Fever in the ICU is a diagnostic dilemma that continues to baffle even the best of critical care physicians. The ultrasound can be a useful tool in this regard, by exploring relatively inaccessible locations for hidden abscesses as in deep abdominal, paranasal sinuses, pleural cavity. Tu et al. demonstrated that ultrasound was a safe and reliable diagnostic aid for empyema.[21] The use of ultrasound for necrotizing fasciitis has been described by Kehrl,[22] while Lichenstein has suggested the possibility of an ultrasound equivalent for fever of unknown origin in the ICU patient.[23]

Raised Intracranial Pressure

The assessment of raised intracranial pressure (ICP) in ICU patients who are sedated and on mechanical ventilation can be facilitated by use of USG. Assessing the diameter of optic nerve sheath (ONSD) by USG guidance can help detect an increase in ICP.[24] The findings of Shirodkar et al. support USG based ONSD estimation as an early test for diagnosing raised ICP.[25] In a meta-analysis, Ohle et al. found a sensitivity and specificity of 95.6% and 92.3% respectively for detection of raised ICP with ocular USG.[26] Shirodkar et al. found a good correlation between USG and MRI enabled measurement of ONSD ($r = 0.02$, $P < 0.001$).[27]

Vascular Access

Ultrasound has gained popularity as a guided interventional tool for vascular access in ICU patients. USG based internal jugular venous (IJV) localization for cannulation was described by Ullman and Stoelting as early as the year 1978.[28] In 1986, Yonei et al. described the in-plane approach of USG based IJV cannulation.[29] Since then, USG-guided central venous cannulation has become a standard of care and is widely practiced. A plethora of literature is available comparing alternative approaches and illustrating the advantage of USG-based technique over the landmark-based methods.[30,31] Gordon et al. demonstrated a success rate of 99.9% for USG-guided IJV cannulation with 87% first pass attempts.[32] In a meta-analysis of 27 trials evaluating the outcome of USG-guided central venous cannulation, it was found that USG-guided catheterization was quicker and safer than landmark-based method.[33] A novel 'medial-oblique' approach for IJV cannulation, described by Dilisio in 2012, is being explored as it has been shown to provide better visualization of anatomy and minimize the risk of arterial puncture[34] as also demonstrated by Baidya et al.[35]

The usefulness in detection of DVT using bedside portable USG is at par with formal duplex sonography with up to 98% agreement with the same.[36] Application of USG for inserting intra-arterial lines for invasive monitoring and insertion of dialysis catheters is also noteworthy. Li et al. observed a 80% success rate in first attempt with USG-based technique as compared to the palpation method (42%).[37]

In a meta-analysis of four trials, Shiloh et al. found that the first attempt success was likely to be more with USG-guided method for arterial cannulation with a 71% improvement (RR, 1.71; 95% CI, 1.25–2.32) as compared with the palpation method.[38] In a meta-analysis of seven trials evaluating the outcome of ultrasound application for hemodialysis catheter insertion, Rabindranath et al. found that the failure risk of catheter placement in the first attempt was reduced significantly with ultrasound guidance, with reduction in risk of inadvertent arterial puncture and subsequent hematoma, as well as time for successful cannulation and the number of attempts required.[39]

POCUS IN MEDICAL EDUCATION

Recognizing that POCUS has the potential to be an important tool for diagnosis and clinical management, some medical schools have embraced the technology to augment anatomy lessons, and as an adjunct to the physical examination. In clinical clerkships, students are increasingly familiarized with ultrasound at the POC, either as a part of their training or incidentally in clinical practice. In graduate medical education, ultrasound training is required and fairly standardized in emergency medicine residency programs in the United States.

LIMITATIONS

The limitations of POCUS performed by a critical care physician are mainly related to the clinician and the particular patient. Patient factors, such as obesity, the presence of edema or subcutaneous emphysema, and suboptimal patient position can reduce image quality and make interpretation difficult. This is especially limiting in performing bedside echocardiography where appropriate patient positioning cannot be achieved and mechanical ventilation frequently hampers image quality. The clinician-related limitations pertain mostly to training and education. This situation may lead to suboptimal care of the patient.

As outlined above, ultrasound has proven to be a modality of immense potential for supporting clinical acumen and guiding the management of ICU patients. While it has established itself as a third eye for an intensivist in management of cardiac and respiratory conditions, it is fast digging its claws into so far unexplored terrain of neurologic monitoring and cardiac arrest situations (Fig. 3). As the primary treating physician in the ICU set up is an intensivist, it is an empowering tool to a well-trained intensivist. While all this is said, the approach has to be goal directed or based on focused protocols. For successful implementation of USG-based strategies, some requirements need to be met. Guidelines must be laid for indications and performance of each examination with adequately trained clinicians who

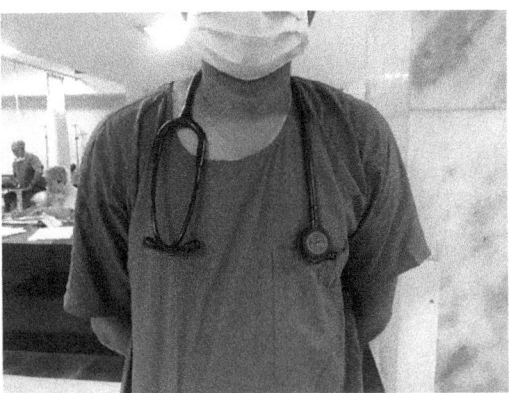 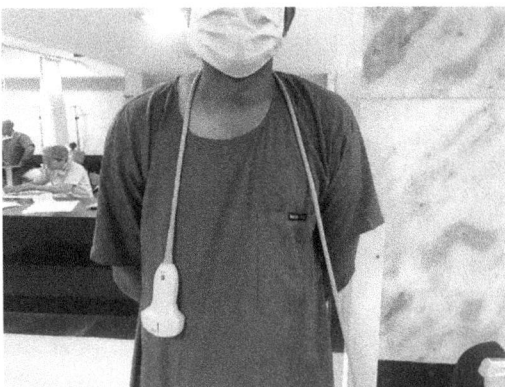

Fig. 3: The new age essential stethoscope: Ultrasound look beyond the visible for the vision.

are performing these examinations. The standardized technical specifications of the USG instruments for each examination should be in place. Proper documentation and archiving of cases should be ascertained. An overseer system should be in place to ensure that the prescribed guidelines are met, in terms of providing accreditation or denoting credentials at institutional or individual level.

KEY POINTS

1. POCUS: Disruptive innovation.
2. It is not a substitute, rather an adjunct to physical examination.
3. Done by nonradiologist.
4. Save time, save radiation, save money, save lives.
5. While we have a good evidence of literature credentialing the efficacy of POCUS in critical care, outcomes-based research is needed for many areas.

REFERENCES

1. Greenbaum LD, Benson CB, Nelson LH, et al. Proceedings of the compact ultrasound conference sponsored by the American Institute of Ultrasound in Medicine. J Ultrasound Med. 2004;23: 1249-54.
2. Solomon SD, Saldana F. Point-of-care ultrasound in medical education—stop listening and look. N Engl J Med. 2014;370(12):1083-5. doi:10.1056/NEJMp1311944.
3. Ramsingh D, Rinehart J, Kain Z, et al. Impact assessment of perioperative point-of-care ultrasound training on anesthesiology residents. Anesthesiology 9. 2015;123:670-82.
4. Ebrahimi A, Yousefifard M, Kazemi HM, et al. Diagnostic accuracy of chest ultrasonography versus chest radiography for identification of pneumothorax: a systematic review and meta-analysis. Tanaffos. 2014;13(4):29-40.
5. Lichtenstein D, Van Hooland S, Elbers P, et al. Ten good reasons to practice ultrasound in critical care. Anaesthesiol Intensive Ther. 2014;46(5):323-35.
6. Lichtenstein D, Goldstein G, Mourgeon E, et al. Comparative diagnostic performances of auscultation, chest radiography and lung ultrasonography in acute respiratory distress syndrome. Anesthesiology. 2004;100:9-15.

7. Mok KL. Make it SIMPLE: enhanced shock management by focused cardiac ultrasound. J Intensive Care. 2016;4(1):51.
8. Lichtenstein DA. Lung ultrasound in the critically ill. Ann Intensive Care. 2014;4:1.
9. Perera P, Mailhot T, Riley D, et al. The RUSH exam: Rapid Ultrasound in Shock in the evaluation of the critically ill. Emerg Med Clin North Am. 2010;28(1):29-56.
10. Atkinson PR, McAuley DJ, Kendall RJ, et al. Abdominal and Cardiac Evaluation with Sonography in Shock (ACES): an approach by emergency physicians for the use of ultrasound in patients with undifferentiated hypotension. Emerg Med J. 2009;26(2):87-91.
11. Rose JS, Bair AE, Mandavia D, et al. The UHP ultrasound protocol: a novel ultrasound approach to the empiric evaluation of the undifferentiated hypotensive patient. Am J Emerg Med. 2001; 19(4):299-302.
12. Jensen MB, Sloth E, Larsen KM, et al. Transthoracic echocardiography for cardiopulmonary monitoring in intensive care. Eur J Anaesthesiol. 2004;21(9):700-7.
13. Ha YR, Toh HC. Clinically integrated multi-organ point-of-care ultrasound for undifferentiated respiratory difficulty, chest pain, or shock: a critical analytic review. J Intensive Care. 2016;4:54.
14. Rabiei H, Rahimi-Movaghar V. Application of ultrasound in pulseless electrical activity (PEA) cardiac arrest. Med J Islam Repub Iran. 2016;30:372.
15. Ozen C, Salcin E, Akoglu H, et al. Assessment of ventricular wall motion with focused echocardiography during cardiac arrest to predict survival. Turk J Emerg Med. 2016;16(1):12-6.
16. Wang XT, Zhao H, Liu DW, et al. The echocardiographic evaluation and its predicted prognosis of acute left cardiac systolic dysfunction in critically ill patients. Zhonghua Nei Ke Za Zhi. 2016; 55(6):430-4.
17. Milne J, Atkinson P, Lewis D, et al. Sonography in Hypotension and Cardiac Arrest (SHoC): Rates of abnormal findings in undifferentiated hypotension and during cardiac arrest as a basis for consensus on a hierarchical point of care ultrasound protocol. Cureus. 2016;8(4):e564.
18. Manasia AR, Nagaraj HM, Kodali RB, et al. Feasibility and potential clinical utility of goal-directed transthoracic echocardiography performed by noncardiologist intensivists using a small handcarried device (SonoHeart) in critically ill-patients. J Cardiothorac Vasc Anesth. 2005; 19(2):155-9.
19. Bagger T, Sloth E, Jakobsen CJ. Left ventricular longitudinal function assessed by speckle tracking ultrasound from a single apical imaging plane. Crit Care Res Pract. 2012;2012:361824.
20. Eisen LA, Davlouros P, Karakitsos D. Left ventricular diastolic dysfunction in the intensive care unit: Trends and Perspectives. Critical Care Research and Practice, vol. 2012, doi:10.1155/2012/964158
21. Tu CY, Hsu WH, Hsia TC, et al. Pleural effusions in febrile medical ICU patients: chest ultrasound study. Chest. 2004;126(4):1274-80.
22. Kehrl T. Point-of-care ultrasound diagnosis of necrotizing fasciitis missed by computed tomography and magnetic resonance imaging. J Emerg Med. 2014;47(2):172-5.
23. Lichtenstein DA. Point-of-care ultrasound: Infection control in the intensive care unit. Crit Care Med. 2007;35(5 Suppl):S262-7.
24. Dubourg J, Javouhey E, Geeraerts T, et al. Ultrasonography of optic nerve sheath diameter for detection of raised intracranial pressure: a systematic review and meta-analysis. Intensive Care Med. 2011;37(7):1059-68.
25. Shirodkar CG, Rao SM, Mutkule DP, et al. Optic nerve sheath diameter as a marker for evaluation and prognostication of intracranial pressure in Indian patients: an observational study. Indian J Crit Care Med. 2014;18(11):728-34.
26. Ohle R, McIsaac SM, Woo MY, et al. Sonography of the optic nerve sheath diameter for detection of raised intracranial pressure compared to computed tomography: a systematic review and meta-analysis. J Ultrasound Med. 2015;34(7):1285-94.
27. Shirodkar CG, Munta K, Rao SM, et al. Correlation of measurement of optic nerve sheath diameter using ultrasound with magnetic resonance imaging. Indian J Crit Care Med. 2015;19(8):466-70.

28. Ullman JI, Stoelting RK. Internal jugular vein location with the ultrasound Doppler blood flow detector. Anesth Analg. 1978;57(1):118.
29. Yonei A, Nonoue T, Sari A. Real-time ultrasonic guidance for percutaneous puncture of the internal jugular vein. Anesthesiology. 1986;64(6):830-1.
30. Brass P, Hellmich M, Kolodziej L, et al. Ultrasound guidance versus anatomical landmarks for subclavian or femoral vein catheterization. Cochrane Database Syst Rev. 2015;1:CD011447.
31. Troianos CA, Jobes DR, Ellison N. Ultrasound-guided cannulation of the internal jugular vein. A prospective, randomized study. Anesth Analg. 1991;72(6):823-6.
32. Gordon AC, Saliken JC, Johns D, et al. US-guided puncture of the internal jugular vein: complications and anatomic considerations. J Vasc Interv Radiol. 1998;9(2):333-8.
33. Hind D, Calvert N, McWilliams R, et al. Ultrasonic locating devices for central venous cannulation: meta-analysis. BMJ. 2003;327(7411):361.
34. Dilisio R, Mittnacht AJ. The "medial-oblique" approach to ultrasound-guided central venous cannulation—maximize the view, minimize the risk. J Cardiothorac Vasc Anesth. 2012;26(6):982-4.
35. Baidya DK, Chandralekha, Darlong V, et al. Comparative sonoanatomy of classic "Short Axis" probe position with a novel "Medial-oblique" probe position for ultrasound-guided internal jugular vein cannulation: A crossover study. J Emerg Med. 2015;48(5):590-6.
36. Boddi M, Barbani F, Abbate R, et al. Reduction in deep vein thrombosis incidence in intensive care after a clinician education program. J Thromb Haemost. 2010;8(1):121-8.
37. Tang L, Wang F, Li Y, et al. Ultrasound guidance for radial artery catheterization: an updated meta-analysis of randomized controlled trials. PLoS ONE. 2014;9(11):e111527. https://doi.org/10.1371/journal.pone.0111527
38. Shiloh AL, Savel RH, Paulin LM, et al. Ultrasound-guided catheterization of the radial artery: a systematic review and meta-analysis of randomized controlled trials. Chest. 2011;139(3):524-9.
39. Rabindranath KS, Kumar E, Shail R, et al. Ultrasound use for the placement of haemodialysis catheters. Cochrane Database Syst Rev. 2011;(11):CD005279.

Chapter 1B

Applications in Critical Care Medicine

Arvind Kumar, Manish Soneja

INTRODUCTION

Sonography is a very important tool for managing patients in intensive care unit. Broadly, it can be grouped into cardiac sonography and noncardiac sonography. A more simplified and clinically more relevant way to understand its use in critical care would be a head to toe ultrasonography (USG). The utility of USG in measuring intracranial pressure (ICP), brain perfusion, and measuring pupillary size now has been well proven.

Besides this catheterization, managing collections in pleural cavity, pneumothorax, pneumonia, and pulmonary edema by lung sonography has shown promising results. In echocardiography it is useful for detecting and managing pericardial effusion and tamponade, assessment of stroke volume, systolic and diastolic function, and cor pulmonale. Focused echocardiographic evaluation in resuscitation (FEER) in emergency ward has increased its optimal use by critical care physicians in major cardiac emergencies. Hemodynamics, volume status assessment, and predicting volume responsiveness through assessment of inferior vena cava (IVC); abdominal USG for managing collections in peritoneal cavity and recesses, hydronephrosis, and bladder catheterization is of much importance. An organized approach in emergency of abdominal sonology is known as the focused assessment with sonography for trauma (FAST). Lower limb quick vascular scans are used for early and timely diagnosing deep venous thrombosis and its proper management.

HEAD AND NECK INCLUDING SUBCLAVIAN VEIN

Optic Nerve Sheath Diameter, Transcranial Doppler, Pupillary Reaction, and Size

Managing ICP is a major concern in critical care wards. Ideally in emergency wards less than 20-25 mm of mercury is recommended in intensive care settings. Invasive (probe or catheter) monitoring of ICP has the risk of infection and hemorrhage. Brain imaging computed tomography (CT) and magnetic resonance (MR) are often used to make decisions before lumbar puncture but its accuracy in knowing ICP is not precise. Various studies worldwide have shown promising usability of optic nerve sheath diameter (ONSD) as a reliable indicator of raised ICP in trauma patient as shown in Figures 1A to D. Considering the variation among geographical distribution an ONSD of 0.45-0.48 cm can be taken as normal in adult and an ONSD more than 0.50 cm has a sensitivity of 86% (79-92%), specificity of 98% (96-99%) for

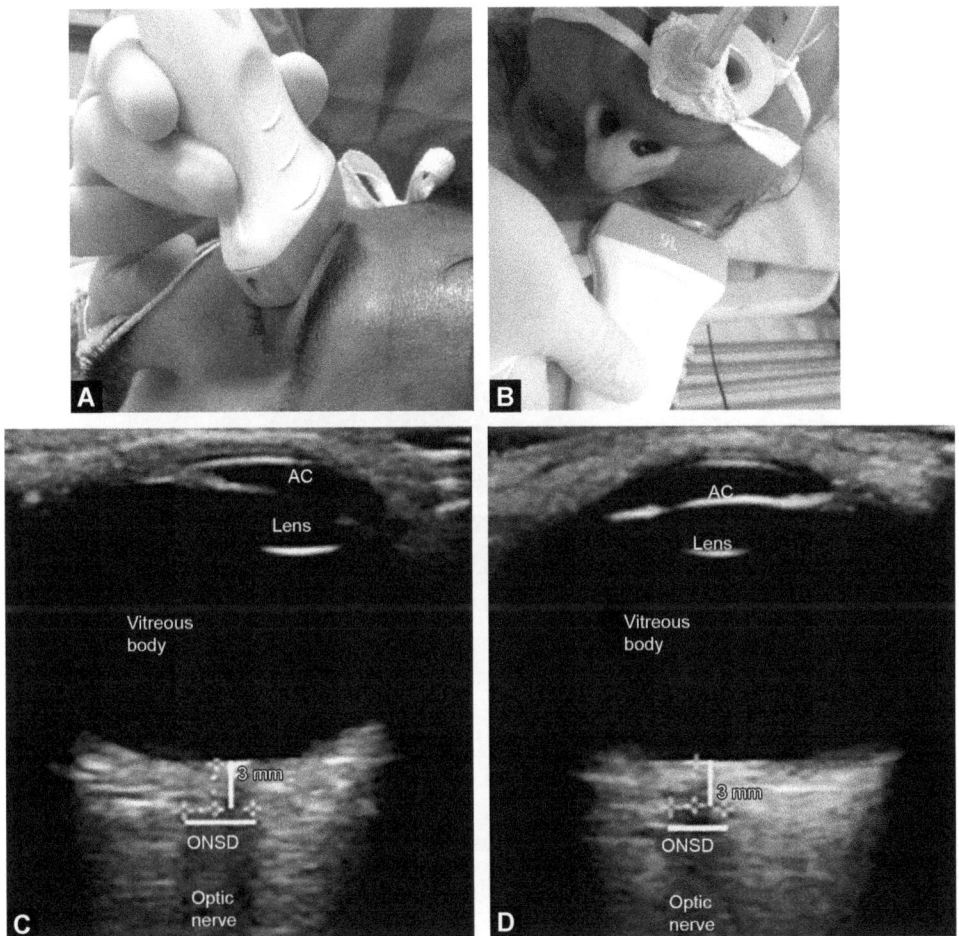

Figs. 1A to D: Sonographic measurement of optic nerve sheath diameter (ONSD) by two different approaches as depicted in the pictures A and B. The corresponding images of A and B have been shown in C (lateral) and D (anterior), respectively.

an ICP more than 20 mm Hg.[1] Transcranial Doppler (TCD) USG can detect brain death by studying the flow pattern in the middle cerebral artery; however, the TCD would need more studies before commenting on its application. Size of pupil and its reaction to light could easily be detected by USG in case of edema of eyelid by a probe of high frequency (10–13 MHz) placed transversely angling upwards. The pupillary reaction can be readily appreciated and size measured with a light shone on the ipsilateral eye (Figs. 2A to C) (for detail, *see* Chapter 11).

Internal Jugular Vein and Subclavian Vein

Critically ill-patients at times need a catheter in a central vein for proper diagnosis and treatment. Catheterization may be done in internal jugular vein (IJV), subclavian, and femoral veins. These procedures are associated with risk of puncture of arteries, hematoma, infection,

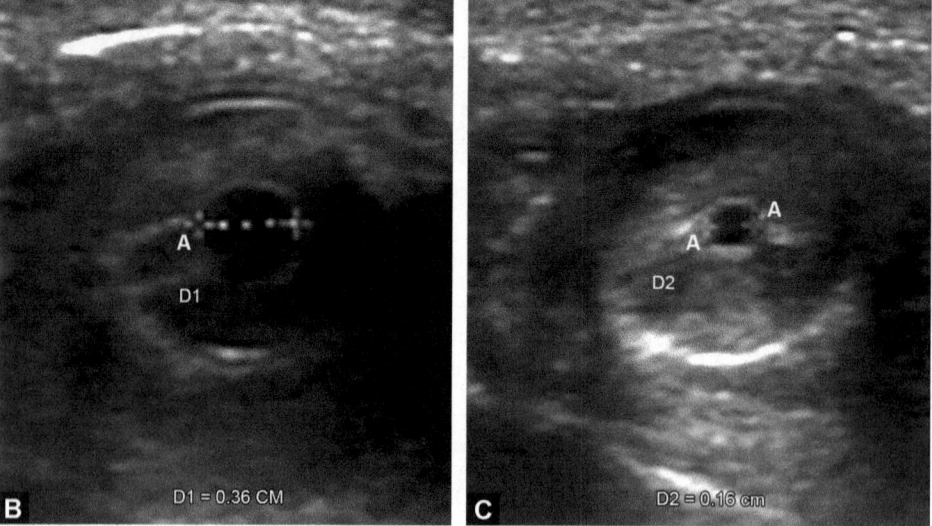

Figs. 2A to C: Probe of the ultrasound for assessing pupillary size (D1) and light reflex (D2).

and thrombosis. USG-guided techniques have been proven beyond doubts to be superior to anatomical landmark technique in catheterization of large vessels. In one of the meta-analysis it showed significant benefit in terms of complication rates, arterial puncture, hematoma formation, and success in venous catheterization in first attempt by 71%, 72%, 73%, and 57%, respectively. It also significantly decreased the time taken for the procedure (Figs. 3A and B)[2] (for detail, *see* Chapter 3).

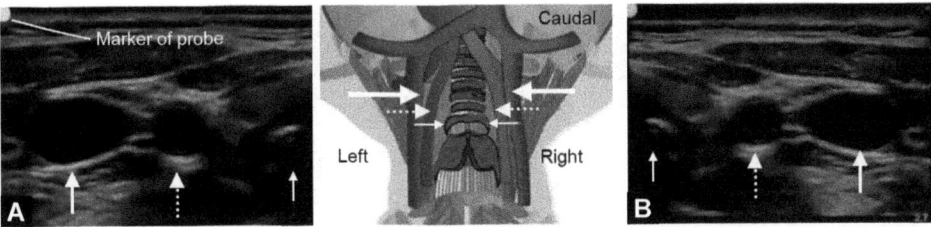

Figs. 3A and B: Ultrasound assessment of jugular vein. Trachea, common carotid artery, and internal jugular vein (IJV) has been shown by thick, thin and dotted arrows, respectively. (A) On left side is showing structures when index position of the probe is facing laterally on left side; and (B) The same is facing medially on right side.

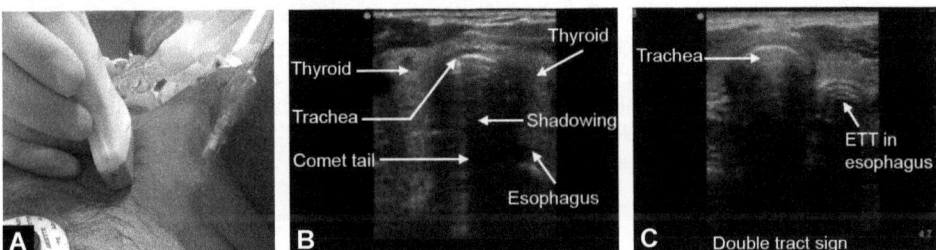

Figs. 4A to C: (A) Transverse probe just above the suprasternal notch; (B) Normally trachea appears as hyperechoic curve with artifact (comet tail) and shadowing. The esophagus is usually seen more distally and on side of the screen as an oval structure with hypoechoic center and a hyperechoic wall; and (C) Intubation of esophagus reveals an adjacent hyperechoic curvilinear structure with comet tail artifact and shadowing, consistent with endotracheal tube in the esophagus. This has been referred to as the "double tract sign".

ULTRASONOGRAPHY OF THE AIRWAY

Postendotracheal intubation sonography of lung could be easily performed to assess lung sliding sign for confirming lung expansion. The same is used to confirm tube within the trachea as shown in Figures 4A to C. During percutaneous dilatational tracheostomy (PDT), prominent vessels could be easily avoided from injury by the use of USG. Under ultrasound guidance prior to PDT, the level of tube could easily be adjusted in relation to cricoid cartilage (for detail, see Chapter 4).

UTILITY OF CHEST SONOGRAPHY IN CRITICAL CARE

Computed tomography is far more superior then chest radiography in lung imaging and interpretation. Because of this in the beginning USG drew less attention for chest imaging. In necessity, intensive care unit (ICU) physicians have pioneered USG of the chest given the benefits of ability to carry out repeated examinations, easy reproducibility, and bedside availability. For pleural sonography, a linear high frequency probe (8–15 MHz) is ideal though a low frequency microcurvilinear probe images the lung well, but resolution is slightly compromised. The lung is evaluated from anterior to posterior in up to down fashion as shown in the Figures 5A and B here in different zones. The same is done till the diaphragm is seen. A rationale sonographic approach of lung is being described here (for detail see Chapter 5).

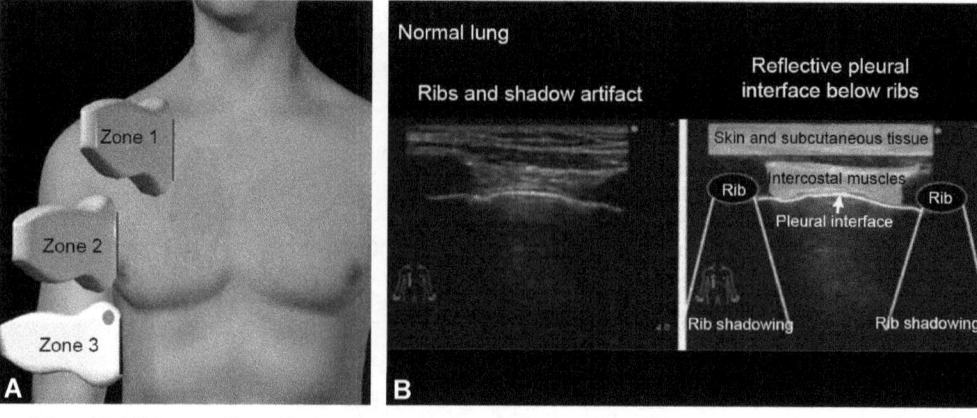

Figs. 5A and B: (A) Zones of lung. Zone 1: Second to third intercostal space (ICS) anterior chest wall for assessment of lung sliding, A and B lines, and their artifacts, Zone 2: Third to fourth ICS anterior axillary line for assessment of lung point, A and B lines, and their artifacts, Zone 3: Following posterior axillary line to passes posterior chest lower down up to diaphragm; and (B) Normal lung.

Normal Lung

The probe with index position towards 12 o'clock (neck of patient) is kept in the intercostal space vertically as shown in the Figure 5A. Grossly lung ultrasound is assessed in three mentioned zones in the Figures 5A and B; part B of the figure shows normal ultrasound image of the lung. An anechoic (artifactual) shadow images (black) is generated by the ribs. A hyperechogenic horizontal line could be well-appreciated in between the two ribs at around more than 0.5 cm away from the top of the screen, which is known as "pleural line" denoting junction between aerated lung and the soft the thoracic wall. The sonographic "image" of the lung is composed of artifacts. This is for the fact that air in lungs stops progression of the ultrasound beam. Two major artifacts arise which can be appreciated: A lines: Hyperechogenic, regularly placed, motionless, and horizontally placed lines represent reverberations from the pleura. These are artifacts of repetition. This artifact pattern can be seen in up to two-thirds of normal lungs. B lines: These are vertical lines originating from the pleura to the ultrasound screen edge. The "comet tail image" is defined as narrow base hyperechogenic, coherent bundle spreading out from the pleural line to the distal edge of the screen. The following similar artifacts should not be considered as B lines. Firstly, vertical comet tail artifacts, which are short, broad, and ill-defined, not reaching the farther end of the monitor, are not classified as B lines. These are labeled as *Z lines*, found in normal lung and in pneumothorax. They are lesser in echogenicity than pleural line and vanish after few centimeters. They are independent in movement in comparison to lung sliding. Second, comet tail artifacts: Counteracted superficially on pleural line. They are seen in emphysema or parietal foreign bodies like shot gun pellet. They are known as E lines.[3] In emergency ward, if a patient is blue what is recommended to be performed is called as bedside lung ultrasound in emergency (BLUE) protocol (Flowchart 1).[4] This using ultrasound algorithm alone has an accuracy of up to 90%.

Flowchart 1: The BLUE protocol,[4] comparing lung ultrasonography results on initial presentation with the final diagnosis by the ICU team. Uncertain diagnoses and rare causes (frequency < 2% proposal for diagnosing various causes of acute respiratory failure by means of bedside lung sonography in addition to relevant vascular assessments, which can achieve an accuracy of up to 90%.

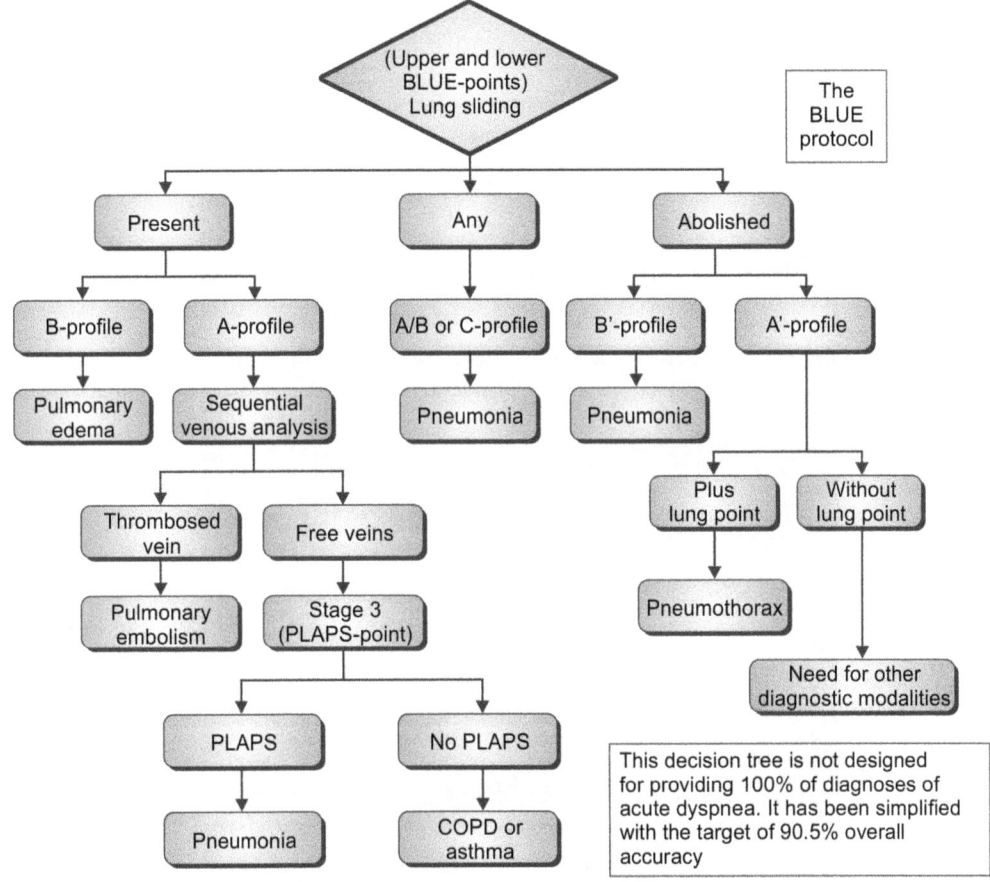

Abbreviations: BLUE: Bedside lung ultrasound in emergency; COPD: Chronic obstructive pulmonary disease; PLAPS: Posterolateral alveolar and/or pleural syndrome.
A profile means predominantly A lines, B profile means predominantly multiple anterior diffuse B lines, A/B profile means A lines and B lines on contralateral side in predominance, and C profile means profile suggesting consolidation.

Pneumothorax (Figs. 6A to D)

Conventionally, pneumothorax is suspected clinically, while confirmation of the same in ICU at bedside could be a tedious issue. Relatively pneumothorax presents with nonspecific signs. But many features, such as decrease in oxygen saturation [peripheral capillary oxygen saturation (SpO_2)], increase in airway resistance, and decrease in compliance of the lung suggest pneumothorax in ICU setting. Supine position, anteroposterior X-rays have poorly sensitivity. It may be false negative in many cases. CT assesses pneumothorax well but procedural delays are unavoidable. Other drawbacks include high cost issue and high radiation exposure. Sonography of lung has proven its utility over other modality in diagnosing pneumothorax (for detail *see* Chapter 5).

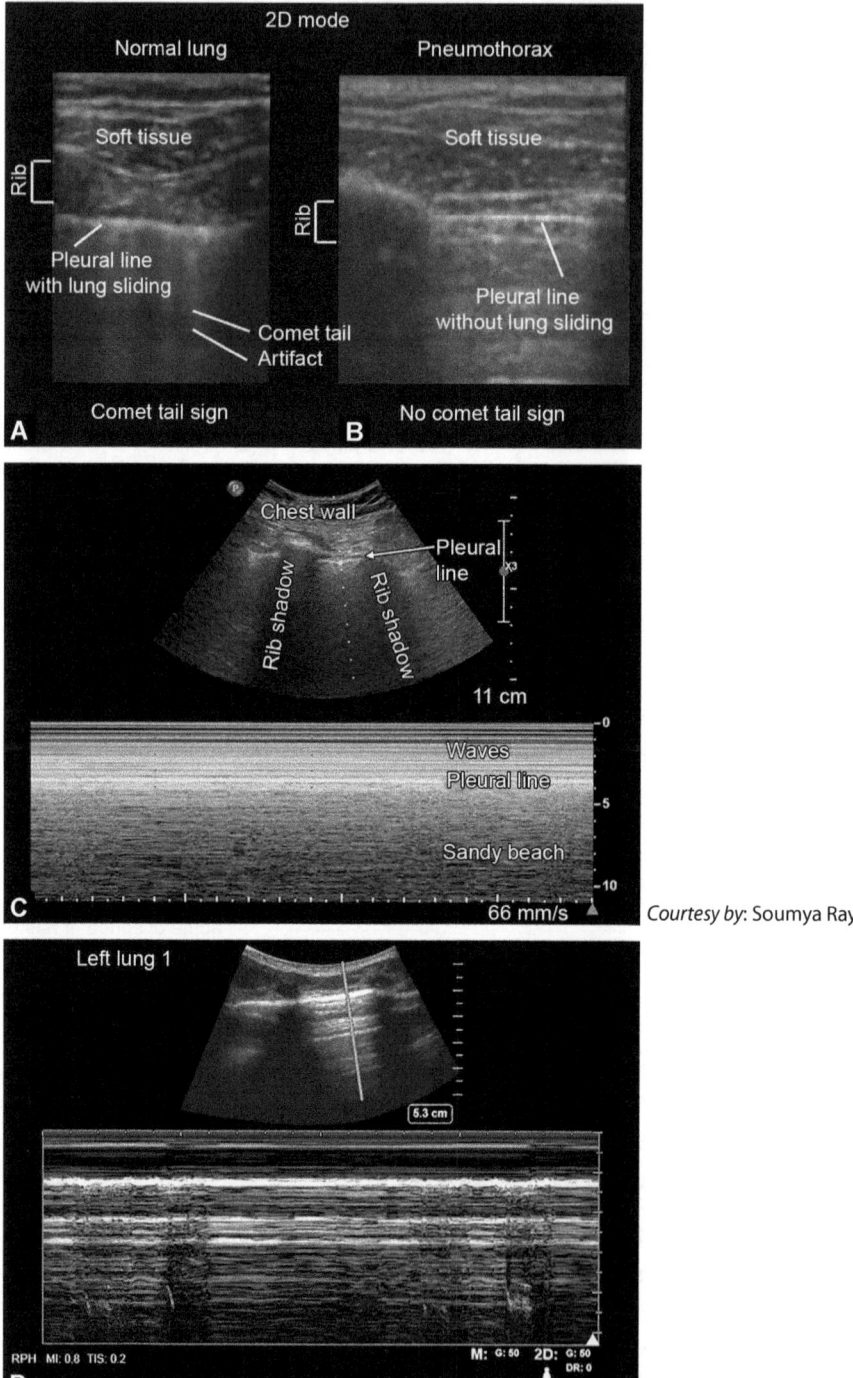

Figs. 6A to D: (A) Recognizing primary ultrasonography (USG) findings of normal lung, lung sliding with pleural line; (B) Recognizing USG findings suggestive of pneumothorax, absence of comet tails, and lung sliding. Corresponding images as seen on M-mode; (C) Seashore sign; and (D) Barcode sign. Apical pneumothorax on left side displaying the barcode sign on M mode.

Absence of pleural sliding is characteristic of pneumothorax because of the collapsed lung secondary to air interspersed between the pleura. As ultrasound penetrates poorly through air visceral pleura cannot be visualized. On M mode, "bar code" or "stratosphere" sign is seen instead of the "seashore" sign. Pneumothorax is effectively ruled out by presence of pleural slide. Pleural adhesions, acute respiratory distress syndrome (ARDS), endobronchial intubation, and emphysematous bullae may show similar effects as of absent pleural sliding. Lung point is the most specific sign of pneumothorax on lung sonography. "Lung point" is the junction of visceral and parietal pleura during inspiration or at expansion of lung parenchyma. Later pleural slide can be appreciated partly after a particular time in inspiration. So, it is easy to imagine that the size of pneumothorax can be assessed by the position of lung point and the appearance of pleural sliding in inspiration.[5] Pneumothorax in polytrauma has been extensively studied by ultrasound for the diagnosis and is proven to be better than radiography of chest. The sensitivity and specificity of chest sonography for diagnosing pneumothorax is more than 90% when CT is considered for comparison.

Pleural Effusion

Chest radiographs are considered an average modality for detection of pleural effusions; parenchymal opacities by this modality may even be inferior to basic clinical examination like auscultation. Sonography of chest diagnose, quantifies and guides well in draining pleural fluid (Fig. 7A). Pleural effusion in majority of cases is small but at times they could be large and may necessitate therapeutic drainage where USG plays a safe and pivotal role. The lung is seen floating like a "jelly fish" in cases of large fluid collection. Effusions may be septate and echogenic, if exudative, pyothorax and hemothorax appear hyperechoic on sonography. The effusive volume could be calculated by measuring interpleural (visceral and parietal) distance in midexpiration at the base, by the formula;[6] effusive volume (mL) = the above measurement (mm) × 20. Drainage of pleural effusion by the Seldinger technique using pig-tail catheter is reliably performed under USG guidance avoiding injury to major organs close proximity. A pig-tailed catheterization would be sufficient for majority of transudative effusions (for detail *see* Chapter 5).

Consolidation

The sonographic use for consolidation has gained popularity over years. It is seen as tissue like structure in echogenicity closely resembling liver and labeled as hepatization (Fig. 7B). Lung parenchyma at the side of consolidation looks like shredded paper. This appearance is recognized as shred sign on sonography which lies deep to pleura. Air bronchograms could be also seen in the affected parenchyma, appearing as hyperechoic, linear areas, which appears brighter with inspiration with air passing through them. Taking "shred sign" as a tool for diagnosis of pneumonia, Lichtenstein et al. showed a sensitivity and specificity of 90% and 98%, respectively for establishing consolidation by sonography in a prospective study (for detail *see* Chapter 5).[7]

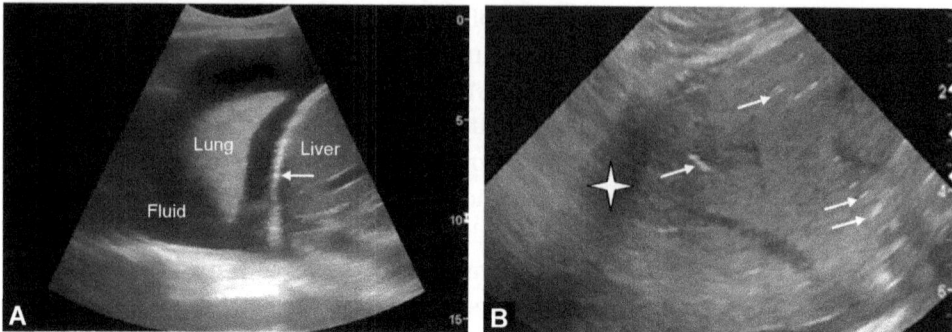

Figs. 7A and B: (A) Sonography of lung where lung is seen small and a floating jelly fish is well-appreciated. Arrow pointing to diaphragm; and (B) Consolidation of lung showing echogenicity (arrows).

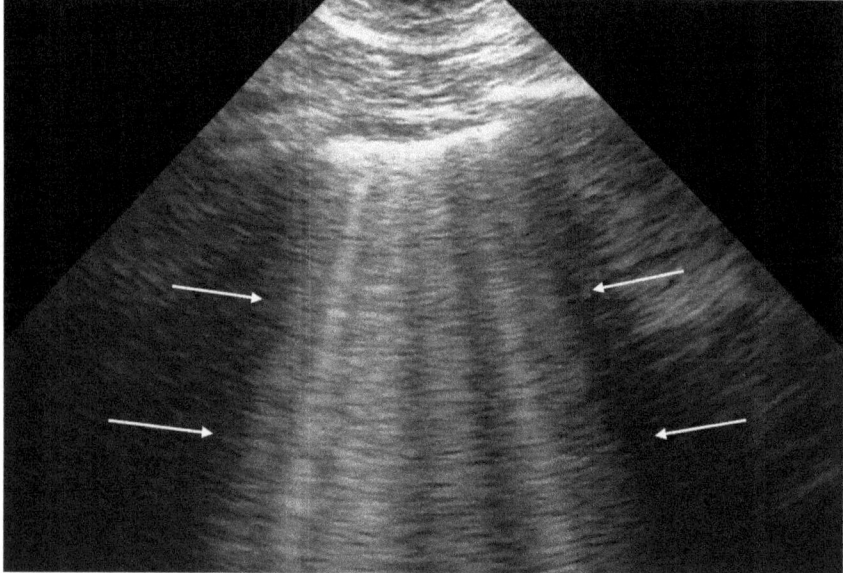

Fig. 8: Pulmonary edema, several "B" lines spreading out from the pleura (arrows).

Pulmonary Edema

Ultrasound waves bounces off from the layers of fluid-air, which leads to formation of "B" lines in the lung parenchyma. These B lines start from the pleural line and move with the pleura as it slides. These lines spread out to reach the distal end of the frame. They are known as "lung comets". But "B" lines only in excess of three is considered significant and reflects fluid in the lung and in turn reflects consolidation (Fig. 8). The comet tail sign has a sensitivity and specificity of around 90% in the diagnosis of consolidation. The number of lung comets is correlated with extravascular lung water. The sonographic "lung comet" has an association with New York Heart Association (NYHA) class, left ventricular (LV) systolic and diastolic function. The changes are similar in ARDS but they are not homogenous. They could also be seen in lung contusion, fibrosis, and atelectasis (for detail, *see* Chapter 5).

ECHOCARDIOGRAPHY IN THE CRITICAL CARE WARD

Echocardiography performed by physicians is valuable for diagnosis and management of critically ill patients. Referral to specialist may entail delays in extracting precise and focus information required in echocardiography which is a major point of concern in critical care wards, though an expert echocardiography can be considered at a later stage whenever needed, if required.[8] Focus assessed transthoracic echo is well-proven in 97% of critically ill-patients for obtaining relevant hemodynamic parameters using standard echocardiographic windows for optimizing care.[9]

BASIC VIEWS IN CRITICAL CARE (FIGS. 9A TO E)

Basic echocardiographic views mostly used in critical care setting are depicted in Figures 9A to E. They are parasternal long axis (PLAX), parasternal short axis view, apical four and five

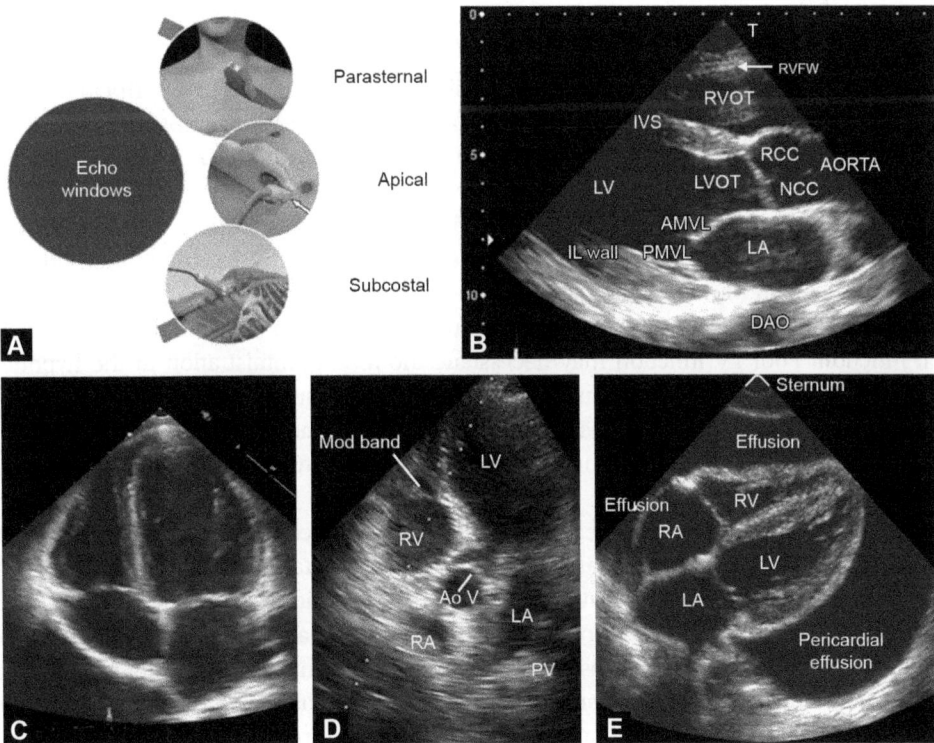

Figs. 9A to E: Depicting various basic echocardiac views useful in critical care setting. (A) Position of probes for obtaining basic views; (B) Parasternal long axis (PLAX) view, right ventricular free wall (RVFW), right ventricular outflow tract (RVOT), and interventricular septum (IVS); (C) Apical four-chamber view; (D) Apical five-chamber view; and (E) Subcostal view.
Abbreviations: LVOT: Left ventricular outflow tract; RCC: Right coronary cusp; AMVL: Anterior mitral valve leaflet; PMVL: Posterior mitral valve leaflet; NCC: Non coronary cusp; DAO: Discending aorta; LA: Left atrium; IL wall: Inferolateral wall; RA: Right atrium; RV: Right ventricular; LA: Left atrium; LV: Left ventricular; PV: Popliteal vein; AoV: Aortic valve; RVFW: Right ventricular free wall.

chamber views, and subcostal views. Apical views are made by keeping the probe at apex beat and the easiest way out is first to palpate and then proceed. At apex with index position at 3 o'clock or the pointer facing towards left shoulder with judicious horizontal movement of the probe in the intercostal space is sufficient to make an apical four chamber view. By tilting the probe slightly upward will make a five-chamber view which is good for assessing volume sample at left ventricular outflow tract (LVOT). The PLAX view are acquired by keeping the probe at second to fifth intercostal space along left border of the sternum and the marker pointing to 11 o'clock or right shoulder. For parasternal short axis views, a clockwise rotation towards left shoulder with pointer at cranial side elicits aortic short axis view. Left lateral positioning of the patient improves the view. At the same point where aortic short axis is acquired tilting the probe downwards and different cuts of short axis is acquired like mitral and papillary level views. In case of difficulties in making abovementioned views which is a common scenario in critical care setup subcostal view is easy to make and a useful tool to gather relevant information. In some patients, this may be the only accessible echo window (for detail see Chapter 7).

ASSESSMENT OF VENTRICULAR FUNCTION

Quick assessment of LV global function is possible and of significance in critical care setup. "Eyeball" estimates of LV ejection fraction (EF) are effective and acceptable mode to estimate LV function. This is comparable to quantitative measures.[10] Moreover, at times quantitative method may not be truly depicting the picture. For example, in case of vasodilation, sepsis the EF may be overcalculated due to unloading of the left ventricle resulting from low systemic vascular resistance. Quantitative measurement of EF by Teichholz method is done in the PLAX view on M mode by measuring the LV internal diameters in systole and diastole which is just segmental representation of LV function and can be wrongly extrapolated for global function. Poor LV function may necessitate inotrope administration in the hypotensive patient. So, diagnose is of prime importance, once fluid resuscitation is optimized. It favorably resembles stroke volume measurements using a pulmonary artery catheterization. It must be understood that frequent sonography of heart is essential for assessing the progress of the patients with treatment in ICU.

The area of right ventricle is measured from apical four-chamber view. Right ventricular (RV) to LV ratio in diameter or area can effectively suggests dilatation of right ventricle. Normally it is less than 0.6, if it is 0.6–1.0 dilatation is considered to be moderate and critical if more than 1.0.[11] As with the LV, the RV function could be evaluated visually. The interventricular septum (IVS) should be evaluated critically as it gives many vital information pertaining to critical care. In case of acute pulmonary embolism (PE), ARDS, or high ventilation pressures, IVS deviates towards left and takes a "D" shape which could be well-appreciated in PLAX view. Here the loss of usual shapes of ventricles can well be appreciated (for detail see Chapter 7).

Pulmonary Embolism

Acute PE may be difficult to diagnose due to nonspecific clinical signs and may present with dramatic clinical deterioration; contrast-enhanced computed tomography (CECT) is a sensitive tool for PE, though at times it is very tedious process to transfer a sick patient from ICU

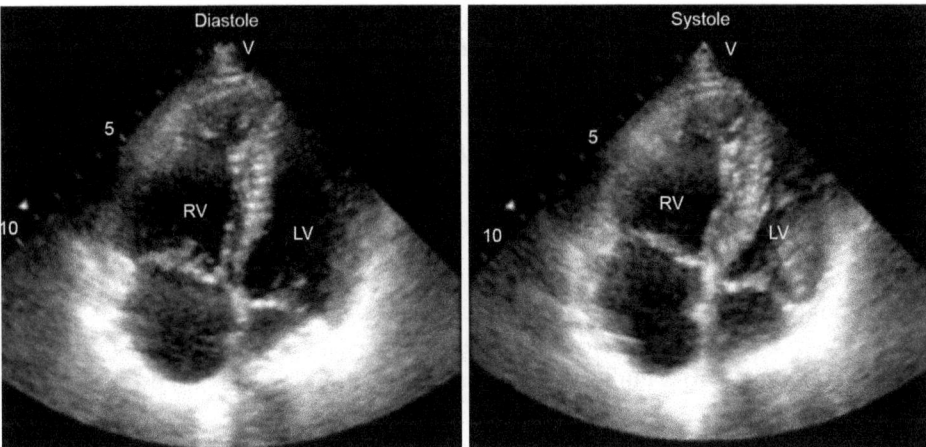

Fig. 10: Right ventricular dilation. The ratio of RV and LV could be objectively calculated and can be labeled as dilation and dysfunction (normal value is less than 0.6).
Abbreviations: LV: Left ventricular; RV: Right ventricular.

to the radiology department. Echocardiography is one of the best tools for bedside diagnosis of massive PE. Although not specific, RV dilatation and dysfunction are characteristic finding of acute PE. RV dysfunction, restricted to the midseptum with apical sparing is particularly seen in PE (Fig. 10). This is a differentiating point compared to other causes of RV dilatation and dysfunction (for detail *see* Chapter 7).[12]

Pericardial Effusion

It is defined as abnormal fluid in between visceral and parietal pericardial space. Transthoracic echocardiography helps in easy diagnosis, where the fluid is seen as black nonechogenic space around the heart. For the visualization of pericardial fluid, if one is not able to get the conventional windows, the subcostal view usually offers a good alternative. Cardiac tamponade, a life-threatening emergency has specific features seen on echocardiography. The clinical diagnosis of tamponade can be done on echocardiography (*see* Fig. 9E), which shows diastolic collapse of the right atrium (RA) and RV; in early diastole, there is collapse of RA, while later RV collapse occurs. Effusion is hemodynamically significant, if RA collapse is longer and objectively lasts for more than one-third of the R-R interval. After a diagnosis is made, pericardial effusion may be safely drained with sonographic guidance (for detail *see* Chapter 7).

MEASUREMENT OF CARDIAC OUTPUT

According to hydraulic formula, flow across an orifice is equal to the product of cross-sectional area and the velocity of flow across the orifice. Flow through the LVOT per heart beat (stroke volume) may be calculated using this formula. However, since velocity across the LVOT varies with time throughout systole, all the velocities need to be integrated as the velocity time integral (VTI), across the Doppler spectrum. Pulse wave Doppler is applied and the resulting waveform is traced with the sample volume placed at the LVOT on an apical five-chamber view. The built-in software package calculates VTI. The LVOT diameter is measured at the point of attachment of the aortic leaflets, on the PLAX view when the valve is fully open in systole. LVOT area = $\pi \times r^2$ or, $\pi \times (d/2)^2$, stroke volume = $\pi \times d2/4 \times VTI$, and cardiac output = stroke volume × heart rate, variation of VTI with respiration may be indicative of volume responsiveness and has been validated in patients undergoing coronary artery surgery, under closed chest conditions (for detail *see* Chapter 8).

MEASUREMENT OF CHAMBER PRESSURES

Pressure inside cardiac chambers is calculated with Doppler echocardiography. Using the Bernoulli equation, pressure gradients are calculated by measurement of velocity of flow: Pressure gradient = 4 (velocity)^2. To calculate pulmonary artery systolic pressure or RV pressure, the pressure gradient across the RV and the RA is measured from the tricuspid regurgitation jet. Pressure gradient across the tricuspid valve = RV systolic pressure − RA pressure. RV systolic pressure (equal to pulmonary artery systolic pressure) = pressure gradient + RA pressure [central venous pressure (CVP)]. Similarly, left atrium (LA) pressure can be measured from the mitral regurgitation jet. Pressure gradient across the mitral valve = LV systolic pressure − LA pressure. LA pressure = LV systolic pressure (systolic BP) − pressure gradient across mitral valve (for detail, *see* Chapter 7).

ASSESSMENT OF HYPOVOLEMIA

Diagnosing hypovolemia and timely intervention with intravenous fluid rush could be instrumental in lifesaving at the same time untitrated and overzealous fluid resuscitation can endanger life. Fluid resuscitation should be monitored and done according to "volume responsiveness" and hypovolemia assessment. Other static indicators of volume status are pulmonary capillary wedge pressures and CVPs. They do not give accurate assessment. Pulmonary capillary wedge pressures and CVPs do not give proper assessment. Variation in size of IVC throughout respiration can be assessed at patient's bedside (Fig. 11). The IVC is focused at the level of epigastrium in long axis and is assessed in M mode. In a subject with spontaneous respiration, an increase in IVC diameter of 2 cm is suggestive of high right-sided pressure. The increase is more in patients with spontaneous ventilation compared to mechanical ventilation. In case of patient of septic shock on mechanical ventilation, a variation in size of IVC of more than 12% is indicative of increased cardiac output with volume overload.[13] In septic shock patients, IVC diameter corresponds well with CVP, intrathoracic blood volume index, and extravascular lung water index which are measured by transpulmonary thermodilution technique. Close approximation of LV walls at the level of papillary muscle suggests significant hypovolemia (for detail *see* Chapter 8).

THE HYPOTENSIVE PATIENT

In the hemodynamically unstable patient a systematic ultrasonographic approach can help in diagnosis and management as outlined in Flowchart 2. Echocardiography is the first step, looking specifically for LV systolic function and cardiac tamponade. A quick look at the LV function, gives a good estimate of EF, comparable to other methods. Acute massive PE is suggested by dilatation of the RA and RV, in certain clinical context. If one suspects

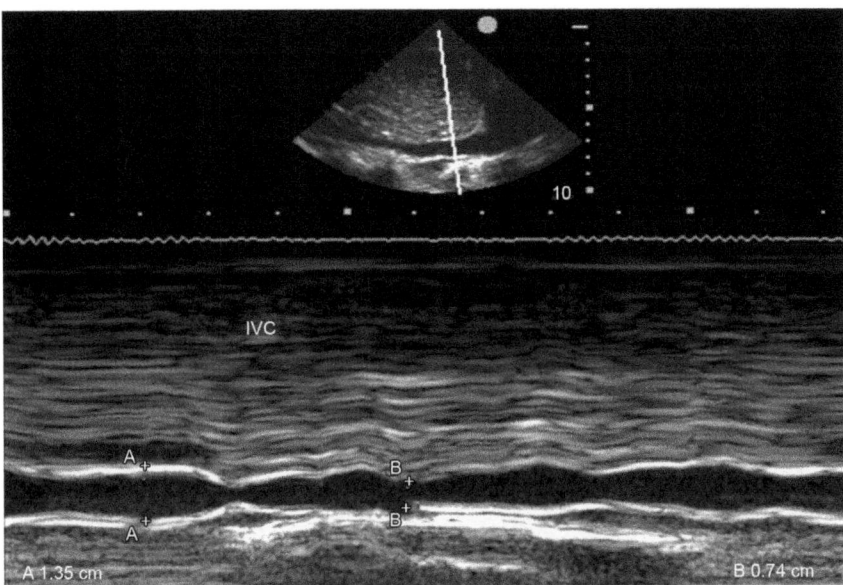

Fig. 11: Inferior vena cava (IVC) tracing in M mode showing marked variation of IVC diameter (A in expiration = 1.35 cm, and B in inspiration = 0.75 cm), the upper portion shows same in B mode.

hypovolemia, respiratory variation in IVC diameter should be assessed. Variation in IVC diameter with spontaneous breathing may be seen even in normovolemic patients; in these conditions for diagnosis, a passive leg raising test is done and the LV outflow tract VTI is measured before and after this test. In this test both the lower limbs are elevated to an angle of 45° from the semirecumbent position, for 2 minutes. If the difference in VTI is more than 12–15%, it indicates volume responsiveness. If in ultrasound of lung, intercostal space shows less than three B lines, it suggests that patient may be given a fluid challenge. If no variation in IVC diameter or VTI is seen in a patient with abnormal B lines (more than three per intercostal space), it is considered to be a case of excessive vasodilatation (for detail, see Chapter 8).

DEEP VEIN THROMBOSIS

A specific and sensitive test for the diagnosing lower limb deep vein thrombosis (DVTs) is compression sonography. A trained bedside clinician may perform modified two-point compression. Compression sonography of the lower limb is done using 5–12 MHz frequency linear probe. For diagnosis of DVT, the patient is placed supine, with the leg externally rotated, and the transducer is first placed distal to the inguinal ligament with pointer towards the patient's right. As the probe is slide up and down, one can identify common femoral vein and the saphenous vein junction. Just lateral to the vein, one can identify the common femoral artery (Fig. 12). If the vein compresses completely at the time when artery gets mildly deformed with the application of firm pressure, a clot can be ruled out. A clot within the vein is indicated by lack of compressibility. To examine the common femoral vein, the saphenofemoral junction is focused and a segment 2 cm above and below it is examined. Next, the common femoral

Flowchart 2: Approach to hypotensive patient.

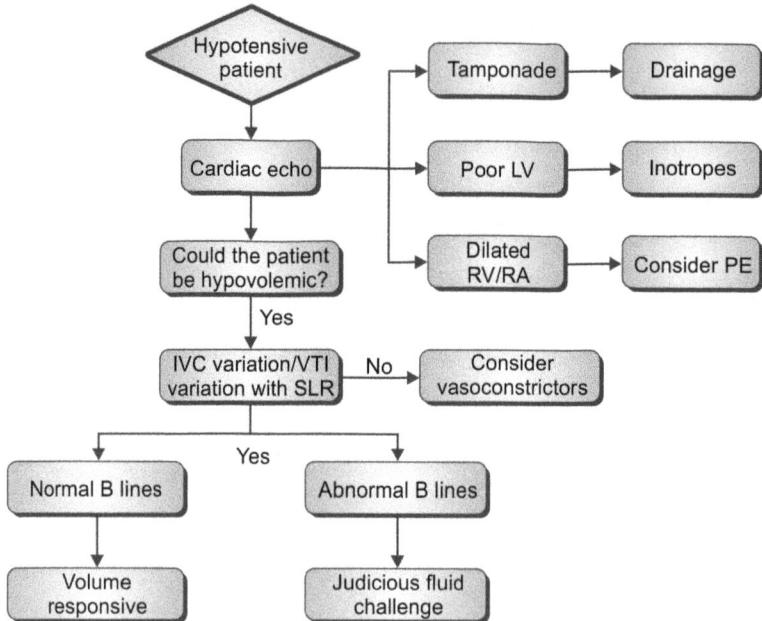

Abbreviations: IVC: Inferior vena cava; LV: Left ventricular; RA: Right atrium; RV: Right ventricular; VTI: Velocity time integral; PE: Pulmonary embolism; SLR: Straight leg raise.

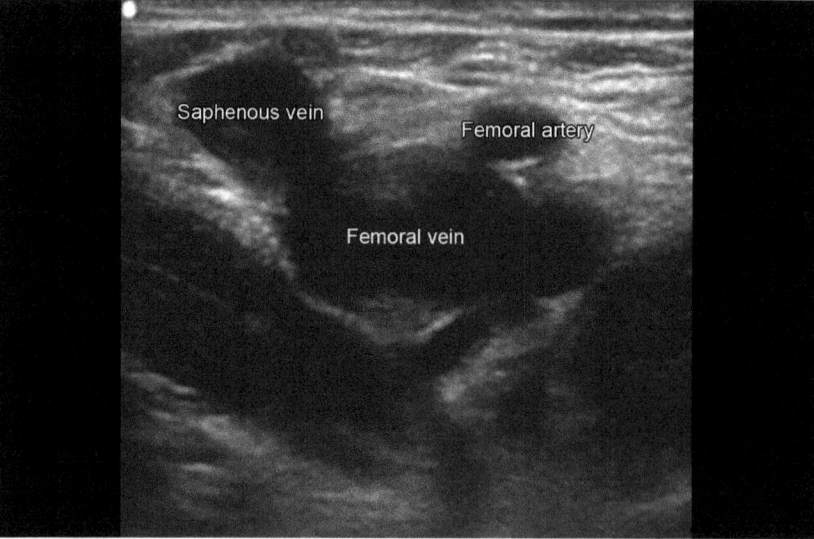

Fig. 12: Normal sonography of femoral and saphenous vein including femoral artery, appears as mickey mouse sign, with pressure on probe the collapsibility of femoral vein could be appreciated in case of deep vein thrombosis (DVT).

vein bifurcation into superficial and deep femoral veins is focused and compressed to exclude any clot. Popliteal vein (PV) is examined in a flexed knee with the probe in the popliteal fossa, and marker towards the right side of patient. Here by sliding the probe proximally and distally, one can identify the popliteal artery located anterior to the popliteal vein. A segment of 2 cm of the vein is examined until it divides into peroneal vein and the superficial and deep tibial veins. Transverse (short axis) view is the best for examination of veins. If a sagittal view is used, it increases the number of false-positive results as when pressure is applied with probe in sagittal view, it may easily slide off the vein. The clots, especially in acute cases, can be seen as hyperechoic areas in the lumen of the vein, but in few cases the only sign of a clot is absence of complete compressibility. Color flow Doppler can identify a clot as a filling defect. To aid in diagnosis, one can squeeze the calf which helps to increase the venous flow and thus fills the lumen which suggests the absence of a clot. The sensitivity of diagnosing DVT is not improved by the use of color Doppler. Tomkowski et al. studied the efficacy of compression sonography in diagnosing DVT compared to ascending venography in 160 acutely ill patients. They concluded that while the negative predictive value was high when compression sonography was used, i.e. 98% [95% confidence interval (CI) = 95-99%], the incidence of distal DVT was underestimated with a positive predictive value of 75% (CI = 30-95%)[14] (for detail, see Chapter 15C).

ABDOMINAL ULTRASOUND
Focused Assessment with Sonography in Trauma (Figs. 13A to H)

In cases of trauma causing blunt injury, hemoperitoneum needs to be identified. In spite of the fact that CT scan is very reliable, it carries the risk as the patient needs to be transferred from emergency to the CT room, which may be hazardous. Peritoneal lavage can be used for diagnosis in this condition; but it increases the incidence of negative laparotomies being done. Moreover, checking for ongoing bleed is another big challenge.

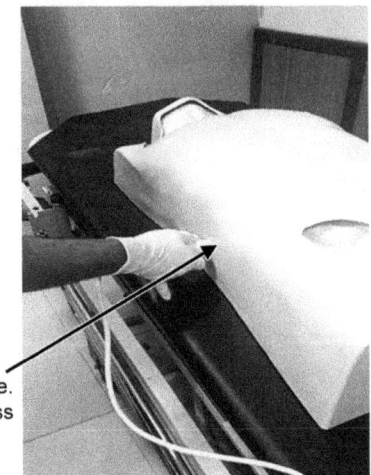

Probe is placed in coronal plane. Visualizing hepatorenal recess

Courtesy by: Kapil Dev Soni

Fig. 13A: Right upper quadrant view.

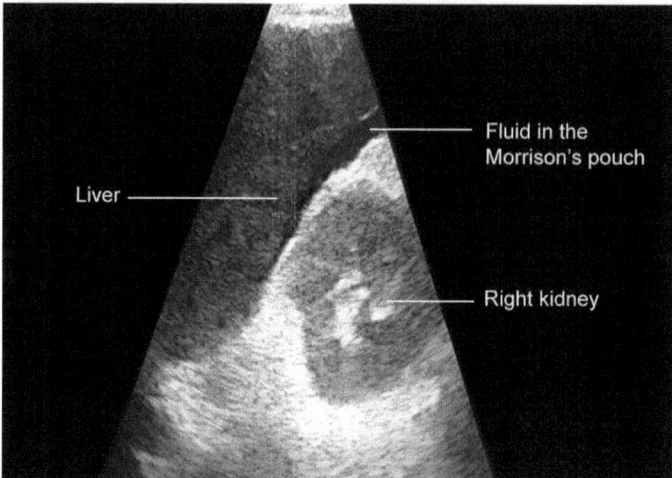

Fig. 13B: Fast positive in right upper quadrant view.

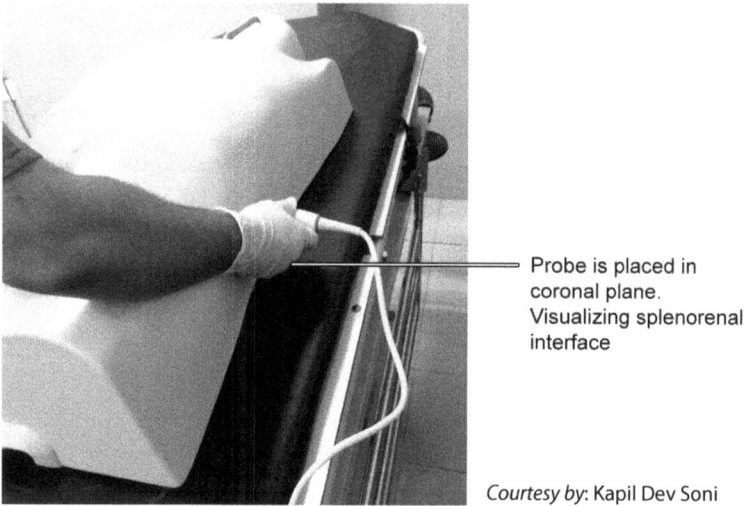

Fig. 13C: Left upper quadrant view.

Focused assessment with sonography in trauma can be used to diagnose free fluid in pericardial and peritoneal cavity, with the advantage that patient's position need not be changed. FAST examination is best done by 3.5–5 MHz curvilinear phased array probe. To examine the right upper quadrant, the probe needs to be placed along the thoracoabdominal line where we can look for free fluid in the Morrison's pouch at the interface of liver and the kidney. To examine the left upper quadrant, probe is placed along the posterior axillary line, lateral to the spleen where one can look for free fluid in the splenorenal pouch. In the pelvis, bladder

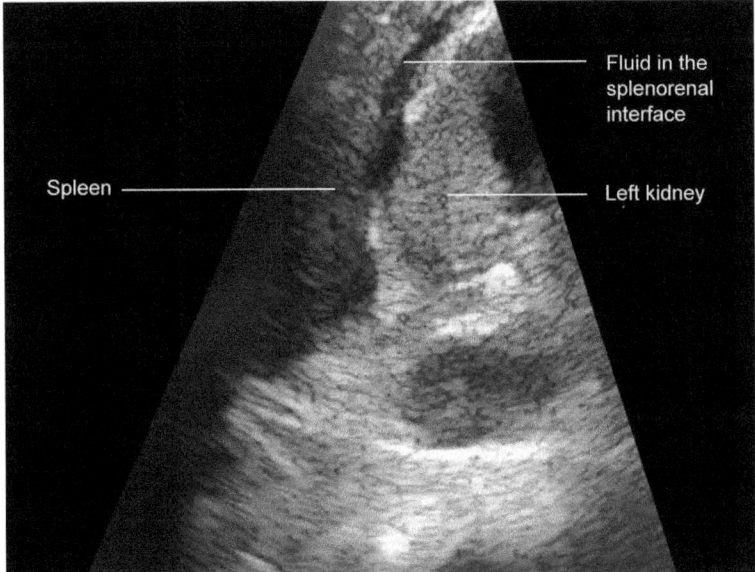

Fig. 13D: Fast positive in left upper quadrant view.

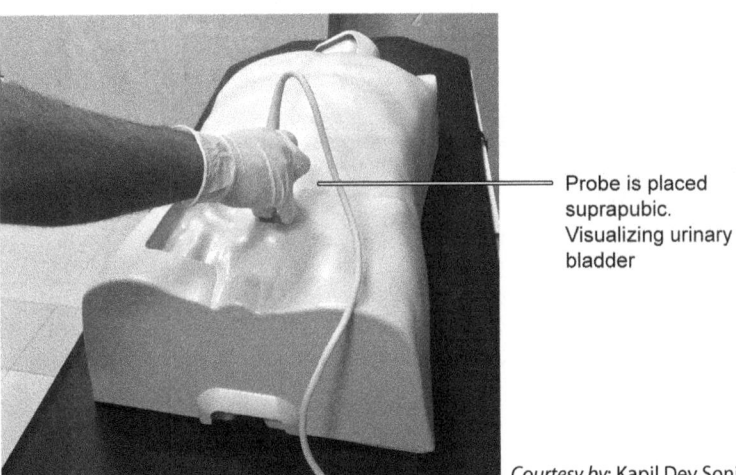

Fig. 13E: Suprapubic view.

can be examined by placing the probe just above the symphysis pubis transversely. If free fluid is seen in pouch of Douglas in females or rectovesical pouch in males, its volume is quantified by imaging of the right and left lower quadrants along with the paracolic gutter. An extended FAST examination of bilateral hemithorax can be used to look for hemothorax and the anterior chest can be viewed for pneumothorax. In unstable patients as an alternative to CT scans, FAST

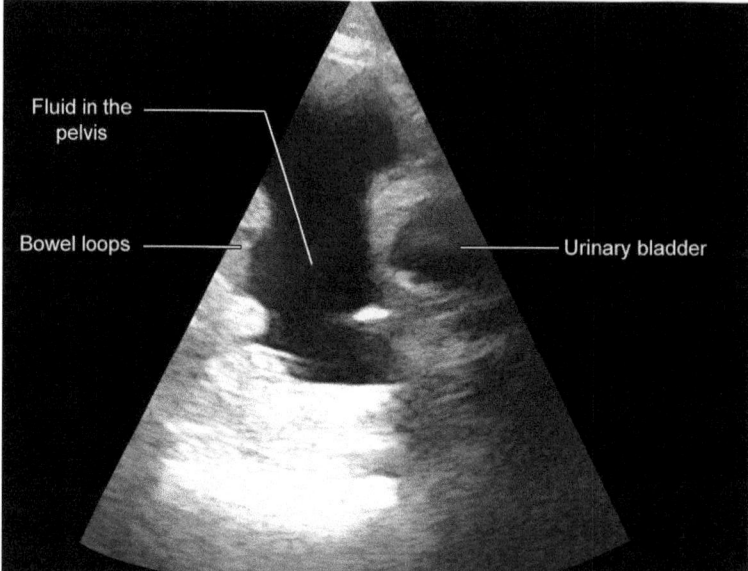

Fig. 13F: Fast positive in suprapubic view.

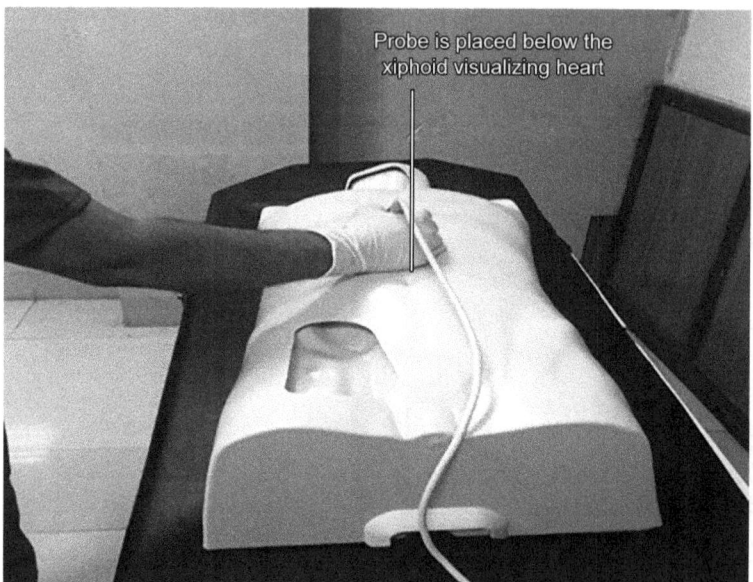

Fig. 13G: Subxiphoid view.

examination can be used as sensitive modality. It has specificity and accuracy of 98–100%. Apart from being quick and noninvasive, it is a good screening modality as patient does not need to be shifted and repeat examinations are feasible (for detail, *see* Chapter 12).

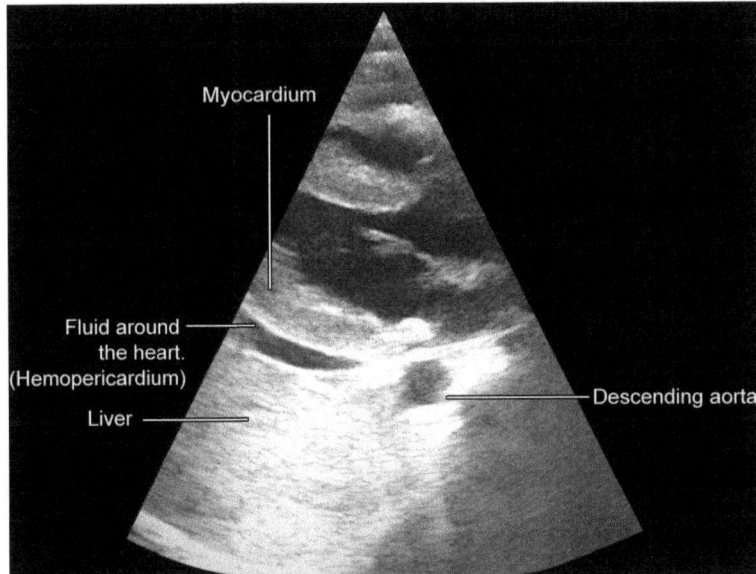

Courtesy by: Kapil Dev Soni

Fig. 13H: Fast positive in subxiphoid view.

REFERENCES

1. Rajajee V, Vanaman M, Fletcher JJ, et al. Optic nerve ultrasound for the detection of raised intracranial pressure. Neurocrit Care. 2011;15(3):506-15.
2. Brass P, Hellmich M, Kolodziej L, et al. Ultrasound guidance versus anatomical landmarks for subclavian or femoral vein catheterization. Cochrane Database Syst Rev. 2015;1:CD011447.
3. ICU Sonography. (2009). ICU Sonology tutorials. [online] Available from www.criticalecho.com/content/icu-echo-tutorials. [Accessed November, 2017].
4. Lichtenstein DA, Mezière GA. Relevance of lung ultrasound in the diagnosis of acute respiratory failure: the BLUE protocol. Chest. 2008;134(1):117-25.
5. Chacko J, Brar G. Bedside ultrasonography: Applications in critical care: Part I. Indian J Crit Care Med. 2014;18(5):301-9.
6. Balik M, Plasil P, Waldauf P, et al. Ultrasound estimation of volume of pleural fluid in mechanically ventilated patients. Intensive Care Med. 2006;32(2):318.
7. Lichtenstein DA, Lascols N, Mezière G, et al. Ultrasound diagnosis of alveolar consolidation in the critically ill. Intensive Care Med. 2004;30(2):276-81.
8. Stanko LK, Jacobsohn E, Tam JW, et al. (2005). Transthoracic Echocardiography: Impact on Diagnosis and Management in Tertiary Care Intensive Care Units. [online] Available from www.prismic-io.s3.amazonaws.com/hemodynamik%2F1a8d55d6-11ae-48aa-86f6-5923967ee912_transthoracic+echocardiography-+impact+on+diagnosis+and+management+in+tertiary+care+intensive+care+units.pdf. [Accessed November, 2017].
9. Jensen MB, Sloth E, Larsen KM, et al. Transthoracic echocardiography for cardiopulmonary monitoring in intensive care. Eur J Anaesthesiol. 2004;21(9):700-7.
10. Hope MD, de la Pena E, Yang PC, et al. A visual approach for the accurate determination of echocardiographic left ventricular ejection fraction by medical students. J Am Soc Echocardiogr. 2003;16(8):824-31.
11. Slama M, Maizel J. Echocardiographic measurement of ventricular function. Curr Opin Crit Care. 2006;12(3):241-8.

12. Kurzyna M, Torbicki A, Pruszczyk P, et al. Disturbed right ventricular ejection pattern as a new Doppler echocardiographic sign of acute pulmonary embolism. Am J Cardiol. 2002;90(5):507-11.
13. Feissel M, Michard F, Faller JP, et al. The respiratory variation in inferior vena cava diameter as a guide to fluid therapy. Intensive Care Med. 2004;30(9):1834-7.
14. Tomkowski WZ, Davidson BL, Wisniewska J, et al. Accuracy of compression ultrasound in screening for deep venous thrombosis in acutely ill medical patients. Thromb Haemost. 2007;97(2):191-4.

Chapter 1C

Should Intensivist do the Ultrasonography: Guidelines, Qualification, Experience and Training?

Shailaja Behera, Puneet Khanna

INTRODUCTION

At present ultrasound has become indispensable to the intensivist. It is portable [point-of-care (POC)], so critically ill-patient need not be shifted, helps in rapid diagnosis and appropriate management, and most importantly conducting procedures more safely. All over the world there are hundreds of workshops organized to train the intensivist and physicians taking care of the critically ill-patients. But critical care ultrasonography has a broad ambit. POC ultrasound has a definite role in the quick assessment of cardiac, pulmonary, hemodynamic, vascular, neurologic, and gastrointestinal status. Here we will enlighten about the guidelines for ultrasonography in critically ill-patients.

TRAINING AND ACCREDITATION

In 2008, the first document on training and accreditation of echocardiography in critical care was developed by an international group of experts and was published by the World Interactive Network Focused on Critical Ultrasound (WINFOCUS).[1]

Subsequently in 2009, American College of Chest Physicians (ACCP) and La Societe de Reanimation de Langue Francaise (SRLF) formed a working group published a consensus statement about competency for performing critical care ultrasonography.[2]

Subsequently in 2011, European Society of Intensive Care Medicine (ESICM) proposed training guidelines and standardization of competency assessment for critical care ultrasonography.[3] Followed by in the year 2014, the same group published consensus statement on the standards for advanced echocardiography.[4]

The guidelines as given by the ESICM expert group *(Consensus reached)*:
- Every critical care physician should have training in critical care echocardiography (CCE) and general critical care ultrasound (GCCUS) with optional component of advanced level echocardiography.

Recommendations for Training in General Critical Care Ultrasound and Basic Critical Care Echocardiography

Structure of the course should meet specific learning goals as described in the ACCP/SRLF competence statement.[2]

The American College of Chest Physicians/SRLF[2] recommendations in GCCUS includes the following ultrasonography:
- *Lung*
- *Pleural*
- *Vascular: Guidance of vascular access and diagnosis of venous thrombosis*
- *Abdominal.*

Basic CCE includes competence in assessment of:
- *Left ventricular size and regional wall motion abnormality*
- *Right ventricular size and contractility*
- *Pericardial fluid/tamponade*
- *Inferior vena cava size and respiratory variability for fluid responsiveness*
- *Severe valvular incompetence (color Doppler).*

Lectures and didactic cases with image-based training of least 10 hours *training,* each in GCCUS[5] and basic CCE[6,7] is *mandatory.*

A *blend of* internet learning and lecture *helps the trainees to acquire better skill in ultrasonography.*[1,8]

There was no consensus on the number of examinations to be performed by a trainee for competency especially in GCCUS but for basic CCE review of literature suggests that *at least* 30 fully supervised transthoracic echocardiogram (TTE) is *essential* for competency in image acquisition *in basic CCE*[7,9-11] *but for GCCUS no consensus has been reached in the number of examinations to be performed.*

Initial practical training should be done in healthy volunteers to teach standard views and spatial orientation of acquired images with different transducer manipulation.

A logbook containing the scanning activity should be maintained by a trainee.

A locally qualified physician who regularly performs GCCUS and basic CCE in intensive care unit (ICU) *should supervise the practical training of a trainee.*

Transesophageal echocardiography (TEE) can *be taught to trainee as an optional component* of basic CCE if resources are available.[12]

Certification in basic CCE and GCCUS is not mandatory *still few countries have it.*

Training in Advanced Critical Care Echocardiography

Intensivists must be trained in basic CCE before going for advanced CCE.

A structured course as described in ACCP/SRLF competence statement,[2] *with at least* 40 hours *divided between didactic cases with image-based training and lectures is required for teaching.*

As for basic CCE and GCCUS both internet-based learning and formal lectures should be used.

Transesophageal echocardiography competency is must for advanced CCE. *An advanced trainee requires around* 150 *fully supervised TTE studies and* 50 *independent TEE studies for competency in image acquisition and interpretation.*[13,14]

CONCLUSION

Care of critically ill-patients is unthinkable without knowledge of use of ultrasound. So competence in basic CCE and GCCUS is must for intensivists. It is the responsibility of international and national critical care societies to provide a pragmatic approach to train the intensivists in critical care ultrasonography for better care of seriously ill-patients.

REFERENCES

1. Price S, Via G, Sloth E, et al. Echocardiography practice, training and accreditation in the intensive care: document for the World Interactive Network Focused on Critical Ultrasound (WINFOCUS). Cardiovasc Ultrasound. 2008;6:49.
2. Mayo PH, Beaulieu Y, Doelken P, et al. American College of Chest Physicians/La Société de Réanimation de Langue Française statement on competence in critical care ultrasonography. Chest. 2009;135(4):1050-60.
3. Expert Round Table on Ultrasound in ICU. International expert statement on training standards for critical care ultrasonography. Intensive Care Med. 2011;37(7):1077-83.
4. Expert Round Table on Echocardiography in ICU. International consensus statement on training standards for advanced critical care echocardiography. Intensive Care Med. 2014;40(5): 654-66.
5. Chalumeau-Lemoine L, Baudel JL, Das V, et al. Results of short-term training of naive physicians in focused general ultrasonography in an intensive-care unit. Intensive Care Med. 2009;35(10): 1767-71.
6. Manasia AR, Nagaraj HM, Kodali RB, et al. Feasibility and potential clinical utility of goal-directed transthoracic echocardiography performed by noncardiologist intensivists using a small hand-carried device (Sono Heart) in critically ill-patients. J Cardiothorac Vasc Anesth. 2005;19(2):155-9.
7. Vignon P, Dugard A, Abraham J, et al. Focused training for goal-oriented hand-held echocardiography performed by noncardiologist residents in the intensive care unit. Intensive Care Med. 2007;33(10):1795-9.
8. Breitkreutz R, Uddin S, Steiger H, et al. Focused echocardiography entry level: new concept of a 1-day training course. Minerva Anestesiol. 2009;75(5):285-92.
9. Benjamin E, Griffin K, Leibowitz AB, et al. Goal-directed transesophageal echocardiography performed by intensivists to assess left ventricular function: comparison with pulmonary artery catheterization. J Cardiothorac Vasc Anesth. 1998;12(1):10-5.
10. Melamed R, Sprenkle MD, Ulstad VK, et al. Assessment of left ventricular function by intensivists using hand-held echocardiography. Chest. 2009;135(6):1416-20.
11. Vignon P, Mücke F, Bellec F, et al. Basic critical care echocardiography: validation of a curriculum dedicated to noncardiologist residents. Crit Care Med. 2011;39(4):636-42.
12. Charron C, Prat G, Caille V, et al. Validation of a skills assessment scoring system for transesophageal echocardiographic monitoring of hemodynamics. Intensive Care Med. 2007;33(10):1712-8.
13. Ryan T, Armstrong WF, Khandheria BK, et al. Task force 4: training in echocardiography endorsed by the American Society of Echocardiography. J Am Coll Cardiol. 2008;51(3):361-7.
14. Quiñones MA, Douglas PS, Foster E, et al. ACC/AHA clinical competence statement on echocardiography: a report of the American College of Cardiology/American Heart Association/American College of Physicians-American Society of Internal Medicine Task Force on Clinical Competence. J Am Coll Cardiol. 2003;41(4):687-708.

2A

Basic Physics of Ultrasound and the Doppler Phenomenon

Devasenathipathy Kandasamy, Ananya Panda

OVERVIEW

- Introduction to ultrasonography (USG)
- Properties of USG including propagation, reflection, refraction, and attenuation of sound waves
- Basic ultrasound instrumentation
- Safety considerations.

INTRODUCTION

Ultrasound refers to sound waves above the range of audible frequency, namely above 20 kHz. As compared to audible sound waves, which spread across the room, ultrasound waves have a shorter wavelength and can be formed into a narrow beam and used for imaging. This chapter covers the properties of ultrasound waves, basic ultrasound instrumentation, and certain important safety considerations.

PROPERTIES OF ULTRASOUND

Ultrasound is mechanical energy that propagates through matter as a pressure wave producing alternating compression and rarefaction. Thus, unlike light and X-rays, ultrasound requires a material medium for propagation. The velocity by which ultrasound propagates through a particular medium is called *propagation velocity* which in a given material, for imaging purposes, is a constant, independent of the frequency or wavelength of the sound.

The propagation velocity depends on the material through which sound travels. The greater the density, lower the velocity. The greater the compressibility of the medium, the lower is the velocity of sound. The velocity also depends on the temperature of the tissue, which for practical purposes is considered constant. The typical figures of sound in different materials in the body are given in Table 1. Accordingly, air has a lower density but is more compressible than soft tissue or water; hence sound has a lower propagation velocity in air.

The knowledge of propagation velocity is important in distance or depth measurement in ultrasound. Assuming that the velocity (v) through which sound travels through soft tissue is known, i.e. 1,540 m/sec and the time (t) in which sound travels to the tissue at depth (d) and its echo returns back to the transducer is known, then depth (d) can be calculated as $d = v \times t/2$. The time is halved as it includes time to both reach the tissue at depth d and return back to the transducer.

Acoustic impedance is another important concept in ultrasound imaging, which determines the behavior of sound waves when it encounters interfaces between different tissues. As sound waves encounter an interface, part of it is reflected and part of it is transmitted. The proportion of sound waves reflected back versus transmitted forward depends on the difference in acoustic impedance between the two tissues. Acoustic impedance (Z) is the product of the density (ρ) of the material and the velocity (v) of sound in it, thus $Z = \rho v$. The greater the difference in acoustic impedance, greater is the reflected fraction. The lesser is the difference in acoustic impedance, greater is the fraction transmitted. If the acoustic impedances of two tissues, Z_1 and Z_2 are same, there is 100% transmission and no reflection and the two mediums are said to be acoustically matched. Some examples of acoustic impedances of different human tissues are given in Table 1 and the typical reflection factors are given in Table 2.

For example, at an interface between bone ($Z = 5 \times 10^6$) and soft tissue ($Z = 1.5 \times 10^6$), the reflected fraction is 30% and 70% of energy is transmitted. However, it is generally not possible to image through bone due to another property called attenuation. At an interface with air/gas, nearly 100% of sound is reflected. Thus gas-filled organs cast a shadow and organs underneath cannot be imaged. Neither can normal lung be imaged nor can gas-filled bowel lumen. Similarly, if there is air between ultrasound transducer probe and human body, sound cannot be transmitted, hence we apply a coupling jelly between the transducer probe and the patient's body.

Table 1: Velocity of sound and acoustic impedance of different materials.

Materials	Velocity (ms^{-1})	Acoustic impedance (Z) (kgm^{-2}s^{-1})
Air	330	430
Fat	1,450	1.38×10^6
Water	1,480	1.48×10^6
Average soft tissue	1,540	1.50×10^6
Liver	1,550	1.64×10^6
Kidney	1,560	1.62×10^6
Muscle	1,580	1.70×10^6
Bone	3,200–4,000	5.3×10^6
Lead zirconate titanate (transducer material)	4,000	30×10^6

Table 2: Typical reflection factors.

Interface	Reflection percentage
Air tissue	99.9
Soft tissue lead zirconate titanate (PZT)	80
Bone muscle	30
Fat muscle	1
Blood muscle	0.1
Liver muscle	0.01

Reflection factor (R) given by $R = (Z_1-Z_2)^2/(Z_1-Z_2)^2$

Reflection

Apart from differences in acoustic impedances, the way ultrasound is reflected also depends on the size and surface features of the reflecting interface and the angle of beam insonation. If the reflecting surface is large and smooth, it acts as a mirror and is called a *specular reflector*. Some examples of specular reflectors include diaphragm, wall of urine-filled bladder, and endometrial stripe. Also, because scanners will detect only those reflected waves that return back to the transducer, a specular reflector will return only those echoes which insonate perpendicular to its surface. Sound waves which insonate at an angle less than 90° will be reflected away from the transducer and will not be detected.

Most of the reflection that occurs in the body is due to rough surfaces and smaller interfaces within solid organs. These interfaces within an organ have sizes much smaller than the wavelength of incident sound and echoes from these interfaces are scattered in all directions. This is termed as *diffuse reflection*. The constructive and destructive effects of these widely scattered sound waves are responsible for producing ultrasound *speckle* or the characteristic *echotexture* of various solid organs.

Refraction

When sound travels from one medium with certain propagation velocity (c_1) into another medium with a different propagation velocity (c_2), it undergoes a change of direction and this is governed by Snell's law: $\sin \theta_1 / \sin \theta_2 = c_1/c_2$ where θ_1 and θ_2 are angles of incidence and refraction, respectively. Refraction is an important cause of misregistration artifact as an object that may be shown at a different position from where it actually is, because the transducer assumes that sound waves travel in the same direction and the echoes originate from the same straight path of travel. This can be rectified by moving or changing the position of the probe to increase the insonation angle, such that waves strike more perpendicularly to the reflecting surface.

Attenuation

As sound waves travel through a particular medium, it undergoes a loss in its intensity and is attenuated exponentially with the depth of travel. This *attenuation* occurs due to a combination of *absorption* (and conversion to heat due to frictional/viscous forces in the medium), *reflection*, and *scatter*. The attenuation is measured in decibels (dB) as it is a logarithmic scale and allows a wide range of intensity ratios to be expressed in a compact manner. The attenuation of sound is important as it determines the depth of tissue that can be imaged and affects transducer selection and image optimization. Attenuation is directly proportional to frequency of sound and thus higher frequency transducers (linear transducers) are used to image more superficial structures (e.g. neck/breast) while lower frequency transducers (curved transducers) are used to image deeper structures (e.g. abdomen/pelvis). The attenuation is also very high in bone (15 dB) and air (40 dB) as compared to average soft tissue (0.7 dB), hence ultrasound is not used to image these structures. On other hand, attenuation through water is very low and hence water-filled structures can be used to enhance ultrasound penetration, e.g. using a urine-filled bladder to visualize uterus.

BASIC ULTRASOUND INSTRUMENTATION

The basic components of ultrasound scanners include: (a) Transmitter or pulser to energize the transducer, (b) Ultrasound transducer or the probe, (c) Receiver and processor to detect and amplify the backscattered energy and prepare them for display, (d) Display that shows ultrasound image in a format that can be recognized and interpreted, and (e) Picture archiving and communications systems (PACS). These components are briefly described as follows:

- *Transmitter/pulser:* This energizes the transducer with short bursts of high-voltage pulse and this is called pulsed ultrasound. Pulsed ultrasound is routine practice in most of clinical imaging. The transmitter also controls the time interval between two pulses and this is called pulse repetition frequency (PRF). The PRF determines the depth of tissue from which unambiguous data can be collected. Routine diagnostic imaging usually uses PRF between 1 kHz and 10 kHz.
- *Transducer:* Ultrasound transducers basically convert electrical energy to mechanical energy and vice versa. Transducers are able to do this because they are made of piezoelectric materials, which start vibrating when stimulated by an electric pulse. They are also capable of generating electric voltage when they are struck with returning sound waves. Thus, after being energized by pulsed electrical voltage from the transmitter/pulser, the electrical energy is converted to mechanical/sound energy and on return of the sound waves to the transducer, the transducer reconverts the received sound/mechanical energy to electrical energy, which is used for image formation. Common piezoelectric materials used are plastic polyvinylidene difluoride (PVDF) and lead zirconate titanate (PZT). However, these substances can lose their piezoelectric effect when heated beyond a certain temperature called Curie temperature (350°C for PZT). Hence, transducer probes should not be autoclaved or immersed in boiling water.

When energized by pulsed electrical voltage, transducers produce a spectrum of sound waves having a range of frequencies. The range of frequencies is called bandwidth. Shorter is the pulse of ultrasound produced by the transducer, greater is the bandwidth. Broad-bandwidth imaging is preferred as it also increases the bandwidth of frequencies received back by the transducer and thus decreases artifacts.

Modern transducers are composed of arrays of thin piezoelectric elements, which are sequentially fired to produce the beam and steer the beam in different directions for real-time imaging. These arrays can have various configurations, such as linear, curved, sector, and annular. *Linear transducers* are high frequency transducers with frequency range of 7–14 MHz. These are used for imaging superficial structures, such as breast, neck, scrotum, vascular structures, and for pediatric abdomen. *Curvilinear* and *sector transducers* are lower frequency transducers between 3 MHz and 5 MHz. Curvilinear probes are most commonly used probes for adult abdominal imaging and obstetric imaging. Sector probes have a small footprint and can be used to scan through small acoustic windows, such as intercostal spaces and cranium through fontanelle.

- *Receiver:* This detects and amplifies the weak electric voltages produced by the transducer face when struck by the returning sound waves. The receiver preferentially amplifies echoes returning from deeper tissues as they are more attenuated, to produce an image of uniform brightness and this is called *time-gain compensation* (TGC). While automatic

TGC settings are given for various applications in modern scanners, this can be also adjusted manually to optimize the displayed image.

The receiver also performs a *dynamic range compensation* to ensure optimum gray-scale image display. The returning echoes can have a wide range of amplitude, up to factor of $1:10^{12}$, i.e. 120 dB. However, since the entire range of amplitudes cannot be displayed on the ultrasound screen, the receiver compresses and remaps the data to a range to 35–40 dB to ensure that the image can be displayed on the screen. However, it is recommended to have as wide as a range as possible for a given scanner, to determine subtle differences in echogenicity between various tissues.

- *Image display system:* Currently, ultrasound is predominantly used as a real-time brightness or B-mode scanning, wherein images are displayed as gray scale on a screen and the brightness of the image depends on the intensity of returning waves. Thus, on a black background, signals with the maximum intensity are displayed white, absent signals are shown as black, and signals of intermediate intensity are shown as varying shades of gray. Current image displays can have 2^8, i.e. 256 shades of gray for every pixel. These images shown on the screen can be further saved, archived, and printed or kept in digital storage in the PACS system.

Other modes of imaging include A-mode or amplitude mode and M-mode or motion mode, all of which have been described in more detail in the next chapter.

SAFETY CONSIDERATIONS

While ultrasound wave is nonionizing and is not associated with radiation concerns of X-ray and computed tomography (CT) scan, it has its own set of safety considerations, related to the intensity or the power of the sound waves. While diagnostic ultrasound is considered safe, it is recommended that the time—averaged intensity of ultrasound beam should not exceed 100 mW/cm^2 and total sound energy should not exceed 50 J/cm^2. This is to avoid potential risks of local heating, tissue cavitation, acoustic streaming of cellular contents in the direction of beam affecting cell membrane permeability, and mechanical damage to cell membranes. Two indices are used as quality control checks. Thermal index (TI) is the ratio of power required to that emitted to raise the temperature of the body by 1°C. Mechanical index (MI) is a measure of maximum amplitude of the pressure wave and is defined as peak rarefaction pressure divided by square root of ultrasound frequency. Both these indices are displayed on the display screen and should be generally kept below 0.5. If TI increases above 1, then exposure time should be reduced. This is an important consideration which should be taken into account.

FURTHER READING

1. Christopher RB. Physics of ultrasound. In: Rumack CM, Wilson SR, Charboneau JW, Levine D (Eds). Diagnostic Ultrasound, 4th edition. Philadelphia: Elsevier; 2010. pp. 2-33.
2. Fowkles JB, Holland CK. Biologic effects and safety. In: Rumack CM, Wilson SR, Charboneau JW, Levine D (Eds). Diagnostic Ultrasound, 4th edition. Philadelphia: Elsevier; 2010. pp. 34-52.
3. Williams J. Imaging with ultrasound. In: Allisy-Roberts P, Williams J (Eds). Farr's Physics for Medical Imaging, 2nd edition. Philadelphia: Elsevier; 2007. pp. 147-68.

Chapter 2B

Modes of Ultrasonography

Devasenathipathy Kandasamy, Ananya Panda

OVERVIEW

- Modes of ultrasonography (USG) (A-mode, B-mode, and M-mode)
- *B-mode imaging:* Image acquisition and optimization
- Special imaging modes
- Artifacts.

INTRODUCTION

Ultrasound imaging involves the use of sound waves above the frequency of 20 KHz for imaging. Typical frequency ranges of ultrasound used in diagnostic imaging are 3–15 MHz. As briefly mentioned in Chapter 1, there are three modes of ultrasound imaging, namely A-mode, B-mode, and M-mode, of which B-mode is the most commonly used mode of imaging. This chapter discusses the three modes in more detail with chief emphasis on B-mode imaging, real-time imaging, and image optimization. The last section in this chapter briefly discusses ultrasound image artifacts.

A-MODE (AMPLITUDE-MODE) IMAGING

The simplest and the oldest mode of imaging, which is now practically limited to examination of eye. The A-mode is illustrated in Figures 1A to C. The probe is held stationary against the patient (Fig. 1A) and is pulsed with a voltage of few hundred volts lasting for few nanoseconds. Ultrasound pulse travels at a different speed in the body and it takes time (t) to reach interface a (Fig. 1B). Some of the sound energy is reflected back from a, along the path as an echo pulse which further takes time (t) to return back to the probe. The probe now acts as receiver, generates a small signal voltage that after amplification, produces a short vertical trace proportional to the echo strength (Fig. 1B). The other interfaces b and c also produce blips at B and C, respectively. The locations of the blips along the trace denote the *depths* of the corresponding interfaces along the beam. A time-calibrated scale can be superimposed on the trace with a sequence of marker corresponding to 1 cm depth in the body. Since sound gets attenuated with distance, the echoes returning from the deeper parts of the body are weaker. Thus, an interface located deeper produces a weaker signal than an exactly *identical* interface located more superficially. The receiver thereby performs time-gain compensation (TGC) to preferentially amplify echoes returning later from deeper structures, so that *all identical interfaces located at different depths produce similar signal* (Fig. 1C).

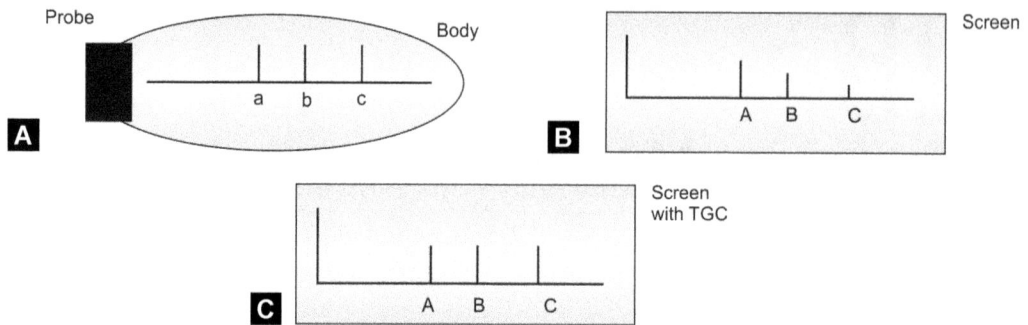

Figs. 1A to C: A-mode imaging. (A) Section through transducer and patient; (B) Trace on screen without time-gain compensation (TGC); and (C) Trace on screen with TGC.

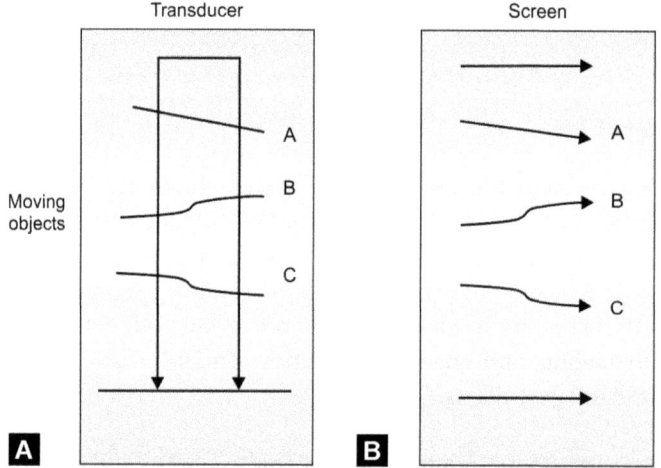

Figs. 2A and B: M-mode imaging. (A) Section through transducer patient; and (B) Appearance on monitor screen.

M-MODE (MOTION-MODE) IMAGING

This is used for imaging of rapidly moving structures, such as cardiac valves, cardiac walls, fetal heart beats, and also for detecting pneumothorax. In this mode the gray scale image is frozen on the screen and is used to direct the beam from the transducer along a line of interest. Echoes from the moving structures are displayed as a horizontal moving line on the screen with brightness of the line proportional to the intensity of the reflected sound wave and interspacing between the lines depending upon the relative depth between these structures along the line of interest (Figs. 2 and 3). Quantitative analyses, such as calculation of heart rate and flow across the cardiac valve are also possible in this mode.

B-MODE (BRIGHTNESS-MODE) IMAGING

This is the commonly used mode in day-to-day practice. In this mode, the gray scale image is actually made up of hundreds of tiny dots, each dot whose brightness is proportional to

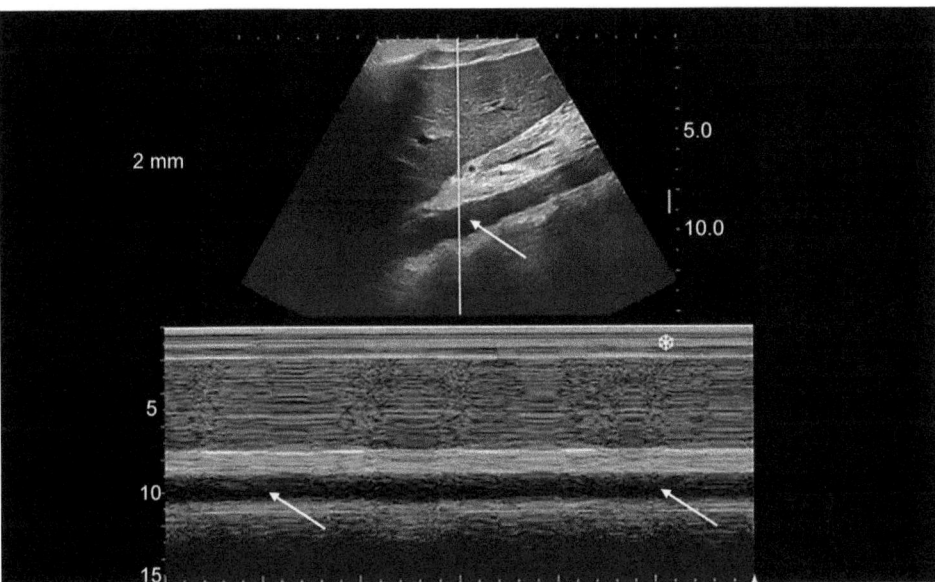

Fig. 3: M-mode imaging of inferior vena cava (arrows) shows lines of varying gray scale intensities and depths along the vertical line placed on the B-mode image.

the intensity of the sound wave, and which combine together to form one image. The basic steps in image formation are described as follows.

The transducer is pulsed at regular short intervals and the ultrasound pencil beam so formed scans back and forth across a two-dimensional section of the patient in either a linear, curvilinear, or sector pattern depending upon the shape of the transducer selected. Figures 4A and B show a simplified version of the scan lines traveled by each ultrasound pulse to produce the image. The points of interest that lies approximately perpendicular to the sound beam produce echoes, which are received by the transducer to form the image. The returning echoes are displayed on the screen as tiny bright dots, with brightness corresponding to the amplitude of the signal at each of the interfaces encountered. Each time the transducer is pulsed, the ultrasound beam takes a new scan line through the patient and the trace starts at the point on the screen corresponding to the skin and travels in the same direction as the ultrasound wave to produce the image.

Real-time imaging is possible because the image is automatically scanned in a succession of frames rapid enough to demonstrate motion of tissues and scan through organs. The rapid scanning can be achieved by two ways, by *mechanical or electronic scanning*. Of the two, mechanical scanners are older, cheaper, and more prone to breakdown. Most modern scanners currently use electronic scanning to achieve rapid real-time scanning.

In *mechanical scanning*, the transducer is either mechanically oscillated or if there is an array of transducers, the transducers are rotated rapidly and the one nearest to the body surface is energized. In either case, the ultrasound beam sweeps through a section of the body, each sweep producing one image frame. The speed of oscillation or rotation can be varied to increase frame rate rapid enough to capture images in real time.

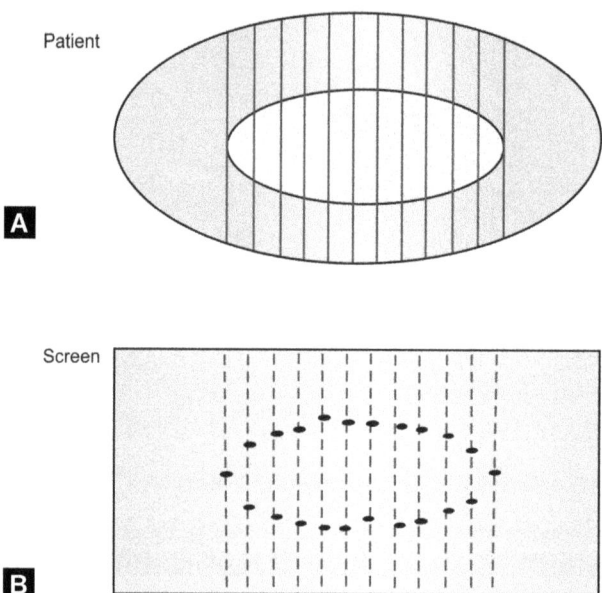

Figs. 4A and B: B-mode imaging. (A) Linear scan through patient; and (B) Appearance on monitor screen.

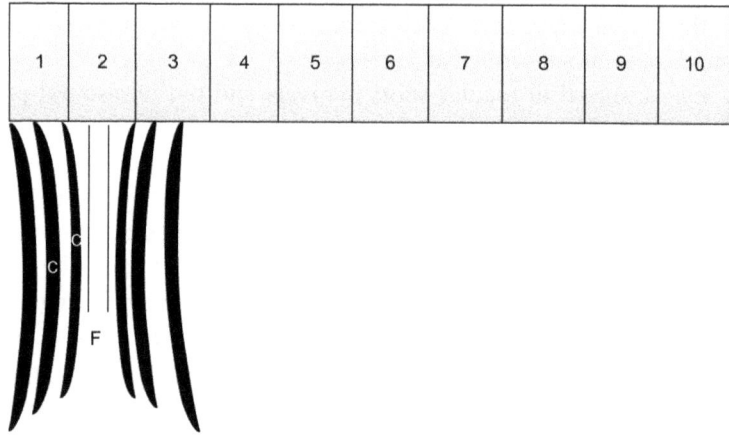

Fig. 5: Cross-section through linear transducer array and ultrasound beam with electronic focusing at focal point F.

In electronic scanning, the transducer is made up of large number (approximately 128) of separate narrow piezoelectric strips, each about a wavelength in width. These are energized in overlapping groups, so that a well-defined ultrasound beam is formed with a focus point F (Fig. 5). The timing of the applied pulses can be changed to change the focal depth F. The greater the time delays between energizing subsequent pulses, shorter is focal length. Modern scanners also provide option of multiple zone focusing.

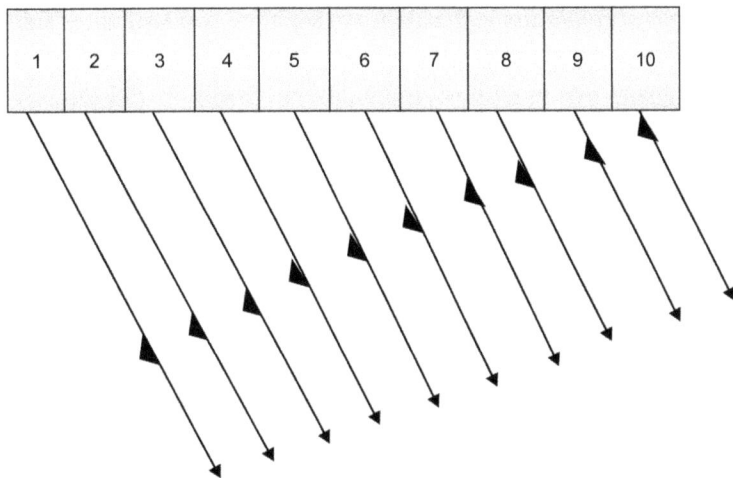

Fig. 6: Steered beam produced by phased array with beam reinforcement in one direction.

The direction of beam can also be steered electronically. For example, if all the elements are energized at the same time, the beam travels forward. If they are energized one after the other in a sequence, they reinforce in one direction and the beam swings to that direction (Fig. 6). Thus, by slightly changing the time delays between subsequent pulses, the entire beam can be made to sweep across a section of the patient.

A few factors determine the effectiveness and quality of real-time imaging. These are as follows:
- *Scan line density:* This is the number of scan lines that pass from the transducer to the patient. These depend on the number of piezoelectric elements in the transducer. To obtain high-resolution images, each frame should be made up of sufficient scan lines, usually 100 lines/frame would suffice.
- *Frame rate:* To get the effect of real-time imaging, large number of frames should be scanned per second. The frame rate depends on number of lines per frame and the pulse repetition frequency (PRF). Thus, *frame rate × lines/frame = PRF*. For example, at 30 frames/second and scan line density of 100 lines/frame, the PRF is 3 kHz.
- *Depth of view:* Each pulse should have sufficient time to make the return journey from the deepest structure before the next pulse is generated. Decreasing the PRF increases the depth of view. Thus, it is not possible to have high scan line density, and high PRF and also produce image with large depth of view. As *depth of view × (scan lines/frame) × (frame rate) is a constant,* there is a trade-off between PRF and depth of view.

IMAGE RESOLUTION

Image quality depends on spatial resolution, temporal resolution, contrast, and absence of artifacts. *Spatial resolution* has three components. *Axial or depth resolution* depends on the ability to separate two objects along the same scan line. The axial resolution is half the pulse length. The shorter the pulse length or higher the frequency of the transducer, better is the axial resolution. *Lateral resolution* is the ability to separate two structures side by side at the same depth. This depends on beam width being narrower than the gap between two objects. Since the beam is narrowest at its focal point, lateral resolution is best at the focus point. Lateral resolution is increased by using a smaller transducer and by beam focusing. *Azimuthal or elevation resolution* is the resolution in a plane perpendicular to the transducer or ultrasound beam and determines slice thickness. Greater the slice thickness worse is the resolution in the elevation plane. Axial resolution is always better and greater than the lateral and elevation resolution.

Temporal resolution is the ability to rapidly scan moving structures and increases with frame rate and PRF. *Contrast resolution* is the ability to differentiate subtle differences in echogenicity and can be improved by adjusting dynamic range and using special modes, such as tissue harmonic imaging (THI) and spatial compound imaging. Freedom from artifacts can be achieved by knowing how and when to adjust settings and also on the operator's skill and technique.

IMAGE OPTIMIZATION
Basics of Setting up Images

The patient should be comfortable. The body part to be scanned should be exposed fully and optimally positioned. For example, extending the neck for thyroid scanning and placing a pillow below the shoulders for patient comfort. Enough coupling jelly should be used to ensure transmission of ultrasound waves to and fro from the transducer to the body. Advantage should be taken of optimum acoustic windows whenever possible, for example, using a full bladder to scan the pelvic structures or using stomach distended with water to visualize the pancreas.

Optimal transducer selection is important. As described earlier, linear high frequency transducers are useful for superficial structures, vascular, musculoskeletal applications, and pediatric patients. While the resolution is high the depth of penetration is limited. Curvilinear transducers have lower frequency but greater depth of penetration and are used for general purpose imaging while sector transducers have smaller footprint and are useful when there is a small window to place the transducer, such as intercostal spaces or fontanelle.

It is recommended to use the various factory presets given in a scanner as various ultrasound settings that are already optimized for the particular region of interest and images can further be optimized by manually adjusting settings, such as overall gain and TGC and by avoiding artifacts.

The depth should be adjusted, such that the screen or the field of view (FOV) covers just enough of the area of interest. For example, if scanning the breast, adjust the depth to just show the breast tissue, underlying pectoralis muscles, and cross-section of the ribs. The lungs should not be included in the FOV for breast examination (Figs. 7A and B). Similarly, the focus zone should be adjusted such that the beam is focused at the depth/zone of interest. For example, when visualizing gallbladder for gallstones, the focal point should be adjusted to the gallbladder to enable detection of tiny stones.

The overall gain or brightness should be adjusted, so that the image is neither too bright (or white) nor too dark (or black). Further optimization of image is possible by using TGC, so that

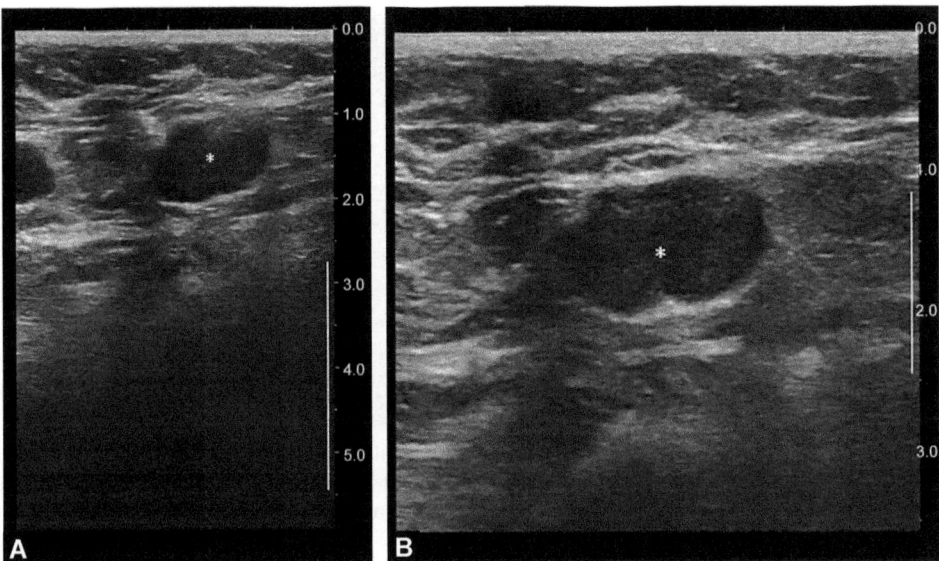

Figs. 7A and B: B-mode ultrasound image. (A) Ultrasound image of breast shows a fibroadenoma (*) in the upper part of the image. The lower part of the image is showing the chest wall and underlying lung which is not adding any information rather reduces the image resolution in the area of interest; and (B) The ultrasound image is obtained after depth correction which shows the area of interest better.

a smooth gray scale image with uniform brightness throughout is obtained on the screen. TGC can either be set to auto or adjusted manually with the array of knobs provided on the setting panel.

Apart from adjusting gain or brightness, contrast can also be adjusted by adjusting the dynamic range. Thus, for a given scanner, the dynamic range should be as wide as possible, to appreciate subtle difference in echogenicity.

Ultrasound scanners also provide facility for zoom. Zoom is of two types: (1) *Read-zoom*, and (2) *Write-zoom*. In read-zoom, the frozen image is magnified similar to a magnifying glass. So, if the image is zoomed too much, the resolution actually decreases, as the image becomes more pixelated. On the other hand, in write-zoom, the image is zoomed while doing the scan. As the smaller area of interest is represented in a larger area, one pixel can represent one echo signal. Thus, magnification occurs without a loss in definition in write-zoom. Thus, it is always preferable to do write-zoom especially while measuring small structures, such as common bile duct or the nasal bone in 1st trimester fetal ultrasound.

Special modes of scanning are included in most modern scanners to improve image resolution and decrease artifacts. Among these, we will discuss two commonly used modes, namely (1) Tissue harmonic imaging, and (2) Spatial compounding.

Tissue Harmonic Imaging

As sound waves propagate through tissue, it becomes distorted and breaks down into its frequency components called harmonics. These changes increase with depth and are more

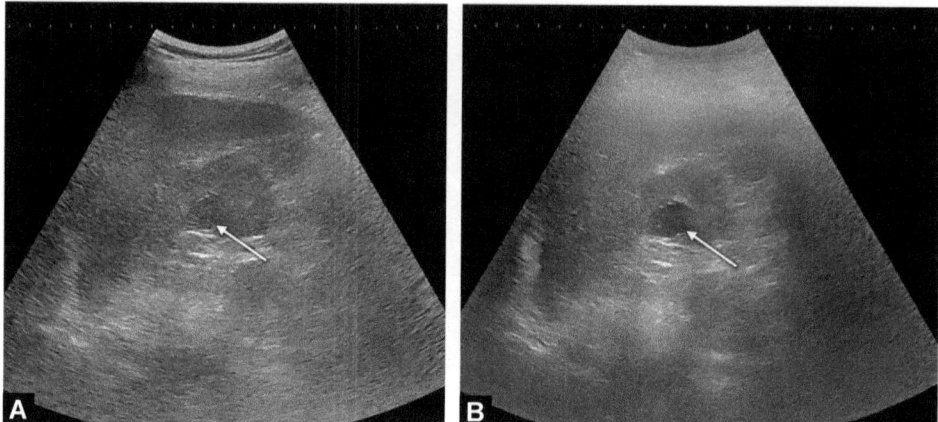

Figs. 8A and B: (A) Ultrasound image of right kidney without tissue harmonic imaging shows an ill-defined hypoechoic structure in kidney (arrow). Based on this imaging it is very difficult to differentiate a cystic lesion from a solid lesion; and (B) Obtained with tissue harmonic imaging clearly shows the lesion to be well-defined and cystic in nature (arrow).

marked in obese patients. Harmonics refer to frequencies that are integral multiples of the true frequency of the transmitted pulse. Thus, for a pulse of 2 MHz, the first harmonic or the fundamental frequency is 2 MHz while second harmonic is 4 MHz, followed by 6 MHz and so on. It has been shown that the second harmonic reaches a useful magnitude on the tissue and if it can be isolated from other frequencies, it can be used to form images instead of the fundamental frequency. Most modern scanners are equipped with methods to isolate second harmonic and use THI. There is improved axial resolution and better lateral resolution with THI.

The advantages of using second harmonic for image formation are as follows:
- Harmonics are generated after the sound waves have passed a few centimeters into the patient. Thus, the echoes generated from harmonics do not show distortion and scattering artifacts arising from the body surface. This is especially useful in obese patients who have thick subcutaneous tissues.
- Reverberation artifacts are decreased as low-amplitude echoes do not produce harmonics or contribute to image formation in THI.
- Penetration can be increased by using a low fundamental frequency while resolution increases by imaging at higher harmonic frequency.
- Low-contrast lesions and fluid-filled lesions, such as cysts are better visualized, as there is reduced acoustic noise (Figs. 8A and B).

Spatial Compounding

One of the causes of decreased ultrasound contrast is ultrasound speckle, which occurs due to scattering of ultrasound beam by multiple tiny reflectors, which are smaller than the wavelength of the insonating sound wave. Speckle makes the image more grainy and decreases overall contrast. Spatial compounding generates the image by insonating the target from multiple different angles. Since data from multiple scan lines are added together to generate the image, the overall image becomes smoother while speckle noise, which is random often, gets canceled out. Spatial compounding thus shows reduced speckle and acoustic noise and improves edge definition and contrast. Drawbacks include decreased posterior acoustic enhancement or shadowing which are useful in identification of cysts and calculi, respectively. The overall aiming is also slower as multiple ultrasound beams are used to insonate the same tissue and data processing and image formation takes longer.

ULTRASOUND ARTIFACTS

Ultrasound is a highly operator-dependent technique. As a sonologist, it is important to identify various ultrasound artifacts to avoid erroneous interpretation. At the same time, certain artifacts can help in diagnosis, such as posterior acoustic enhancement denotes a cyst while posterior acoustic shadowing denotes a calculus.

Artifacts can be broadly divided into four categories, i.e. **(1) Resolution-related artifacts, (2) Attenuation-related artifacts, (3) Propagation-related artifacts, and (4) Ultrasound beam-related artifacts**.

Resolution-related Artifacts

Speckle, as discussed earlier, is due to interference form sound waves scattered from various small structures. Speckle decreases resolution and can be decreased by either using compound imaging or by changing the orientation of probe to insonate the target organ from a different angle.

Attenuation-related Artifacts

Sound waves often get progressively attenuated while passing through tissues. However, while passing through a fluid-filled structure, there is increased forward transmission. This is called *posterior acoustic enhancement* and manifests as a bright band extending downward from the posterior margin of a cyst. If the beam encounters a focal object, that attenuates the sound more than the adjacent tissue, then there is a focal dark band or a shadow extending from the highly attenuating structure, such as a calculus or bowel gas. This is called *posterior acoustic shadowing* (Fig. 9).

Propagation-related Artifacts

Propagation-related artifacts arise due to altered propagation velocities in different tissues and due to multiple echoes generated from tissues. Examples include reverberation, comet-tail, ring-down, side-lobe, and mirror-image artifacts.

Reverberation

In the presence of two parallel highly reflective surfaces, the echoes generated from the primary ultrasound beam may be repeatedly reflected back and forth before returning to

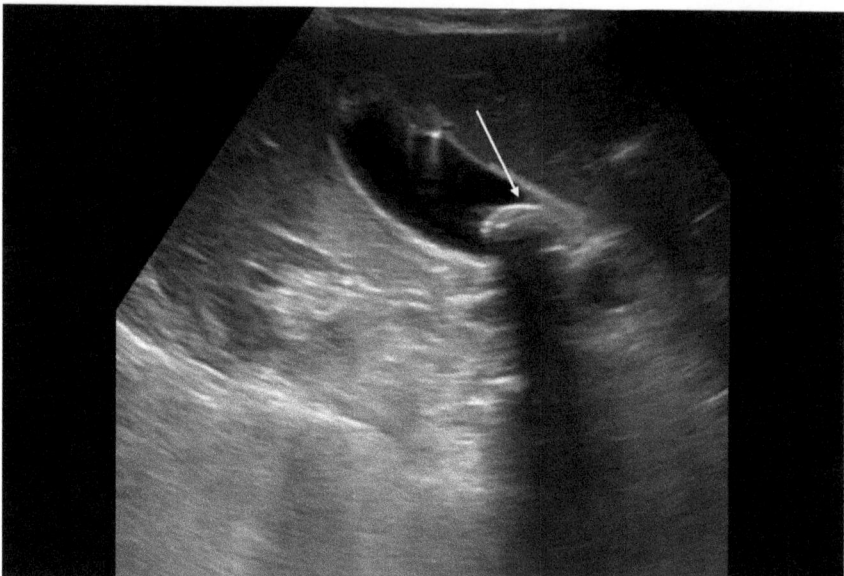

Fig. 9: Ultrasound image of gallbladder showing an hyperechoic structure (arrow) with posterior acoustic shadowing suggestive of gallstone.

the transducer for detection. These are seen as multiple evenly spaced lines on the screen. For example, the lines seen within the distended urinary bladder mimicking a bladder mass or cystitis. This can be resolved by changing the scanning angle.

Comet-tail Artifact

Comet-tail artifact is a form of reverberation artifact. In this, the two reflecting surfaces are very closely spaced and thus the delayed echoes are also very closely spaced, such that they appear as dense tapering echoes distal to a strongly-reflecting surface; for example, behind metallic objects. This artifact is also used in the diagnosis of abnormalities, such as gallbladder adenomyomatosis (Fig. 10).

Mirror-image Artifact

This occurs with specular reflectors, which act as a mirror, and thus for example, structures in the liver can appear to lie in the lung behind the diaphragm because diaphragm acts as a specular reflector (Fig. 11).

Ring-down Artifact

When a small gas bubble resonates, it emits ultrasound continuously, resulting in a track throughout the scan. This effect can also be caused due to air in the stomach. A diagnostic use of this is to identify pneumobilia after percutaneous transhepatic biliary drainage (PTBD) or biliary stent placement to check for patency of the drain or stent.

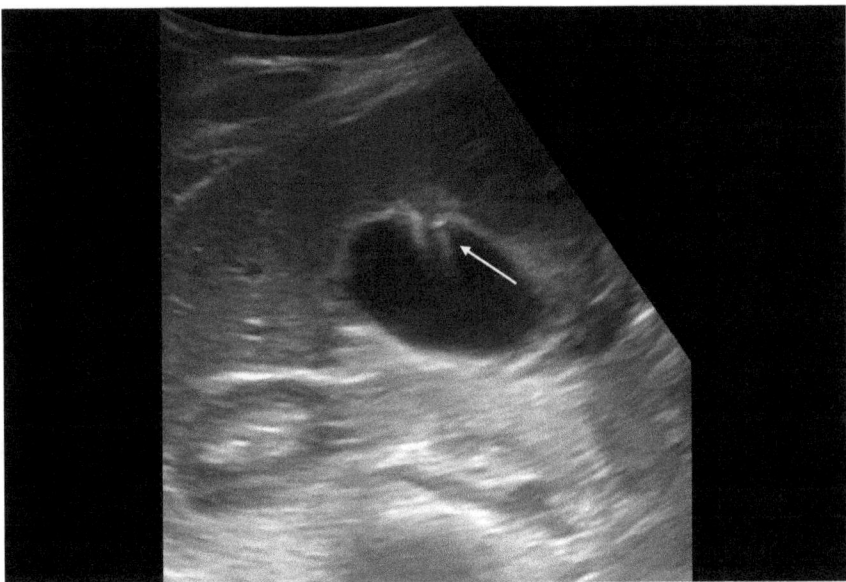

Fig. 10: Ultrasound image of gallbladder showing two linear comet tail-like structures (reverberation artifact) (arrow) arising from the near wall suggestive of adenomyomatosis.

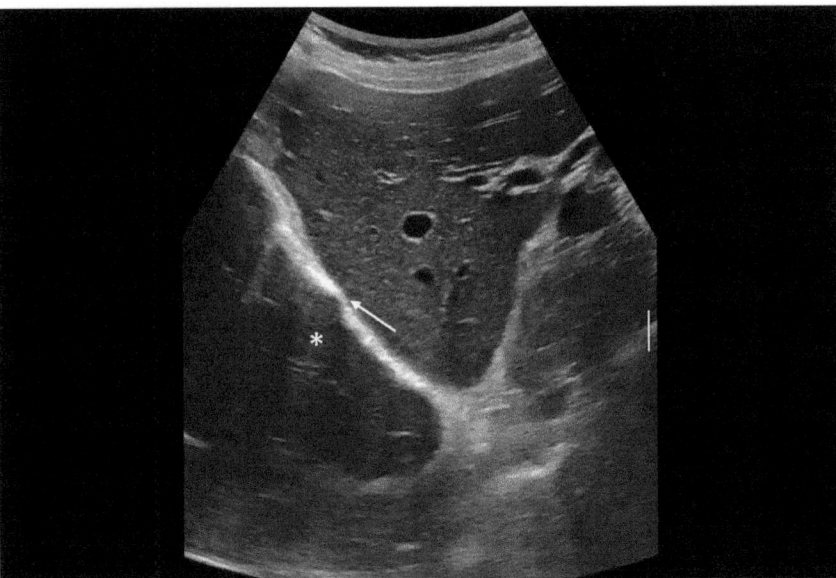

Fig. 11: Ultrasound image of left lobe of liver in sagittal plane shows an echogenic linear structure (arrow) suggestive of diaphragm. Distal to the diaphragm there are structures (*) which are resembling the liver parenchyma like a mirror image suggestive of mirror image artifact.

Ultrasound beam-related Artifacts

Ultrasound beam-related artifacts include beam width and side-lobe artifacts. In beam-width artifact, reflective surfaces located in the widened beam beyond the margin of the transducer can generate echoes which are detected by the transducer and are wrongly represented or displayed on screen as lying within the path of the beam. Side-lobe artifacts are due to echoes from low-amplitude beams that project radially from the main bean and are mostly seen in linear transducers. These can be changed by adjusting the focal zone and by placing the transducer over the center of the object. Side-lobe artifacts can also be eliminated by using THI.

FURTHER READING

1. Hangiandreou N. B-mode US: basic concepts and new technology. Radiographics. 2003;23:1019–33.
2. Feldman MK. US Artifacts. RadioGraphics. 2009; 29:1179–89.
3. Baad M. Clinical significance of US artifacts. RadioGraphics. 2017; 37:1408–23

Chapter 2C

Doppler Imaging

Devasenathipathy Kandasamy, Ananya Panda

OVERVIEW

- Doppler principle
- Modes of Doppler imaging
- Setting up and optimizing Doppler images.

THE DOPPLER PRINCIPLE

Conventional B-mode imaging depends on the reflection of sound waves from stationary interfaces. The Doppler effect is based on the principle that, if the reflecting interface is moving with respect to the sound beam emerging from the transducer, there is a change in the frequency of sound reflected by the moving object. *The change in frequency (Δf) or the Doppler shift is directly proportional to the velocity of the moving interface relative to the transducer.* This is given by the formula:

$$\frac{f - f'}{f} = \frac{2v \times \cos\theta}{c}$$

where f is the frequency of transducer, f' is the frequency of reflected sound wave, f-f' is the change in frequency (Δf), v is the velocity of the moving interface, c is the speed of sound, and θ is the angle of insonation between the ultrasound beam and the moving interface.

Since the only variable in the earlier equation is the velocity of the moving interface (v), the equation can be rearranged to:

$$v = \frac{\Delta fc}{2F\cos\theta}$$

The maximum Doppler shift is obtained when $\theta = 0°$ as cos 0 is 1 and as θ approaches 90°, there is no relative frequency shift as cos 90 = 0. As cos θ changes rapidly, as θ increases beyond 60°, such that small changes in θ more than 60°, leads to large variation in cos θ, it is recommended to align probe in such a way that θ (angle of insonation) is less than 60°.

MODES OF DOPPLER IMAGING

There are three modes of Doppler imaging, namely (1) Continuous wave Doppler, (2) Pulsed Doppler, and (3) Power Doppler. These are briefly described in subsequent paragraphs.

Continuous Wave Doppler

The probe uses two angled transducers, of which one is the transmitting transducer (T) and the other is the receiving transducer (R). Transmitting transducer continuously emits sound at frequency f while receiving transducer receives sound at frequency f'. The original frequency f is suppressed and the Doppler shift (f-f') is extracted electronically. Pulsatile blood flow involves a range of velocities and produces a spectrum of frequencies, which can be displayed on the screen. The Doppler signal can also be heard through a loudspeaker. The higher the pitch, the greater is the velocity, the harsher the sound, greater is the turbulence.

However, in a continuous wave Doppler, as both sound is transmitted and received continuously, it is not possible to locate the depth or the position of the moving reflector or distinguish between the flow in two overlapping vessels at different depths in the path of the beam.

Currently, continuous wave Doppler is limited to small handheld portable Doppler machines for detection of presence and velocity of flow without any other information.

Pulsed Doppler

This is the most widely used mode that combines B-mode and Doppler imaging and provides information regarding, location, velocity and direction of flow, and the color information is superimposed on gray scale image.

In this mode, instead of continuous ultrasound emission, pulsed Doppler devices emit ultrasound waves in brief pulses. Usage of pulses of sound permits the use of the time interval between the transmission of pulse and return of the echo as a means of determining the depth from which the Doppler shift arises. A range gate can also be set to accept only those echoes that arrive within a specified time interval, so that the final spectral curve can only have signals from the selected sampling volume.

The pulsed Doppler can be superimposed with a B-mode scanning and this is called *duplex scanning*. By superimposing these two modes, the operator can precisely control the location, size, and steer angle of the sample volume from which velocity information needs to be obtained. Color Doppler flow imaging (CDFI) further enhances duplex imaging by providing color overlay to the velocity information. Thus, flow towards and away from the transducer can be depicted in blue and red color, respectively (this color scheme can also be inverted). The intensity of the color increases with velocity and focal areas of turbulence or stenosis can be seen as color mosaic (Fig. 1).

In a pulsed Doppler, the interval between pulses must be long enough for successive Doppler signals not to overlap. Thus for deeper vessels, a lower pulse repetition frequency (PRF) would have to be set, as sound will take a longer time to go and come back from the deeper located sample volume. A higher PRF can be set for more superficial vessels.

However, low PRF can also cause problem called *aliasing*. If the PRF is too low and the velocity within the vessel is higher than the PRF, aliasing occurs. According to Nyquist equation, *the PRF should be at least twice the maximum Doppler shift frequency produced by the flow*, which is depicted as PRF = 2 Δf. *Thus, the fastest flow that can be accurately measured is the velocity that produces a Doppler shift not more than half of the PRF being used.* A greater flow velocity than this produces aliasing, which is seen as wraparound of the maximum velocity and seen on the other side of the baseline (Figs. 2A and B). Increasing PRF can decrease aliasing, however, this limits the depth from which velocity information can be collected. Thus, it is usually not possible to accurately measure fast velocities in deep vessels. The risk of aliasing

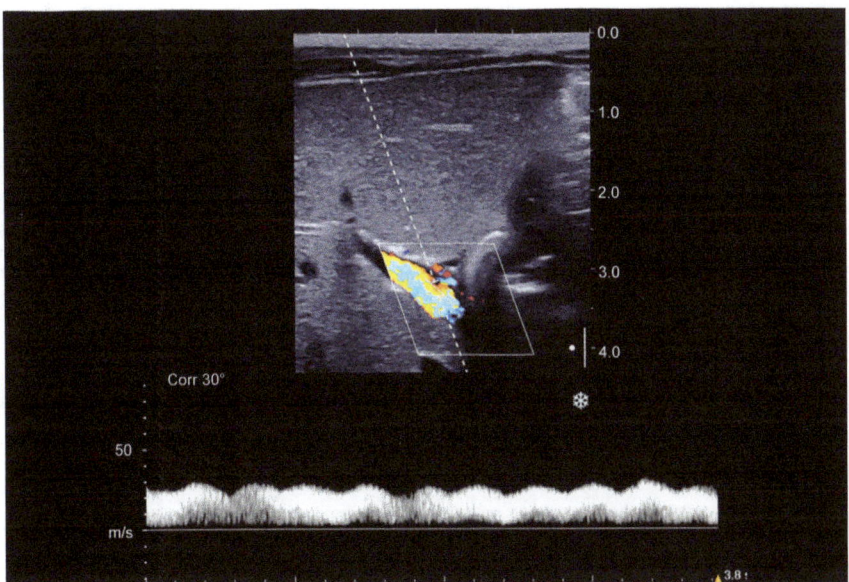

Fig. 1: Color Doppler ultrasound image showing the portal venous flow as colored map over the B-mode image. The obliquely placed box is the color box where any detectable flow is represented in color and the dotted line represents the plane of sample volume. The flow within the sample volume is shown as a graph in the lower part of the image.

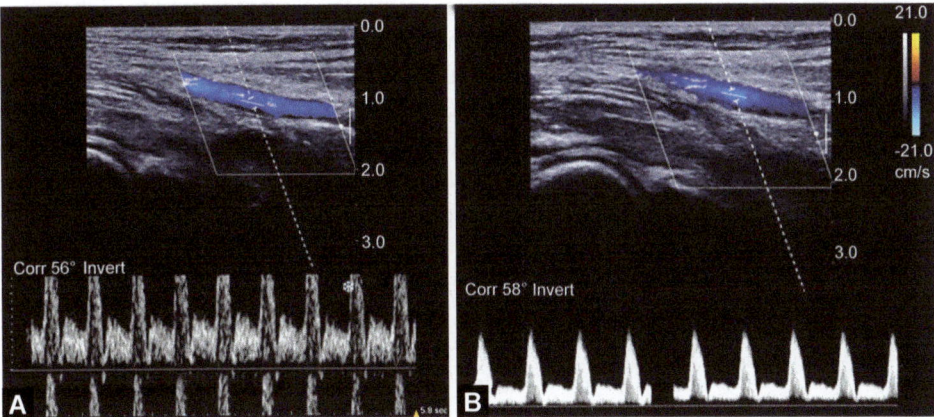

Figs. 2A and B: (A) Ultrasound Doppler image of brachial artery shows color flow in the color box and the graphical representation of the sample volume is showed in the lower part of the image. Part of the flow is shown above the baseline and peak of graph (spectrum) in the systolic phase is shown below the baseline. This phenomenon is called as wrap around or aliasing which can be corrected by increasing the pulse repetition frequency (PRF or scale); and (B) The pulse repetition frequency (PRF) optimized spectrum.

can also be reduced by reducing the Doppler shift by using probe of lower frequency (curvilinear probe over linear probe) or by increasing the angle of insonation (θ). But both these modifications increase errors in velocity measurements.

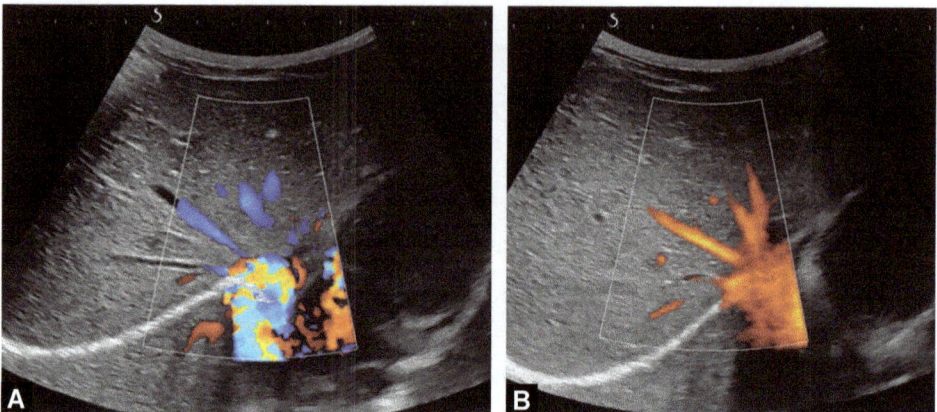

Figs. 3A and B: (A) The hepatic veins in color Doppler mode; and (B) The hepatic veins in power Doppler mode. Power Doppler mode is more sensitive to flow compared to color Doppler mode.

Power Doppler

Power Doppler images map the amplitude of Doppler signal without any indication of direction or velocity. Thus, greater is the volume of flowing red blood cells in a given sample volume, greater is the amplitude of power Doppler signal and is seen as a brighter color (Figs. 3A and B). Power Doppler is chiefly used to distinguish between areas of flow versus no-flow, also, because it is sensitive to even small amount of flow, it is used to pick up smaller velocities or signals weaker than those detected by color Doppler. A practical example of power Doppler is to differentiate between near occlusion and total occlusion in atherosclerotic vessels, wherein near occlusion will show a trickle of color in residual vessel lumen on power Doppler imaging while total occlusion will show a complete color void. However, power Doppler is susceptible to motion and any motion in neighboring tissues or movement of the transducer can produce a false burst of color. Thus, it is best limited to areas where tissue motion is less or absent. Unlike pulsed Doppler, color Doppler is not subjected to aliasing because there is no directional information.

SETTING UP AND OPTIMIZING DOPPLER IMAGES

Doppler imaging is technically challenging because of the various factors that come into play. Erroneous technique can provide potentially misleading velocity and color flow information. The following section gives a brief primer on how to optimize Doppler information.

Gray scale or B-mode imaging should always be optimized before CDFI is applied. The vessel of interest should be shown correctly on the screen, the depth should not be too large and should be just enough to show the required vessel, and a few centimeters of tissue behind it. Too large a depth will lead to unnecessary delay in imaging, and will limit the maximum PRF that can be used, thus leading to aliasing. The entire vessel should be first scanned in B-mode in both longitudinal and transverse sections to look for areas of narrowing, plaque formation, or intimal thickening. Abnormal areas and the adjacent few centimeters of normal

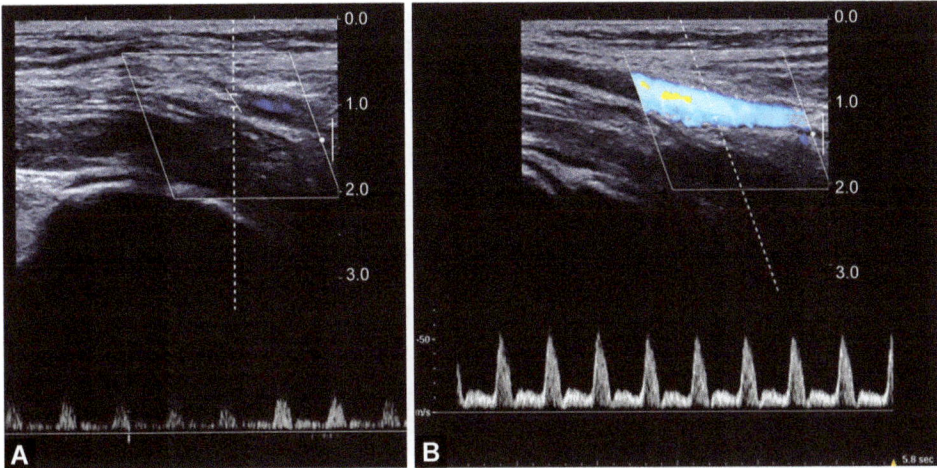

Figs. 4A and B: (A) Color Doppler image of the brachial artery shows that the direction of the sample volume is not optimal and hence the spectrum is erroneous; and (B) Angle of the sample volume is optimized and a good quality spectrum is obtained.

vessel can then be selectively interrogated with color flow and Doppler imaging, thus saving time and maximizing information obtained from the study.

Once the CDFI is activated, it is important to place the sample volume box correctly over the center of the vessel. The sample volume size should be just large enough to cover two-thirds of the vessel lumen and should be placed near the center of the vessel, as the flow is laminar in center of the vessel. Placing the sample gate too close to the wall of the vessel can erroneously depict turbulent flow. The angle of insonation should always be less than 60° as errors increase with angle more than 60°. The angle of insonation can be optimized by angling the sample volume box to the right or left (steering), or by changing the angle of the transducer (Figs. 4A and B).

It is also important to adjust the color gain, so that the color just about fills the vessel lumen without spilling over or bleeding into the adjacent tissue. Too low gain can miss slow flow while too high gain produces noise and can obscure Doppler information.

As discussed earlier, PRF setting should be as high as possible for a given Doppler study, so as not to cause aliasing. The velocity scale can also be adjusted and is related to the PRF; higher the velocity scale, higher is the PRF. Another way of optimizing display is by shifting the baseline up and down to better depict the spectral waveform and to avoid aliasing.

Other settings that can be adjusted include activating wall filters to eliminate "thump" or low frequency high amplitude noise arising from walls of the vessels. While filters are useful in eliminating Doppler noise, if the filter is set too high, it can erroneously eliminate low flow from stenosed vessels. Decreasing the wall filter progressively fills in spectral information near the baseline in the Doppler waveform.

To summarize, Doppler imaging is a specialized ultrasound technique. Nonradiologists can use CDFI to look for patency of vessels prior to central and peripheral line insertions, to evaluate deep vein thrombosis, and to look for position of adjacent vessels before performing any ultrasound-guided intervention to avoid accidental vessel injury. Since Doppler imaging uses longer pulse durations than conventional imaging, it is important not to use Doppler imaging continuously or keep the probe stationary for a long time near soft-tissue/bony interfaces, as that can lead to local heating and potentially adverse thermal bioeffects.

FURTHER READING

1. Merrit CRB. Doppler US: The Basics. RadioGraphics. 1991;11:101-9.
2. Pozniak MA. Spectral and Color Doppler Artefacts. RadioGraphics. 1992;12:35-44.
3. Rubin JM. Spectral Doppler US. RadioGraphics. 1994;14:139-50.

Chapter 3

Ultrasound for Vascular Access

Akhil Kant Singh, Karthic Iyer

HISTORY OF CENTRAL VENOUS ACCESS

In 1929, a 25-year-old, German surgical resident named Werner Forssmann; inserted a 4 Fr ureteric catheter into his left cubital vein, and got an X-ray of the chest to demonstrate the tip of the catheter in the right side of the heart. Later, he injected contrast and visualized the right side of the heart. He published his findings but was met with derision. He was also forced to leave his job and joined the German army as a surgeon.[1]

Although, Forssmann was not particularly encouraged by the response in his own country, Dickinson Richards and Andre Cournand from the US refined Forssmann's technique for testing cardiovascular physiology. For their innovation and research into central venous access, Forssmann, Cournand, and Richards were jointly awarded the Nobel Prize for medicine in 1956.

Catheterization of the central veins has improved by leaps and bounds since the time of the early pioneers. Central venous cannulation is a procedure commonly performed by anesthesiologists and intensivists. The scenario in which one might perform the procedure ranges from controlled conditions like those encountered in the operating rooms; to emergent, life-threatening situations like ongoing cardiopulmonary resuscitation.

Even though a lot of water has flown under the bridge in the form of various developments, since the first attempt at central venous cannulation, the "landmark" technique remains the most commonly attempted method, especially by trainees in developing countries. Studies reporting landmark based approaches to central venous cannulation differ in their results depending on the vessel being accessed, but the consensus is in favor of the usage of ultrasound. Ultrasound-guided approaches have largely supplanted the "landmark" technique as the ideal technique, and for good reason.

- Higher rates of first-pass cannulation, especially in anatomically difficult patients like those who are obese or short necked, and in patients with narrow venous lumen due to hypovolemia.
- Lesser rates of complications like hematoma (especially in patients with coagulation disorders), arterial puncture or pleural injury.
- Identification of vessel thrombosis.

Given its advantages, various organizations including the National Institute for Clinical Excellence, UK, European Federation of Societies for Ultrasound in Medicine and Biology (EFSUMB), Association of Anaesthetists of Great Britain and Ireland (AAGBI)[2] and the American Society of Anesthesiologists[3] have prescribed the regular use of ultrasound guidance while gaining central venous access in their guidelines. In fact, the AAGBI questions the need for attaining competency for all staff in the landmark-based approach in its draft policy on vascular access.[2]

THE EVIDENCE

The medical literature is replete with studies and case reports about ultrasound-guided vascular access. There have been numerous studies about the effect of ultrasound on ease of insertion, number of attempts and complications while performing vascular access procedures.

The evidence is clear regarding the utility of ultrasound in internal jugular vein cannulation in adults. Ultrasound guidance results in improved first attempt success, and overall success rates.[4,5] Whether results are equally applicable to experienced operators as well is unclear, but we think that the use of ultrasound would be beneficial even in this subset. In children again, although the data is limited; the literature supports the use of ultrasound for vascular access in the internal jugular vein.[6]

A Cochrane review of the evidence for ultrasound guidance in subclavian vein cannulation found that it reduces the risk of accidental arterial puncture and hematoma formation. No difference was found in total complications, number of attempts or time taken to insert the guidewire.[7] The catch is that studies reviewed were not of good quality and there was no discrimination between experienced and inexperienced operators. Notwithstanding the poor evidence, we would still recommend that ultrasound be used, particularly by beginners.

For femoral vein cannulations, ultrasound increases first attempt success and decreases total time taken. Ultrasound also has a greater benefit in children[8] and in patients undergoing cardiopulmonary resuscitation.[9,10] Intuitively, beginners derive greater benefit than experienced physicians.[11]

Thus, ultrasound-guided central venous cannulation is an important skill which should be taught to all residents and routinely used in clinical practice.

This chapter will introduce the reader to the basic approaches to central venous cannulation using an ultrasound, and touch briefly upon ultrasound being used for peripheral venous and arterial cannulation.

GENERAL CONSIDERATIONS

In Plane or Out of Plane

Whether one uses the in-plane or out-of-plane technique, is a personal choice. The out-of-plane approach is the more commonly used technique and the ultrasound probe is placed perpendicular to the direction being traversed by the vessel. The needle is visualized as a hyperechoic point in the anechoic circle which is the vein (Fig. 1). In the in-plane technique, the probe is positioned longitudinally, over it. In this view, the vessel is envisioned in the long axis, along with the shaft and tip of the needle as it is advanced (Fig. 2). Herein, lies the advantage of this technique over the out-of-plane approach. It is easy to lose sight of the tip and puncture the posterior wall, if the operator does not fan the probe while using the out-of plane technique. Of course, the main drawback of the in-plane approach is the difficulty encountered while keeping the vessel in the viewing window.

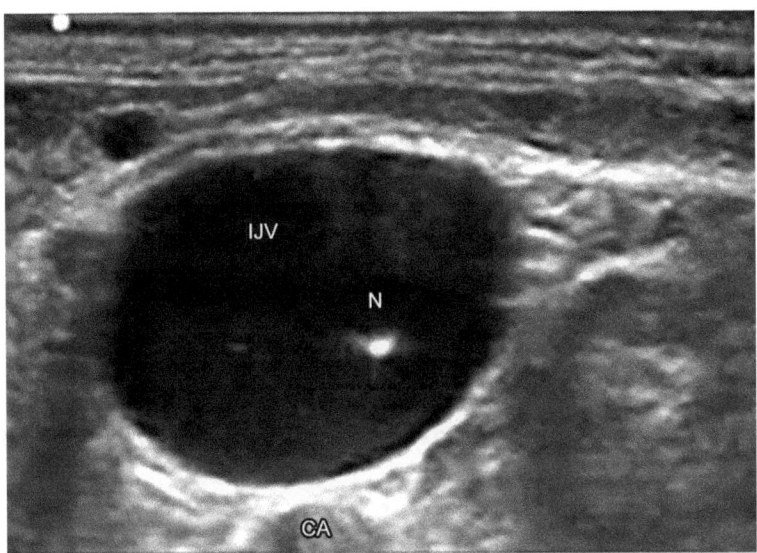

Fig. 1: Needle visualized within the IJV.
Abbreviations: IJV: Internal jugular vein; N: Needle tip; CA: Carotid artery.

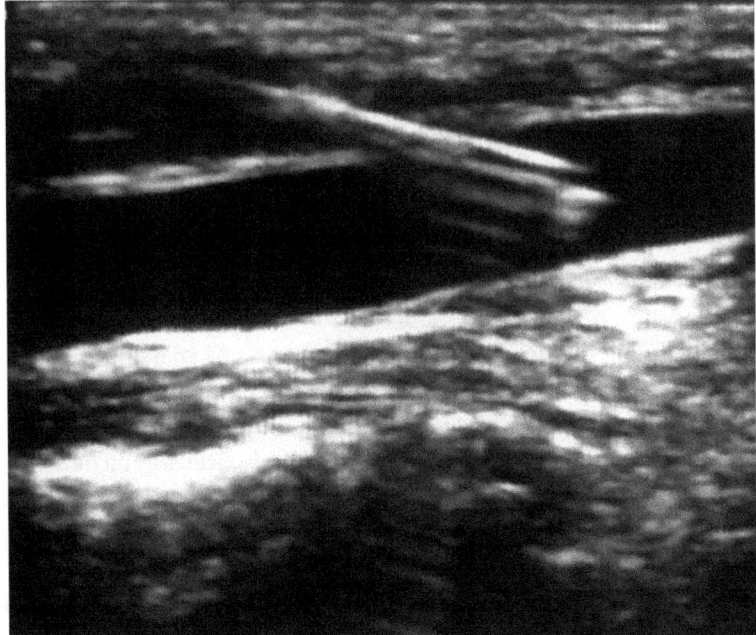

Fig. 2: Needle shaft entering IJV (Longitudinal view).

Static or Dynamic Ultrasound Guidance

Dynamic ultrasound guidance utilizes imaging in real time to visualize the target vessel and the needle pathway. On the other hand, in the static approach the ultrasound is used to assess vessel patency and size, the adjacent structures and to ascertain the site of puncture. The actual vessel puncture is done without ultrasound guidance. The static approach is infrequently used and has been shown to be inferior to the dynamic ultrasound guidance.

Vein or Artery

Correctly differentiating the vein from the artery is, of course the basic step before taking a puncture; but it is not always as straightforward as it sounds. Arterial walls are more hyperechoic, and thus brighter and large arteries are less compressible as compared to veins, it is sometimes difficult to differentiate between the two. In hypovolemic patients, arteries can also be compressed with minimal pressure. If the vein in the area of interest has a thrombus, the operator might find difficulty in compressing the vein, thus confounding the identification. In such cases, color Doppler and/or pulse wave Doppler can be used to aid identification.

One-person or Two-person Technique

Intuitively, a two-person technique is easier for beginners. One person holds the ultrasound probe, while the other inserts the line. This technique allows the person inserting the line to use both her/his hands for performing the procedure but might make image optimization difficult, because the probe control is in the hand of the other person. The other disadvantage is that this technique requires two people which might not always be possible. Experienced operators prefer the one-person technique as it affords better image control. Needless to say, it requires better hand-eye coordination.

THE PROCEDURE

We always find it prudent to "sound" the vessel of interest before actually starting the cleaning and draping process. It gives us an estimate about the depth and caliber of the vessel, its general direction and the relation of the surrounding structures. For example, while obtaining venous access in the neck, turning the head too much toward one side may make the internal jugular vein ride on top of the carotid artery instead of being side-by-side. This increases the risk of arterial puncture, even when the procedure is done under ultrasound guidance. Also, we look for evidence of thrombus in the vessel we are interested in (Fig. 3). This should be done in all patients who have had catheters placed previously (e.g. dialysis-dependent patients). Visualizing the vessel also helps in detecting the presence of valves which might be a cause of difficulty in passing the guidewire.

For placing catheters in the neck veins, it is helpful to place the patient in Trendelenburg position, if feasible. This helps in distending the veins of the necks and avoiding air embolism. In conscious patients, asking the patient to perform a Valsalva maneuver may also help to increase the intrathoracic pressure and consequently, increase the caliber of the veins.

While placing catheters in the neck under ultrasound guidance, we do not place a roll between the scapulae, as is the practice at some places. We do not find it aids the imaging of the vessels or the placement of the catheter.

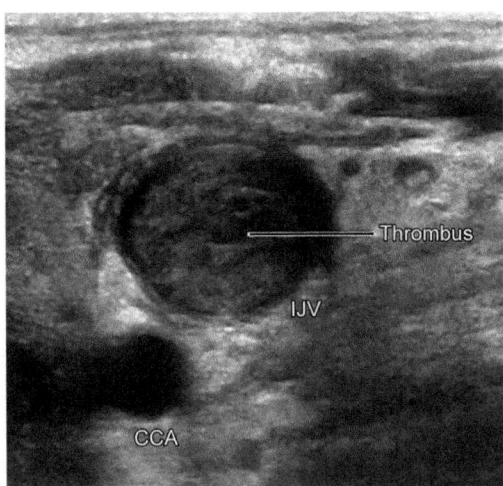

Fig. 3: Transverse view of thrombus within IJV.
Abbreviations: IJV: Internal jugular vein; CCA: Common carotid artery.

For femoral vein cannulation, the patient's leg is externally rotated and slightly abducted.

Maintaining sterility during the procedure is of utmost importance. Using a sterile laparoscopic camera cover is the protocol at our institution. That of course adds to the cost of the procedure in the form of yet another consumable but we find that it ensures asepsis and reduces overall costs by avoiding catheter-associated bloodstream infections. Some operators prefer to use a sterile drape to wrap around the probe and the cord and cover the exposed end of the probe with a transparent dressing like Tegaderm®. Whatever method is used, it is important to place ultrasound gel directly over the transducing surface of the probe and to eliminate air bubbles by smoothing out the camera cover or the transparent dressing. Presence of air bubbles over the probe will result in a suboptimal ultrasound image and will make visualization difficult.

We prefer to use the single operator, out of plane approach. The needle can be positioned close to the probe in an almost vertical orientation or it can be placed a little further away, proximally. If the needle is placed away from the probe, the depth of the vein should be kept in mind and the angle of approach of the needle will have to be adjusted accordingly, so as to "triangulate" the path of the needle toward the center of the vein.

While making the skin puncture, the needle indentation can be visualized on the ultrasound. It is easy to occlude the vessel while pressing down with the probe to maintain image acquisition, especially in small adults and children. Knowing the right amount of pressure to apply comes with experience and is, fortunately, not very difficult. The tip of the needle should be visualized continuously, so as to avoid piercing the posterior wall of the vessel. The needle trajectory should be assessed constantly. If the needle is seen to be going medially or laterally, it should be withdrawn slightly and realigned, so as to puncture the vein.

When the needle enters the vein, blood can be aspirated freely. One may have to insert the needle in a little further because sometimes the endothelium may tent into the lumen. The needle tip should be visualized within the vein at all times. Our practice is to place the probe down and then thread the guidewire in, because we have two hands. For operators blessed with more appendages or those using the two-person technique, it is possible to thread the guidewire under ultrasound guidance with the in-plane technique. Once the guidewire is in place, we confirm its location with the ultrasound. We suggest not just looking for the guidewire in the vein, at the insertion point, but rather as far as possible in to the chest. To achieve this, we slide the probe distally till the clavicle, and then drop the tail to visualize the vein going under. This is done to ensure that the guidewire has not exited the vein.

Subclavian Vein Cannulation

Obtaining vascular access in the subclavian vein is complicated by its anatomy.

The subclavian vein is described as starting from the lateral border of the first rib where the axillary vein continues as the subclavian vein. From here on, it travels over the first rib and under the medial one-third of the clavicle, proceeds to enter the mediastinum and joins with the internal jugular vein behind the sternoclavicular joint; forming the brachiocephalic vein. Most of the subclavian vein runs under the clavicle, and this bony structure makes sono-visualization of the vein difficult, if not impossible. There are three different approaches described in the literature; (1) infraclavicular, (2) through the axillary vein, and (3) supraclavicular. There is some confusion in the literature with regards to the infraclavicular approach and through the axillary vein approach, insofar as the description is concerned. Most studies have used the term "infraclavicular" when they have actually described the axillary vein approach, which is performed in the infraclavicular fossa.

In the infraclavicular approach, it is recommended that a probe with a smaller footprint is used. Lichtenstein[12] and Lanspa et al.[13] have described a technique using a pediatric microconvex probe with cannulation of the proximal axillary vein or subclavian vein. The idea is to get under the clavicle without compressing the vein and visualize the vein in the short axis. We prefer the short axis view as it is difficult to keep the vein centered in the long axis view. The combination of lubrication and a relatively smooth surface makes as the probe is prone to sliding which may result in accidental arterial puncture. The short axis view helps in real time differentiation of the vein and artery while taking the puncture (Fig. 4).

The axillary vein approach to subclavian vein cannulation involves placing the ultrasound probe in the infraclavicular fossa in either the short axis or long axis view. As has been mentioned before, the long axis view allows for the guidewire to be threaded into the vein under vision (Fig. 5).

The supraclavicular approach as described by Mallin et al.[14] comprises using an endocavitary probe (Fig. 6) to picture the confluence of the subclavian vein with internal jugular vein in the supraclavicular fossa. The probe is placed lateral to the clavicular head of the sternocleidomastoid. The small footprint of the probe allows for catheterization to take place in the long axis view. Because the endocavitary probe is usually not available in emergency rooms and operating rooms, a high-frequency linear probe may be used for the purpose, but its larger footprint might make visualization more cumbersome. Another alternative is using a "Hockey Stick" probe (Fig. 7).

With progressive adoption of ultrasound guidance for subclavian vein cannulation, this previously underutilized location is coming back into favor. As long as the operator,

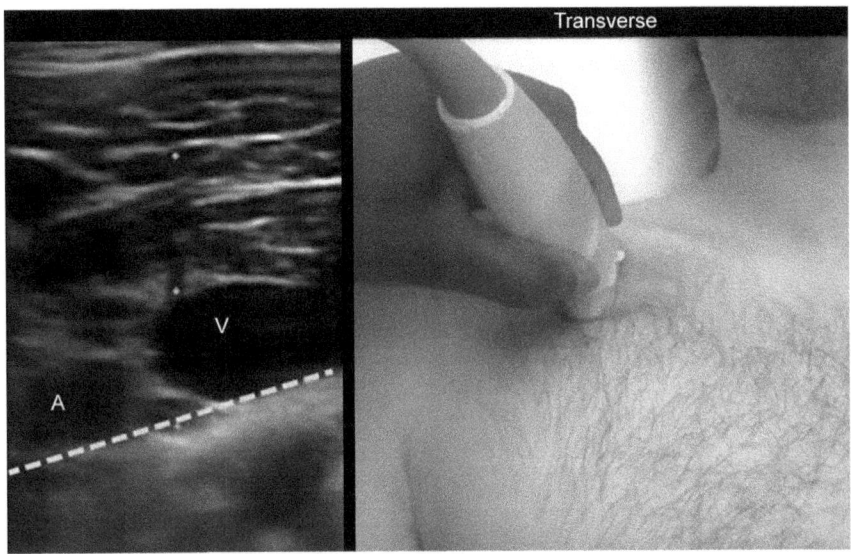

Fig. 4: Transverse view of subclavian vein (V) and subclavian artery (A).

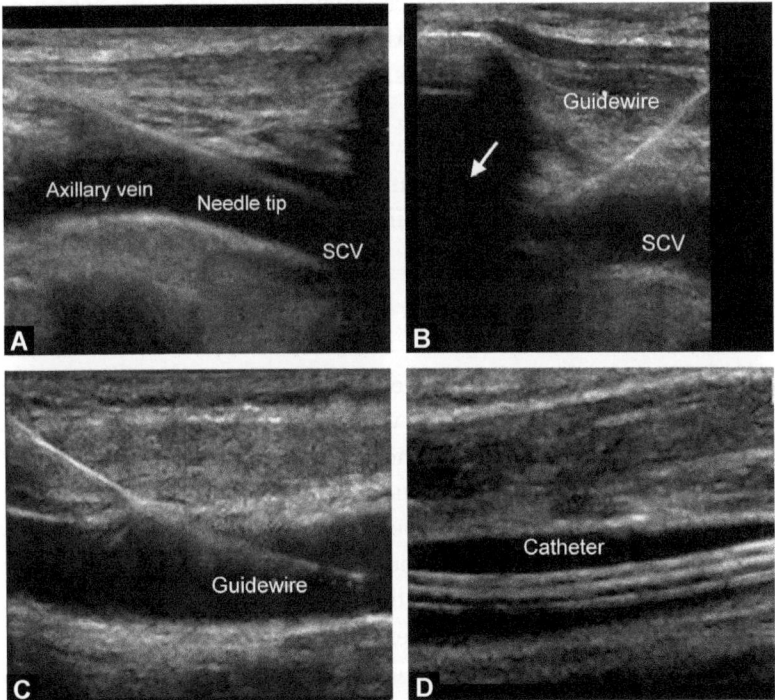

Figs. 5A to D: Longitudinal view of axillary vein puncture (A), guidewire insertion (B and C), catheter in situ (D).
Abbreviation: SCV: Subclavian vein.

especially the novice remains mindful of the surrounding structures; particularly the pleura, the risk of complications is quite low.

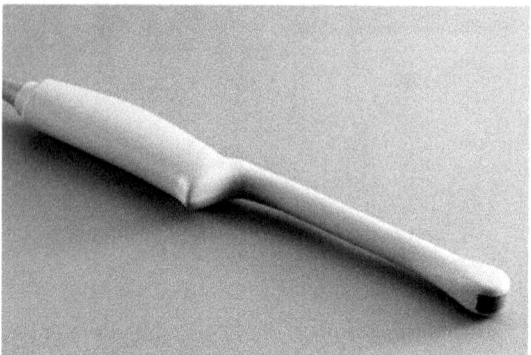

Fig. 6: Endocavitary probe.

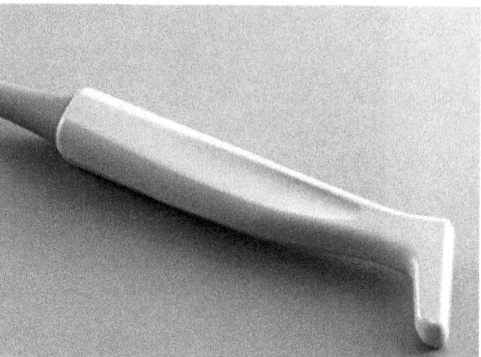

Fig. 7: Hockey stick probe.

Femoral Vein Cannulation

Ultrasound is not commonly used for femoral vein cannulation. Sufficient data is not available to definitely conclude, if reduced complications occur with use of ultrasound, but higher rates of successful first pass cannulations occur with ultrasound. In obese individuals and critically ill-patients with significant edema, where palpation of the femoral artery is difficult or pediatric population where there are higher rates of vessel overlap in the femoral triangle, ultrasound is helpful.

Position of the patient is supine. Mild reverse Trendelenburg position makes the femoral vein more prominent. The ultrasound transducer is placed on the femoral triangle with the pointer facing medially. The vessels are imaged in the short axis view while the needle is inserted out of plane.

External Jugular Vein

External jugular vein cannulation is associated with lesser complications like carotid puncture and pneumothorax. The external jugular vein is usually easily identified as it crosses over the sternocleidomastoid in the posteroinferior direction. Given its narrow caliber, double puncture is common. The vein may be tortuous or contain a valve. These can cause difficulty in the passage of the guidewire. Ultrasound can help delineate the anatomy and detect valves or thrombus in the vein although it does not improve the success rate or the speed of cannulation. Position of the patient is similar to that for internal jugular vein cannulation. The transducer is initially placed transversely over the external jugular vein and then the probe is turned clockwise to trace the longitudinal course of the vein. The needle is inserted in an in-plane technique almost parallel to the skin, as the vein is very superficial. Instead of using the needle present in the central line set, one may cannulate the external jugular vein using a 16G intravenous (IV) cannula and then thread the guidewire through the IV cannula.

Once the guidewire has been placed correctly, the technique is the same as the landmark technique. It is good clinical practice to visualize the catheter in the vein. The catheter can be seen like a "target sign" in the vein lumen in the short axis view. Some institutions also take a print of the captured image of the guidewire in the long axis view and attach it to the patient's file.

AFTER THE PROCEDURE

After placing neck lines or performing subclavian vein cannulation, we always look for pleural sliding to rule out pneumothorax. This is especially important in difficult cannulations or in patients with short necks.

It is also advisable to confirm placement of the catheter by injecting an agitated saline flush through the distal port of the catheter and observing the presence of hyperechoic contrast either in the lumen of the vessel, or the right atrium. The heart can be visualized using a phased array or curvilinear probe, either in the apical four chamber view or the subxiphoid four chamber view.

Peripheral Venous Cannulation

Peripheral venous cannulation can be a challenge in patients who are obese, who have had long hospital stays or in children. Ultrasound aided peripheral venous cannulation does not have as robust a body of evidence as central venous cannulation, but various authors have attempted to describe ease of insertion and first attempt success rate and have come up with differing results. Costantino[15] demonstrated first pass success rate of 46% in an emergency department setting while Ueda et al.[16] reported that they were able to cannulate 96% of their patients in the first attempt. Schoenfeld et al.[17] reported better patient satisfaction scores with ultrasound-guided peripheral venous cannulation. We recommend using ultrasound when no veins are palpated (as might be the case in obese patients) or when previous attempts have failed.

While cannulating peripheral veins, it is important to keep the tip of the needle in the ultrasound view. The target is quite small and one can quite easily lose sight of the tip of the needle with even small movements of the probe. As has been described by Ueda et al.[16] this can be easily learnt through a short teaching module.

Arterial Cannulation

Although arterial cannulation by palpation is the preferred method, ultrasound does have a role as well. Specifically, ultrasound guidance comes into play in the obese, the very young and patients who have low systemic pressures; where difficulty might be encountered in palpating a pulse. Also, when cannulation is to be attempted at a site other than the usual locations (radial, brachial, femoral, axillary or dorsalis pedis) or blind cannulation attempts have been unsuccessful, ultrasound can be of great help. Numerous studies also support this practice and report that the use of ultrasound results in higher first attempt cannulation success rates.[18-20]

Arterial cannulation can be performed using both the long axis and short axis approaches. As has been mentioned before, the short axis view is easier to obtain but it is easy to lose track of the tip of the needle. The long axis allows for continuous visualization of the shaft and tip of the needle but this approach has a steeper learning curve. Another method described is cannulation after surface marking of the artery, better known as static guidance. We do not use this technique as we find the other methods better.

The area of interest for the arterial cannulation should be scanned with the ultrasound and artery and vein differentiated before the site is cleaned and draped. While placing the arterial cannula, asepsis has to be maintained. The linear probe should be covered with a sterile plastic cover after applying ultrasound gel liberally to the transducer surface. If a catheter over needle type of cannula is being used, care has to be taken to progress the needle a few millimeters after the flash of blood is observed to ensure the catheter is inside the lumen as well. If there is resistance in pushing the catheter in, it means it is not in the vessel.

REFERENCES

1. Meyer JA. Werner Forssmann and catheterization of the heart, 1929. Ann Thorac Surg. 1990;49(3):497-9.
2. Bodenham Chair A, Babu S, Bennett J, et al. Association of Anaesthetists of Great Britain and Ireland: Safe vascular access 2016. Anaesthesia. 2016;71(5):573-85.
3. American Society of Anesthesiologists Task Force on Central Venous Access, Rupp SM, Apfelbaum JL, et al. Practice guidelines for central venous access: a report by the American Society of Anesthesiologists Task Force on Central Venous Access. Anesthesiology. 2012;116(3):539-73.
4. Karakitsos D, Labropoulos N, De Groot E, et al. Real-time ultrasound-guided catheterisation of the internal jugular vein: a prospective comparison with the landmark technique in critical care patients. Crit Care. 2006;10(6):R162.
5. Brass P, Hellmich M, Kolodziej L, et al. Ultrasound guidance versus anatomical landmarks for internal jugular vein catheterization. Cochrane Database Syst Rev. 2015;1:CD006962.
6. Lausten-Thomsen U, Merchaoui Z, Dubois C, et al. Ultrasound-Guided Subclavian Vein Cannulation in Low Birth Weight Neonates. Pediatr Crit Care Med. 2017;18(2):172-5.
7. Brass P, Hellmich M, Kolodziej L, et al. Ultrasound guidance versus anatomical landmarks for subclavian or femoral vein catheterization. Cochrane Database Syst Rev. 2015;1:CD011447.
8. Aouad MT, Kanazi GE, Abdallah FW, et al. Femoral vein cannulation performed by residents: a comparison between ultrasound-guided and landmark technique in infants and children undergoing cardiac surgery. Anesth Analg. 2010;111(3):724-8.
9. Hilty WM, Hudson PA, Levitt MA, et al. Real-time ultrasound-guided femoral vein catheterization during cardiopulmonary resuscitation. Ann Emerg Med. 1997;29(3):331-6; discussion 337.
10. Siddik-Sayyid SM, Aouad MT, Ibrahim MH, et al. Femoral arterial cannulation performed by residents: a comparison between ultrasound-guided and palpation technique in infants and children undergoing cardiac surgery. Paediatr Anaesth. 2016;26(8):823-30.
11. Wadman MC, Lomneth CS, Hoffman LH, et al. Assessment of a new model for femoral ultrasound-guided central venous access procedural training: a pilot study. Acad Emerg Med. 2010;17(1):88-92.
12. Lichtenstein D. Cathe´te´risme e´cho-guide´de la veine sous-clavie`re en re´animation. Ann Fr Anesth Reanim. 2000;19:266s.
13. Lanspa MJ, Fair J, Hirshberg EL, et al. Ultrasound-guided subclavian vein cannulation using a micro-convex ultrasound probe. Ann Am Thorac Soc. 2014;11(4):583-6.

14. Mallin M, Louis H, Madsen T. A novel technique for ultrasound-guided supraclavicular subclavian cannulation. Am J Emerg Med. 2010;28(8):966-9.
15. Costantino TG, Parikh AK, Satz WA, et al. Ultrasonography-guided peripheral intravenous access versus traditional approaches in patients with difficult intravenous access. Ann Emerg Med. 2005;46(5):456-61.
16. Ueda K, Hussey P. Dynamic Ultrasound-Guided Short-Axis Needle Tip Navigation Technique for Facilitating Cannulation of Peripheral Veins in Obese Patients. Anesth Analg. 2017;124(3):831-3.
17. Schoenfeld E, Shokoohi H, Boniface K. Ultrasound-guided peripheral intravenous access in the emergency department: patient-centered survey. West J Emerg Med. 2011;12(4):475-7.
18. Gu WJ, Tie HT, Liu JC, et al. Efficacy of ultrasound-guided radial artery catheterization: a systematic review and meta-analysis of randomized controlled trials. Crit Care. 2014;18(3):R93.
19. Gao YB, Yan JH, Gao FQ, et al. Effects of ultrasound-guided radial artery catheterization: an updated meta-analysis. Am J Emerg Med. 2015;33(1):50-5.
20. Shiloh AL, Savel RH, Paulin LM, et al. Ultrasound-guided catheterization of the radial artery: a systematic review and meta-analysis of randomized controlled trials. Chest. 2011;139(3):524-9.

Chapter 4A

Airway Ultrasound

Bikash Ranjan Ray

INTRODUCTION

Sonography for assessment and management of the upper airway has emerged as a useful adjunct to existing clinical methods. Increased availability of ultrasound units in the intensive care unit, operating rooms and emergency department has made it more assessable. This has led to effective use of ultrasound in different aspect of airway management. By combining the knowledge of regional anatomy with the skill of ultrasound technology, one can gather a lot of information regarding different aspects of airway assessment in various scenarios, which helps the intensivist in proper decision making and further management. With advancement in technology and development of better probes with high-resolution imaging, ultrasound has emerged as a vital noninvasive tool for airway management in intensive care practice.

ADVANTAGE OF ULTRASOUND IMAGING OF UPPER AIRWAY

The technique is safe, simple, noninvasive, fast, repeatable, reliable and readily available in intensive care units. It has no claustrophobic effects and the technique is safe to use in pregnancy. As opposed to computed tomography (CT) or magnetic resonance imaging (MRI), it does not require strict immobility and can be used in children and infants. The technique is simple to learn and its competency can be acquired after a short training.

CHALLENGES OF AIRWAY IMAGING

Ultrasound imaging depends upon the transmission and reflection of acoustic waves through a medium. Air does not allow transmission of these waves and hence produces poor image on ultrasound. For this reason, use of ultrasound for airway was not popular for many years. However, the modern ultrasound systems are equipped with multiarray and variable frequency transducers that combines with cross beam imaging facility with better spatial resolution and produces high quality image of the superficial soft tissues and airway structures.

AIRWAY ANATOMY: WHAT DOES ULTRASOUND REVEAL?

Air has been described as the enemy for ultrasonography as the ultrasound beams cannot penetrate air. Hence, whenever air is encountered in the path of ultrasound beams; all of it is reflected back, producing a strong echo, which appears as a bright white line. Thus, during airway scanning the air-mucosa (A-M) interface appears as a bright hyperechoic linear

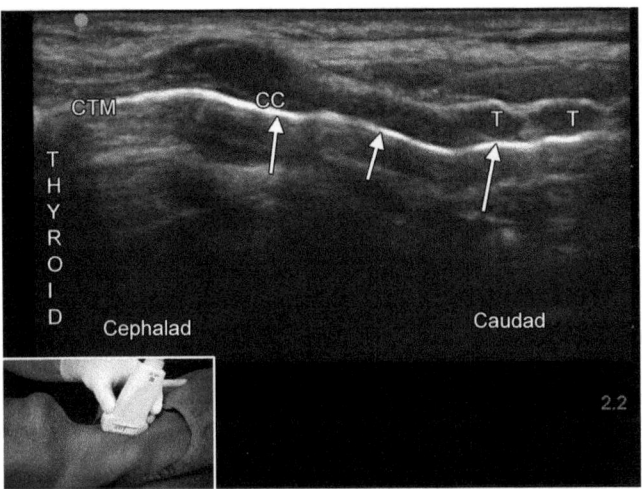

Fig. 1: Sagittal view of the upper airway. (CTM: Cricothyroid membrane; CC: Cricoid cartilage; T: Tracheal rings), in a "string of pearls" appearance. Arrows indicate air mucosal interface.

white line. Anything beyond this white line is considered as artifact (Fig. 1). Hence, only the tissue from the skin to the anterior part of the airway can be visualized distinctly as the intraluminal air prevents visualization of structures beyond the A-M interface.

Acoustic impedance of a tissue is described as the amount of resistance the ultrasound beam encounters while travelling through the tissue. As the ultrasound beam travels, reflection of sound waves occurs at different tissue layers as per the acoustic impedance of the tissues. Difference in impedance is maximum whenever there is a junction of soft tissue with air or bone. Tissues (fat and bone) providing more resistance to the ultrasound beam produce a strong echo and are described as "hyperechoic" whereas tissues, such as fluid or blood provide little resistance to the ultrasound beam, generating little amount of echo and are described as hypoechoic. The hyperechoic structures appear white and the hypoechoic structures appear as black on the screen.

Bone produces a strong echo as the entire ultrasound beam is reflected back from its surface results in an acoustic black shadow, which obscures everything beyond the bone, whereas cartilaginous structures (thyroid, cricoid, and tracheal rings) appear homogeneously hypoechoic as they allow the ultrasound beam to travel due to their high water content. However, cartilages tend to calcify with age and behave like bone. Soft issues like muscles and membranes appear hypoechoic but are heterogeneously striated. Glandular structures (submandibular and thyroid glands), are more homogeneous and have a varying degree of hyperechogenecity compared with nearby soft tissues.

Ultrasound can be used for scanning of the airway from the chin to the mid of the trachea. Airway structures that can be visualized with ultrasound are listed in Box 1. Ultrasound images of the airway (pharynx, larynx) have higher resolution compared with other imaging modalities (CT, MRI), as the structures are superficially located.

> **Box 1:** Structures visualized in ultrasound of airway.
>
> | | | |
> |---|---|---|
> | Tongue | Epiglottis | Thyroid gland |
> | Muscles of floor of tongue | Vocal cords | Esophagus |
> | Hypopharynx | Cricoid cartilage | Pleura |
> | Hyoid bone | Tracheal rings | Carotid and internal jugular vessels |
> | Thyroid cartilage | Cricothyroid membrane | |

ULTRASOUND IMAGING OF AIRWAY

Position of the Patient

Upper airway scanning with ultrasound is performed in supine position with a pillow under the occiput to achieve optimum head extension and neck flexion mimicking the sniffing position for laryngoscopy and intubation.

Ultrasound Transducer

The transducer frequency used for airway ultrasound varies from 2 MHz to 10 MHz. The curvilinear probe (2-5 MHz) is preferred for scanning of the submandibular region as it allows visualization of deeper structures and wider field of view. The linear probe (7.5-10 MHz) is useful for imaging of superficial structures from the skin. Normally, B mode imaging is used for airway imaging. Both M mode and Doppler has limited application in airway imaging, however Doppler may be useful to detect blood vessels within the needle tract, during front of neck procedures. Sagittal, parasagittal and transverse views are used for scanning depending upon the structure of interest.

Sonoanatomy, Image Acquisition and Interpretation of Upper Airway Ultrasound

Imaging of upper airway with ultrasound has been described using transcutaneous and sublingual approaches, however the transcutaneous approach is preferred as it is easy and efficient.

Sublingual Approach

Sublingual approach uses a small footprint, high frequency (2-5 MHz) curved probe placed in the sublingual fossa, intraorally. Imaging can be performed with the probe in both sagittal and transverse positions, allowing a comprehensive assessment of airway structures. This technique has the advantage of better probe-tissue contact compared to transcutaneous approach where it may be difficult to maintain a good probe-skin contact over the uneven and curved surface of the anterior neck. Although simple and efficient, sublingual scanning is less preferable to transcutaneous scanning, as it is often difficult to visualize the epiglottis and laryngeal inlet by the acoustic shadowing cast by the hyoid bone and intraluminal air with the former approach.

Transcutaneous Approach

The transcutaneous approach involves a systematic scanning of the airway from the mentum to the suprasternal notch, along the course of the floor of the mouth. Airway assessment with ultrasound can be classified into scanning of the suprahyoid and infrahyoid regions.

For suprahyoid measurements, the patient is placed in a supine sniffing position and the low frequency curved probe is gently placed in the suprahyoid region both in sagittal and transverse orientation to visualize the tongue, muscles of the floor of the mouth and hyoid bone.

Hyoid Bone (Fig. 2)

Hyoid bone can be visualized in both transverse and sagittal views. In the transverse scanning, it is visible as a hyperechoic inverted U-shaped linear structure with posterior acoustic shadowing. Where as in the sagittal view the hyoid bone has a narrow hyperechoic curved structure with an acoustic shadow.

Tongue and Floor of the Mouth (Fig. 3)

In the sagittal view, the tongue is visualized as a curvilinear hyperechoic structure deeper to the muscles of the floor of the mouth. The intrinsic muscles of the tongue have a striated appearance on sonography. Among extrinsic muscles of the tongue only geniohyoid, genioglossus, and hyoglossus can be visualized, as the mandibular ramus and the mastoid obscure imaging of rest of the muscles. The mylohyoid and geniohyoid muscles appear as linear hypoechoic bands below the skin and subcutaneous fat, extending between the mandible and the hyoid bone. The genioglossus and hyoglossus muscles lay deep to the geniohyoid muscle and are seen running in a fanlike fashion towards the dorsal surface of the tongue.

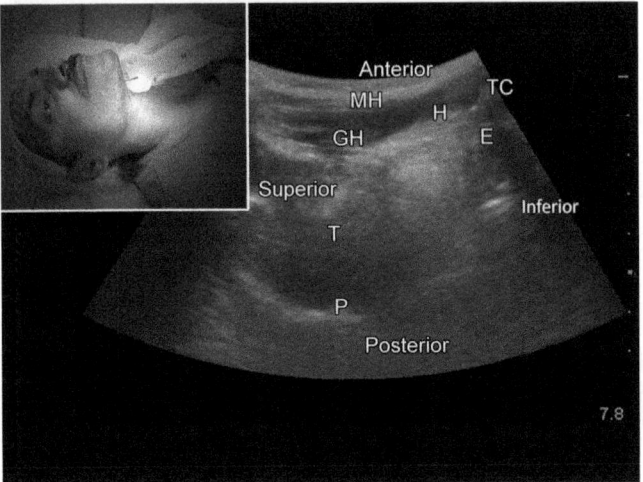

Fig. 2: Extended submandibular sagittal view.
Abbreviations: MH: Mylohyoid; GH: Geniohyoid; H: Hyoid bone; TC: Thyroid cartilage; E: Epiglottis; T: Tongue; P: Palate.

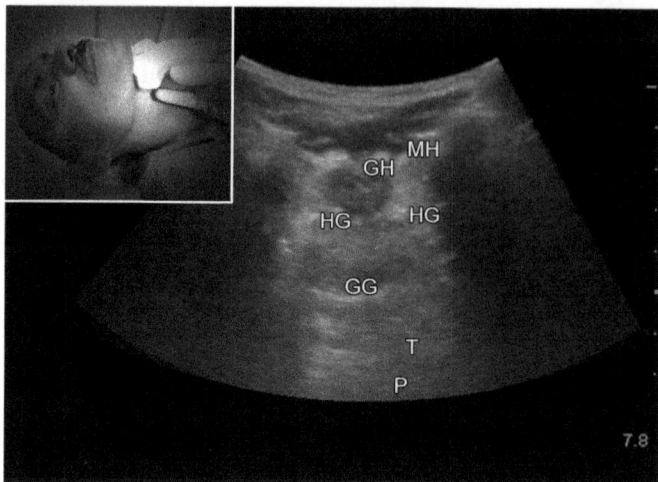

Fig. 3: Transverse submandibular view.
Abbreviations: MH: Mylohyoid; GH: Geniohyoid; HG: Hyoglossus; GG: Genioglossus; T: Tongue; P: Palate.

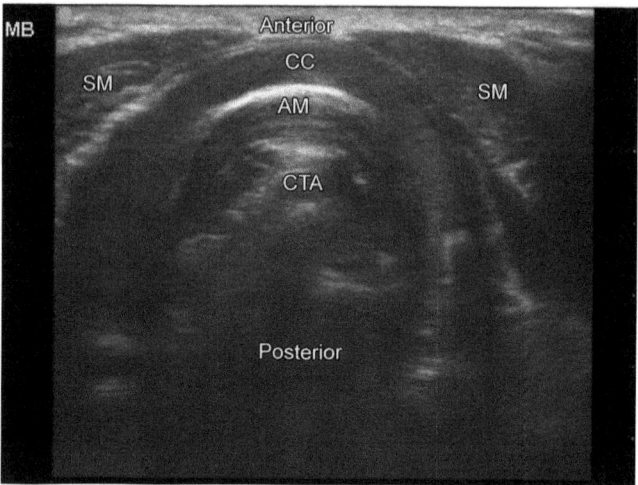

Fig. 4: Transverse view at cricoid cartilage.
Abbreviations: SM: Strap muscles; CC: Cricoid cartilage; AM: Air mucosal interface; CTA: Comet tail artifact/acoustic shadow.

The high-frequency linear probe should be used (both in sagittal and transverse plane) to visualize the structures of the infrahyoid region, which includes thyroid cartilage, the epiglottis, vocal cords, cricoid cartilage, cricothyroid membrane, strap muscles, thyroid gland, tracheal ring cartilages, and esophagus.

Cricoid Cartilage and Cricothyroid Membrane (Figs. 1 and 4)

The infrahyoid scanning usually starts by placing the linear transducer midline of the anterior neck, above the sternal notch in a sagittal plane, with the orientation marker directed towards the patient's head. The tracheal rings appear as a series of hypoechoic structures, which resembles like a "string of beads". The cricoid cartilage is identified as a large oval hypoechoic structure, which appears as a hump at the cephalic end of the "string of beads". The hyperechoic line that is visualized deep to the string of beads represents the air-mucosal interface. The cricothyroid membrane is identified as the transducer is moved cephalad in the sagittal plane. It appears as a hyperechoic band bounded cephalad by the thyroid cartilage and caudally by the cricoid cartilages. In the transverse scanning, the cricoid and the tracheal cartilages appear as inverted U-shaped hypoechoic structures as the transducer is moved caudally from the thyroid cartilage. The cricoid is the first cartilage to appear below the thyroid cartilage, which is the largest compared to other tracheal cartilages.

Thyroid Cartilage and Vocal Cords (Fig. 5)

Thyroid cartilage is visualized as an inverted V-shaped hypoechoic structure while scanning caudally below the hyoid bone in the transverse plane. The vocal cords are best visualized through the thyroid cartilage. The true vocal cords appear as two triangular hypoechoic structures. The echogenic vocal ligaments delineate true cords medially. The false cords lie more cephalad and are more echoic than true cords in appearance. False cords are relatively immobile on phonation in contrast to true cords, which deviate medially.

Epiglottis

The epiglottis is easily identified in the transverse plane, as a hypoechoic curvilinear structure through the thyrohyoid membrane. However, some degree of cephalad or caudal angulation is required for optimal imaging. It may be difficult to visualize the epiglottis in the parasagittal view due to acoustic shadowing by the hyoid bone. In one study, it has been found that the epiglottis could be visualized in 90% of cases in transverse view and 70% in parasagittal view.

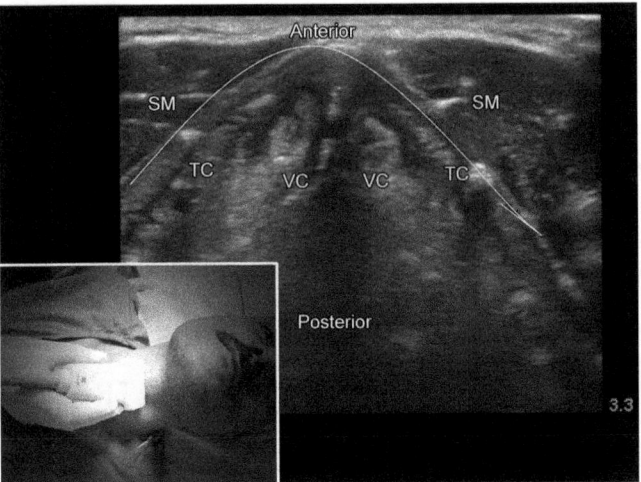

Fig. 5: Midline transverse view at hyoid.
Abbreviations: SM: Strap muscles; TC: Thyroid cartilage; [marked by white line]; VC: Vocal cords.

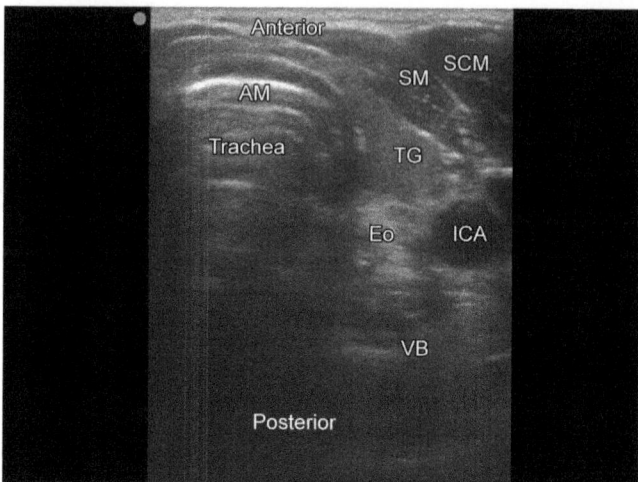

Fig. 6: Oblique transverse view at suprasternal notch.
Abbreviations: AM: Air mucosal interface; SM: Strap muscles; SCM: Sternocleidomastoid; TG: Thyroid gland; ICA: Internal carotid artery; VB: Vertebral body; Eo: Esophagus.

Tongue protrusion and swallowing helps in identification of epiglottis as it becomes prominent as a mobile structure inferior to the base of the tongue. Area anterior to the epiglottis is the pre-epiglottic space, which normally appears as hyperechoic.

Thyroid Gland (Fig. 6)

Transverse scanning at the sternal notch can identify the two lobes and the isthmus of the thyroid gland, which are visualized anterolateral to the trachea. Characteristically, the thyroid parenchyma appears to be more echogenic than the adjacent strap muscle with a homogeneous texture.

Esophagus

Esophagus can be visualized in the transverse view, at the level of the suprasternal notch. As the esophagus is a collapsible tubular structure, its concentric layers and its position (left and posterolateral to the trachea) help in its identification. Proper identification of esophagus is essential for assessment of endotracheal intubation. In awake patient, asking the patient to swallow, which results in visible peristaltic movement inside esophageal lumen, can easily identify it.

CLINICAL APPLICATION OF UPPER AIRWAY ULTRASOUND

Prediction of Difficult Airway

Growing academic interest in the use of ultrasound for identifying difficult airway predictors has led to increasing numbers of papers on this topic in last few years. Submandibular ultrasonography and pretracheal soft tissue thickness has been used with varying results. It may be especially helpful in obese patients as identification of normal anatomical landmarks are difficult and ultrasound helps in identifying anatomical structures and measurements of differential distribution of fat.

Pretracheal soft tissue thickness has been measured with ultrasound using a transverse window at difference levels (Vocal cords, thyroid isthmus, and suprasternal notch, etc.) to predict difficult laryngoscopy in obese patients. However varying and inconsistent results in different studies has failed in finding any definitive predictor of difficult laryngoscopy.

Although pretracheal soft tissue thickness measurement with ultrasound in the infrahyoid region has been used in many studies, there are only few studies using suprahyoid measurements to predict difficult airway. Sonographic suprahyoid parameters that have been studied for prediction of difficult airway includes: the hyomental distance in neutral and extended positions, hyomental distance ratio, tongue cross-sectional area, tongue width, tongue volume, tongue thickness-to oral cavity height ratio, and floor of the mouth muscle cross-sectional area and volume, etc. Among these parameters, the hyomental distance in neutral and extended positions, hyomental distance ratio and tongue volume measurements have shown promising results for prediction of difficult laryngoscopy and difficult intubation.

Preliminary studies have showed promising results, but most of these studies were pilot studies and were limited by small study samples. More systematically designed studies are needed to generate high quality evidence, before this modality can be validated as a routine screening tool for prediction of difficult airway.

Prediction of Appropriate Size Endotracheal Tube, Double Lumen Tube, and Tracheostomy Tube

Selection of proper size endotracheal tube (ETT) is crucial in the scenario of pediatric and neonatal airway management. Ultrasound has been found to be an extremely useful tool for prediction of proper dimension of ETT in these patients' population.

Subglottic transverse diameter measured by ultrasound has an excellent correlation with the outer diameter of proper-sized ETT. The subglottic airway diameter measured with ultrasound also demonstrates a good correlation with the gold standard MRI in children and young adults. It helps in choosing the appropriate size ETT for children and may reduce the number of reintubations with uncuffed tubes, compared to using age-related formulas for selecting ETTs.

Selection of proper size of double lumen tube is essential for lung isolation and one lung ventilation. Ultrasound can be useful in estimation of the diameter of the left mainstem bronchus, and thus the proper size of a left-sided double lumen tube. Outer tracheal diameter measured by ultrasound just above the sternoclavicular joint is strongly correlated with the left main stem bronchus size measured using CT. The ratio between left mainstem bronchus diameter on CT imaging and outer tracheal diameter measured with ultrasound is found to be 0.68 mm. Thus, outer tracheal width measured with ultrasound may be used to calculate the size of the left mainstem bronchus and hence can be used to select the size of a left-sided double-lumen tube.

In tracheostomized children, measurement of the tracheal width and the thickness of soft tissue from skin to trachea with ultrasound, has been used to predict the proper size and shape of a potential replacement tracheostomy tube.

Confirmation of Endotracheal Tube Placement

Definitive airway control while resuscitating critically ill-patients is achieved with endotracheal intubation. After intubation, confirmation of proper position of ETT in the trachea is essential and is usually carried out clinically by auscultation. Various techniques have been described in the literature to verify correct ETT position. However, none of the confirmatory methods has been found to be ideal, to be used in all situations. Waveform capnography with quantitative measurement is considered as the standard of care for confirmation of ETT position but has limited role in cardiac arrest scenarios. It is also affected by low cardiac output, low pulmonary blood flow or use of epinephrine. Upper airway ultrasound has emerged as alternative and adjunctive tool, which can be advantageous in circumstances where capnography may be faulty. Ultrasound can be helpful for confirmation of ETT placement either by direct scanning to identify passage of tube in the trachea or indirectly by looking for lung ventilation, or by combining these techniques.

In order to detect ETT intubation directly, the linear transducer is placed just above the sternal notch to obtain a proper transverse neck window, with the visualization of the esophagus. Successful ETT intubation may be confirmed in real time with the artifacts generated when the tube is introduced inside the trachea. In case of esophageal intubation a new circular structure can be seen with artifacts generated by the presence of the tube inside the esophagus, which resembles the trachea. Indirectly ETT intubation can be confirmed by lung sliding at the plural level or by looking for diaphragmatic movement as a sign of lung ventilation. Lung sliding and diaphragmatic movement can be used for identification of endobronchial intubation. Ultrasound transducer placed at the level of the glottis can identify direct passage of the tracheal tube into the trachea in children, which is characterized by the widening of the gap between the vocal cords.

Static transtracheal scanning at the sternal notch window has been found to be 98.9% sensitive and 94.1% specific for diagnosing esophageal intubation. Upper airway ultrasound correlates well with waveform capnography for detecting proper position of tracheal tube. Meta-analysis of available literature also supports use of ultrasonography for detection of esophageal intubation with a high degree of diagnostic accuracy. Ultrasonography has been suggested as a valuable adjunct for airway management especially for confirmation of ETT placement, in situations where capnography may be unreliable.

Airway ultrasound has three major advantages when used during ETT intubation in critically ill-patients. First, it provides real time imaging of passage of the ETT through the trachea or esophagus, which can be detected before commencing ventilation. Secondly, ultrasound is highly specific for detection of esophageal intubation and is found to be useful when the results of capnography are equivocal. Finally, ultrasound can be used during cardiopulmonary resuscitation (CPR) without interruption of chest compressions. Also, ultrasound is superior to auscultation for confirmation of endotracheal intubation in noisy situations like airlifting, crowdy emergency department, etc.

Localization of Trachea and Cricothyroid Membrane

Ultrasound is helpful in locating trachea in challenging conditions like obesity, short thick neck, neck mass, previous surgery, radiotherapy, etc. It can be used as a prelocation technique for localization of trachea in emergency airway management and awake tracheostomy.

Identification of cricothyroid membrane is crucial in airway management. In all, airway management guidelines, the final solution for failed intubation and ventilation, is always oxygenation through the cricothyroid membrane. It can be achieved either with a small-bore jet-insufflation catheter, or with the larger-bore needle technique, or as a surgical emergency cricothyroidotomy. Yet accuracy of cricothyroid membrane identification through conventional palpatory and surface landmark techniques is only 30%.

Ultrasonography allows reliable and rapid identification of the cricothyroid membrane. Both, transverse and sagittal views can be used for localization of the membrane. Mean time for cricothyroid membrane localization with ultrasound has been found to be 24.3 seconds.

Upper Airway Ultrasound for Percutaneous Dilatational Tracheostomy

Percutaneous dilatational tracheostomy (PDT) is a routinely used procedure in modern intensive care units. Bronchoscopic guidance is recommended for the procedure and is often used. Even though bronchoscopic guidance provides excellent visualization, it has certain disadvantages. It may hamper ventilation leading to hypercarbia and increased intracranial pressure, which may not be tolerated by patients with head and spine injuries. Also, bronchoscopy may itself induce pulmonary hyperventilation and pneumothorax. Ultrasonography guidance is an alternative technique which overcome the limitations of bronchoscopy and provides several advantages for PDT. It helps in real-time localization and visualization of the anterior tracheal wall, identification of tracheal rings and detection of blood vessels in the pretracheal tissue. It also allows clinician to select the proper site for the tracheostomy and helps in estimating appropriate size of tracheostomy tube. One of the studies found that preprocedural ultrasound assessment leads to change in PDT puncture site about one in four patients. US-guided PDT has been found to be successful in cases when bronchoscope-guided PDT has failed.

Many clinical studies have advocated the preprocedural airway ultrasound as it improves safety during PDT by facilitating successful tracheal puncture and wire insertion, and avoiding posterior wall injury. The Traditional landmArk versus ultRasound Guided Evaluation Trial (TARGET) study showed that real-time ultrasound guidance improves success rate of the first PDT attempt and puncture accuracy compared to traditional landmark technique. Similarly, the TRACHUS trial looked at the failure and complication rate of ultrasound-guided PDT in critically ill-patients and suggests it to be as effective as bronchoscopy-guided PDT. Also, when compared with conventional landmark methods, ultrasound-guided PDT has higher success rates and shorter time to successful cannulation.

Prediction of Postextubation Stridor

Air column width at the level of the vocal cords measured with the help of ultrasound, has been found as a potential measurement to predict postextubation stridor. Limited air column width difference before and after ETT cuff deflation has been found to be associated with postextubation stridor. However, results of different studies are not consistent and further researches with larger sample size are required to establish a cutoff value for predicting postextubation stridor.

Miscellaneous Use of Airway Ultrasound in Critically Ill-Patients

Identification of Airway Pathologies Influencing Airway Management

Pathological conditions of airway like subglottic hemangioma, laryngeal stenosis, respiratory papillomatosis and Zenker's diverticulum can be visualized with the help of ultrasound which can influence the airway management technique. Maxillary sinusitis, which may be a source of infection in the intensive care unit can be detected with help of ultrasound. It can also help in decision making before planning for nasal intubation. Ultrasound can be helpful in assessment of epiglottitis, inflammatory mucosal swelling and assessment of vocal cord functions.

Evaluation of Gastric Content

Ultrasound has been found to be effective in quantifying gastric content, both in experimental and clinical studies. It has emerged as a useful and noninvasive bedside technique for estimating the nature of gastric content (empty, clear fluid, thick fluid or solid). The gastric volume can be estimated if only clear fluid is present. Several studies have used ultrasound to assess the gastric fluid volume by measuring the antral cross-sectional area and have found a linear correlation between the antral cross-sectional area and gastric volume. It can help the intensivist for detection of gastric residual volume and rate of gastric emptying in patient on enteral nutrition. Ultrasound also can reliably identify gastric outlet obstruction.

Nerve Blocks Related to Airway

Ultrasound has been used successfully for blocking the superior laryngeal nerve during preparation of awake fiberoptic intubation. The greater horn of the hyoid bone is used as an landmark for identification of the nerve as visualization of the nerve is difficult.

Confirmation of Nasogastric Tube Placement

Combination of airway and abdominal ultrasound can be used for detection of correct placement of nasogastric tube. It has been found to be an effective way for confirmation of nasogastric tube placement with a sensitivity up to 97%.

Diagnosis of Pneumothorax

For detecting or excluding pneumothorax, ultrasound have been found to be as effective as the chest radiograph. It can establish the diagnosis of pneumothorax in patients in whom the pneumothorax is not visible on X-ray and can only be diagnosed by a CT scan. The absence of lung sliding or lung pulsation and presence of prominent A-lines on ultrasonic scan is suggestive of pneumothorax. A "stratosphere sign" on M mode scan and lung point on real time ultrasound are pathognomonic for pneumothorax. The presence of lung sliding can rule out pneumothorax and has a negative predictive value of 100%. Absence of lung sliding alone had a sensitivity of 100% and a specificity of 78%, for the diagnosis of pneumothorax, whereas visualization of the lung point had a sensitivity of 79% and a specificity of 100% for diagnosing pneumothorax. Ultrasound is also the fastest way for diagnosing intraoperative pneumothorax.

FUTURE ASPECTS

With the development of technology in ultrasound, better probes and high-resolution real time imaging, ultrasound will be the preferred and first line of investigation tool for airway assessment and management. 3D and 4D ultrasound will further move the boundaries of airway assessment and availability of smaller portable devices will increase its use.

CONCLUSION

The scope of airway ultrasound in critical care has increased many folds in recent years as it is simple, safe, quick, repeatable, portable and widely available. It also gives real-time dynamic images which are relevant for airway management. A thorough knowledge of the anatomy, sonoanatomy and proper skills for ultrasound scanning can help the intensivist to use ultrasound effectively in various aspects of airway management. Optimal benefit of airway ultrasound can be achieved by using it dynamically in direct conjunction with clinical methods for airway assessment and procedures. With further researches and growing evidences, these airway ultrasound techniques are highly likely to become routine practice in future of airway management.

FURTHER READING

1. Abdallah FW, Yu E, Cholvisudhi P, et al. Is ultrasound a valid and reliable imaging modality for airway evaluation? An observational computed tomographic validation study using submandibular scanning of the mouth and oropharynx. J Ultrasound Med. 2016;36(1):49-59.
2. Andruszkiewicz P, Wojtczak J, Sobczyk D, et al. Effectiveness and validity of sonographic upper airway evaluation to predict difficult laryngoscopy. J Ultrasound Med. 2016;35(10):2243-52.
3. Bajracharya RG, Truong AT, Truong DT, et al. Ultrasound-assisted evaluation of the airway in clinical anesthesia practice: past, present and future. Int J Anesthesiol Pain Med. 2015;1:1-2.
4. Bryson PC, Leight WD, Zdanski CJ, et al. High-resolution ultrasound in the evaluation of pediatric recurrent respiratory papillomatosis. Arch Otolaryngol Head Neck Surg. 2009;135:250-3.
5. Ding LW, Wang HC, Wu HD, et al. Laryngeal ultrasound: a useful method in predicting post-extubation stridor. A pilot study. Eur Respir J. 2006;27:384-9.

6. Elliott DS, Baker PA, Scott MR, et al. Accuracy of surface landmark identification for cannula cricothyroidotomy. Anaesthesia. 2010;65:889-94.
7. Emshoff R, Bertram S, Kreczy A. Topographic variations in anatomical structures of the anterior neck of children: an ultrasonographic study. Oral Surg Oral Med Oral Pathol Oral Radiol Endod. 1999;87:429-36.
8. Ezri T, Gewurtz G, Sessler DI, et al. Prediction of difficult laryngoscopy in obese patients by ultrasound quantification of anterior neck soft tissue. Anaesthesia. 2003;58:1111-4.
9. Friedman EM. Role of ultrasound in the assessment of vocal cord function in infants and children. Ann Otol Rhinol Laryngol. 1997;106:199-209.
10. Garel C, Contencin P, Polonovski JM, et al. Laryngeal ultrasonography in infants and children: a new way of investigating. Normal and pathological findings. Int J Pediatr Otorhinolaryngol. 1992; 23:107-15.
11. Hardee PS, Ng SY, Cashman M. Ultrasound imaging in the preoperative estimation of the size of tracheostomy tube required in specialised operations in children. Br J Oral Maxillofac Surg. 2003; 41:312-6.
12. Hatfield A, Bodenham A. Portable ultrasonic scanning of the anterior neck before percutaneous dilatational tracheostomy. Anaesthesia. 1999;54:660-3.
13. Hatfield A, Bodenham A. Ultrasound: an emerging role in anaesthesia and intensive care. Br J Anaesth. 1999;83:789-800.
14. Hui CM, Tsui BC. Sublingual ultrasound as an assessment method for predicting difficult intubation: a pilot study. Anaesthesia. 2014;69(4):314-19.
15. Jacoby J, Smith G, Eberhardt M, et al. Bedside ultrasound to determine prandial status. Am J Emerg Med. 2003;21:216-9.
16. Jiang JR, Tsai TH, Jerng JS, et al. Ultrasonographic evaluation of liver/spleen movements and extubation outcome. Chest. 2004;126:179-85.
17. Kajeker P, Mendonac C, Gaur V. Role of ultrasound in airway assessment and management. IJUTPC. 2010;1:97-100.
18. Komatsu R, Sengupta P, Wadhwa A, et al. Ultrasound quantification of anterior soft tissue thickness fails to predict difficult laryngoscopy in obese patients. Anaesth Intensive Care. 2007;35:32-7.
19. Kristensen MS. Ultrasonography in the management airway. Acta Anaesthesiol Scand. 2011;55:1155-1173.
20. Kundra P, Mishra SK, Ramesh A. Ultrasound of Airway. Indian J Anaesth. 2011;55:456-62.
21. Lahav Y, Rosenzweig E, Heyman Z, et al. Tongue base ultrasound: a diagnostic tool for predicting obstructive sleep apnea. Ann Otol Rhinol Laryngol. 2009;118:179-84.
22. Lakhal K, Delplace X, Cottier JP, et al. The feasibility of ultrasound to assess subglottic diameter. Anesth Analg. 2007;104:611-4.
23. Lichtenstein D, Biderman P, Meziere G, et al. Sinusogram, a real time ultrasound sound of maxillary sinusitis. Intensive Care Med. 1998;24:1057-61.
24. Lichtenstein DA, Menu Y. A bedside ultrasound sign ruling out pneumothorax in the critically ill. Lung sliding. Chest. 1995;108:1345-8.
25. Lichtenstein DA, Meziere G, Lascols N, et al. Ultrasound diagnosis of occult pneumothorax. Crit Care Med. 2005;33:1231-8.
26. Liu KH, Chu WC, To KW, et al. Sonographic measurement of lateral parapharyngeal wall thickness in patients with obstructive sleep apnea. Sleep. 2007;30:1503-8.
27. Lixin J, Bing H, Zhigang W, et al. Sonographic diagnosis features of Zenker diverticulum. Eur J Radiol. 2011;80:e13-9.
28. Mallin M, Curtis K, Dawson M, et al. Accuracy of ultrasound- guided marking of the cricothyroid membrane before simulated failed intubation. Am J Emerg Med. 2014;32(1):61-3.
29. Manikandan S, Neema PK, Rathod RC. Ultrasound-guided bilateral superior laryngeal nerve block to aid awake endotracheal intubation in a patient with cervical spine disease for emergency surgery. Anaesth Intensive Care. 2010;38:946-8.

30. Milling TJ, Jones M, Khan T, et al. Transtracheal 2-d ultrasound for identification of esophageal intubation. J Emerg Med. 2007;32:409-14.
31. Muhammad JK, Patton DW, Evans RM, et al. Percutaneous dilatational tracheostomy under ultrasound guidance. Br J Oral Maxillofac Surg. 1999;37:309-11.
32. Munir N, Hughes D, Sadera G, et al. Ultrasound guided localisation of trachea for surgical tracheostomy. Eur Arch Otorhinolaryngol. 2010;267:477-9.
33. Muslu B, Sert H, Kaya A, et al. Use of sonography for rapid identification of esophageal and tracheal intubations in adult patients. J Ultrasound Med. 2011;30:671-6.
34. Nicholls SE, Sweeney TW, Ferre RM, et al. Bedside sonography by emergency physicians for the rapid identification of landmarks relevant to cricothyrotomy. Am J Emerg Med. 2008;26:852-6.
35. Perlas A, Chan VW, Lupu CM, et al. Ultrasound assessment of gastric content and volume. Anesthesiology. 2009;111:82-9.
36. Rajajee V, Williamson CA, West BT. Impact of real-time ultrasound guidance on complications of percutaneous dilatational tracheostomy: a propensity score analysis. Crit Care. 2015;19:198.
37. Sartori S, Tombesi P. Emerging roles for transthoracic ultrasonography in pleuropulmonary pathology. World J Radiol. 2010;2:83-90.
38. Shibasaki M, Nakajima Y, Ishii S, et al. Prediction of pediatric endotracheal tube size by ultrasonography. Anesthesiology. 2010;113:819-24.
39. Shiver SA, Blaivas M. Gastric outlet obstruction secondary to linitis plastica of the stomach as shown on transabdominal sonography. J Ultrasound Med. 2004;23:989-92.
40. Singh M, Chin KJ, Vincent W, et al. Use of sonography for airway assessment: an observational study. J Ultrasound Med. 2010;29:79-85.
41. Sustic A, Miletic D, Protic A, et al. Can ultrasound be useful for predicting the size of a left doublelumen bronchial tube? Tracheal width as measured by ultrasonography versus computed tomography. J Clin Anesth. 2008;20:247-52.
42. Sustic A. Role of ultrasound in the airway management of critically ill-patients. Crit Care Med. 2007;35:S173-7.
43. Sustić A, Kovac D, Zgaljardić Z, et al. Ultrasound-guided percutaneous dilatational tracheostomy:a safe method to avoid cranial misplacement of the tracheostomy tube. Intensive Care Med. 2000;26:1379-81.
44. Sustić A, Zupan Z, Antoncić I. Ultrasound-guided percutaneous dilatational tracheostomy with laryngeal mask airway control in a morbidly obese patient. J Clin Anesth. 2004;16:121-3.
45. Teoh WH, Kristensen MS. Utility of ultrasound in airway management. Trends in Anaesthesia and Critical Care. 2014;4(4):84-90.
46. Tsui BC, Hui CM. Challenges in sublingual airway ultrasound interpretation. Can J Anaesth. 2009; 56:393-4.
47. Vigneau C, Baudel JL, Guidet B, et al. Sonography as an alternative to radiography for nasogastric feeding tube location. Intensive Care Med. 2005;31:1570-2.
48. Wojtczak, Jacek A. Submandibular sonography. J Ultrasound Med. 2012;31(4):523-8.

4B

Ultrasound-guided Percutaneous Dilatational Tracheostomy and Cricothyrotomy

Sulagna Bhattacharjee, Souvik Maitra

INTRODUCTION

Percutaneous dilatational tracheostomy (PCT) is a commonly performed bedside intervention in a critical care unit[1] and it is expected that the number of PCTs performed will increase as the demand for critical care services increases. A review of the hospital record database in North America found that the number of tracheostomy performed for prolonged mechanical respiratory support has increased nearly 200% from 1993 to 2002.[2] Another population-based study from Canada predicted that there would be around 80% increase in the number of patients requiring mechanical ventilation from 2000 to 2016.[3] There is some evidence that early tracheostomy may decrease incidence of pneumonia in ventilated patients when compared to late or no tracheostomy.[4] However, there is no level I evidence that early tracheostomy decreases intensive care mortality.[4,5] However, a few retrospective studies have found that early tracheostomy may be beneficial.[6,7] Tracheostomy can be performed either by open surgical technique or by percutaneous dilatational technique (PCT), and both the can be performed at the bedside in an ICU.[8,9] Recent evidence says that there is no difference in these two techniques in terms of procedure related mortality or serious life-threatening events.[10] However, wound infection may be reduced by 75% and postoperative scarring is also reduced with percutaneous technique.[10] An international survey in 2015 found that around 41.6% of all tracheostomies are performed by single-step dilatational technique.[11] This survey has also reported that though in 70% of cases bronchoscopy was used, only in around 21% cases, ultrasound was used to assist cannula placement and dilator advancement, or guidewire advancement.

Though PCT is classically performed with real-time fiber optic bronchoscopy guidance,[12] other methods, such as Ambu aScope[13] or ultrasound[14] has been used. There are studies where PCT has been performed even without guidance.[15,16] Safety profile and procedural complication rate of ultrasonography (USG) guided PCT is similar to bronchoscopy-guided PCT.[14,17] However, ultrasound guided tracheal puncture significantly increases rate of appropriate puncture and first puncture rate when compared with landmark-guided puncture.[18]

Ciaglia et al. in 1985 first described PCT where trachea was punctured by a needle at midline after shallow incision.[19] A 'J tipped' flexible metal guidewire was inserted through the needle and trachea was dilated over the guidewire.

ULTRASONOGRAPHIC ANATOMY OF NECK

Thyroid cartilage, cricoid cartilage, tracheal rings and air-mucosal surface in the trachea are the anterior midline structures seen in either longitudinal or transverse scan of neck (Figs. 1 and 2). Cartilages have a hypoechoic (dark) sonographic appearance and on the other hand, air-mucosal interface appears to be a bright (hyperechoic) line just below the tracheal rings, as the sound waves are strongly reflected due a significant difference in acoustic impedance of air and mucosal membrane. Posterior wall of the trachea is not visualized in USG, because ultrasound beams cannot pass through the air column in the trachea. The carotid arteries are seen as circular hypoechoic pulsatile structures lateral to the trachea and the thyroid gland is seen as a homogeneous structure anterior and lateral to the trachea. Identification of the vascular structures is of particular importance as it has been shown to reduce vascular complications.[20] Anterior jugular vein can be found near the midline and then defined as "vulnerable" as there is a probability of bleeding during procedure. Anterior jugular veins located near the midline and larger than 4 mm in diameter are considered to be at high risk for bleeding.[21] There are reports where anterior jugular vein was dissected and ligated under direct vision.

TECHNIQUES

A number of authors have described techniques of ultrasound-guided PCT, both in prospective and retrospective studies.[18,22-24] Initial studies have used ultrasound for scaning of the anterior neck structures, and they observed that preprocedure scanning of the anterior neck structures may decrease the incidence of vascular complication.

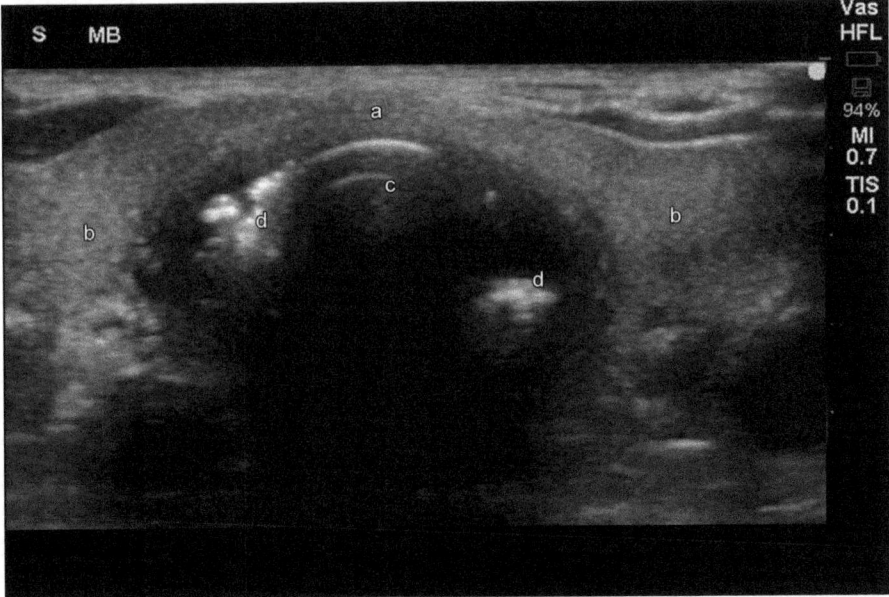

Fig. 1: Short axis out of plane ultrasonic view of tracheal in the midline (a- thyroid isthmus, b- lateral lobe of thyroid gland, c- tracheal lumen, d- cuff of endotracheal tube filled with saline).

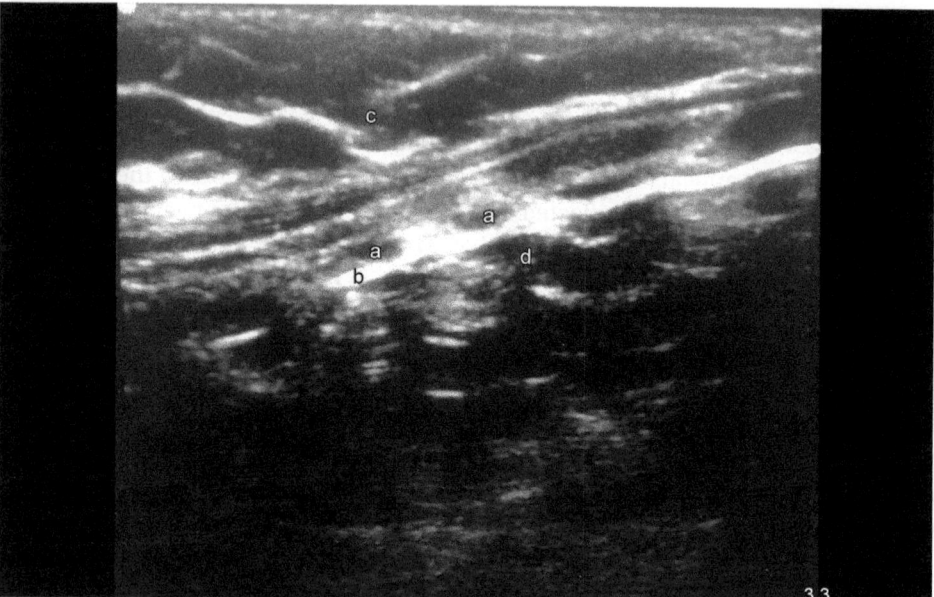

Fig. 2: Longitudinal in-plane view of the trachea, demonstrating ring and air-mucosal surface in cross-section (a- tracheal ring, b- air-mucosal surface, c- pre-tracheal tissue, d- tracheal lumen).

Subsequently, tracheal wall puncture and threading of guidewire was visualized in real-time USG. PCT can be performed either in intubated patients or in patients with laryngeal mask airway. One study reported that preprocedure ultrasound scanning changed the position of intended puncture site in 24% patients.[25]

The first step of this procedure is the scanning of the trachea, paratracheal soft tissue, identification of midline and any vascular structure that may lie in the intended site of tracheal puncture. A linear array probe of 6–13 MHz or 5–10 MHz set in musculoskeletal pre-set is best suited for scanning and real-time puncture. The patients should be positioned in neck extension except in cases of cervical spine injury. Immediate vascular complications can be avoided by identification of the "at risk" arteries or veins.[26] After preprocedure scanning, the endotracheal tube is to be withdrawn under ultrasound visualization till the tip of the endotracheal tube lies below the cricoid cartilage.[18] However, endotracheal tube can also be withdrawn under direct bronchoscopy guidance or under direct laryngoscopy. Where direct laryngoscopy is used, the cuff of the endotracheal tube is kept at the level of glottis.[27]

Next, the anterior tracheal wall is punctured by a needle attached with a saline filled syringe in the midline between first and fifth tracheal ring and the passage of the needle through skin, subcutaneous tissue and anterior tracheal wall is visualized in ultrasound. Ideal puncture point is in the anterior quadrant of the trachea between 1 o'clock and 11 o'clock position.[28] Both a transverse axis out-of-plane (Fig. 3) and a long-axis technique[23] (Fig. 4) can be utilized. Once the anterior tracheal wall is punctured, and the tip of the needle is visualized inside the trachea (Fig. 5), a guidewire is inserted. It is seen passing through the needle and

Ultrasound-guided Percutaneous Dilatational Tracheostomy and Cricothyrotomy 85

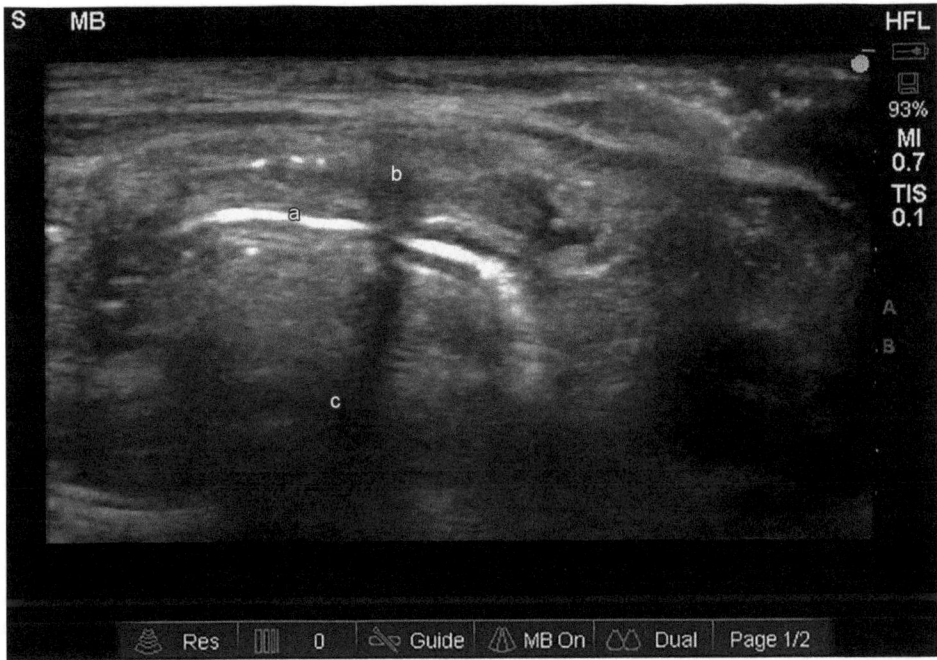

Fig. 3: Short axis out of plane ultrasonic view of trachea in the midline with showing the puncture of anterior tracheal wall with needle (a- air-mucosal surface, b- indentation of the puncture needle, c- tracheal lumen).

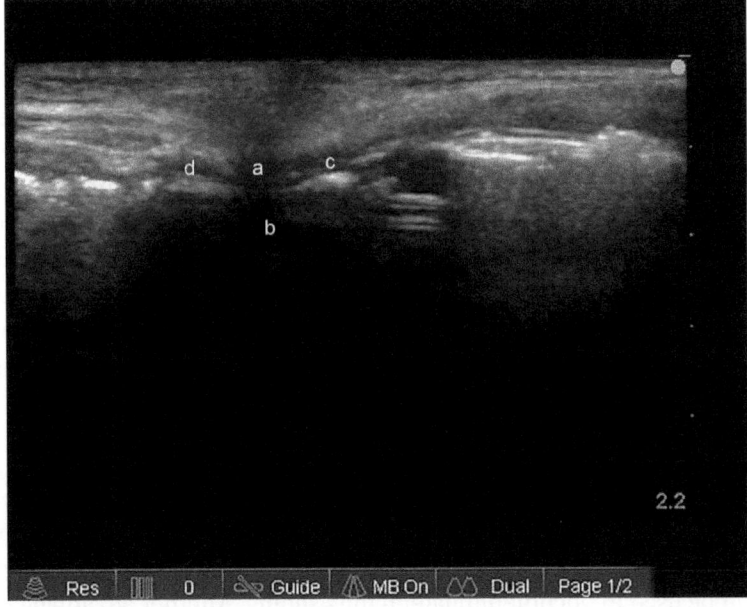

Fig. 4: Ultrasonic view of anterior trachea in the long axis showing the puncture of anterior tracheal wall with needle (a- indentation of the needle, b- tracheal lumen, c, d- tracheal rings).

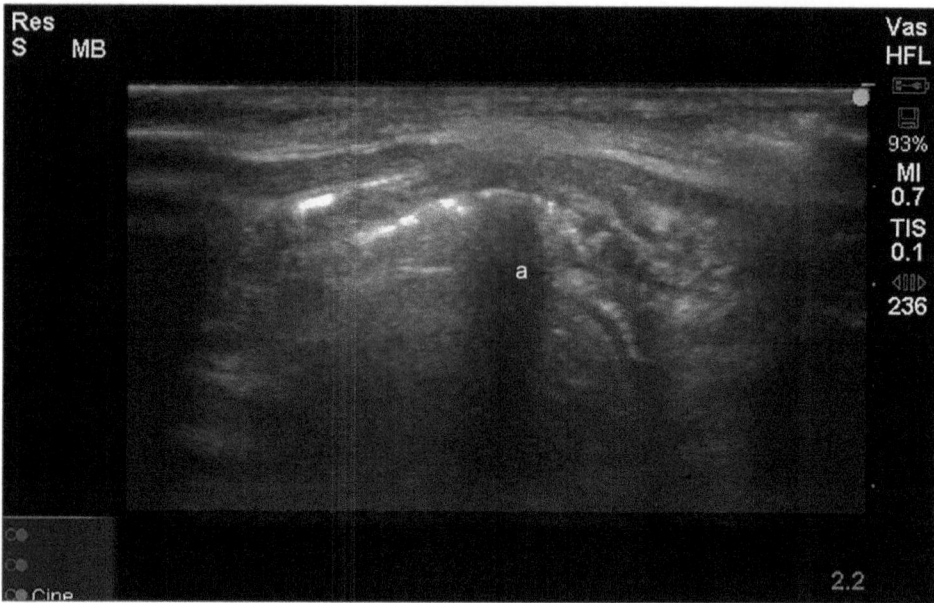

Fig. 5: Short axis out of plane ultrasonic view of trachea with the shadow of tracheal dilator in side tracheal lumen (a- shadow of tracheal dilator inside the tracheal lumen).

going caudally. More importantly, the puncture point should be kept between the first and the fifth tracheal rings, rather than any specific intertracheal ring space. The thyroid isthmus may lie in the path of needle puncture. Avoiding tracheal puncture below third tracheal ring can minimize this problem.[28]

In the long-axis view, the needle shaft and guidewire are seen as hyperechoic lines and in transverse view, they are seen as hyperechoic dots. No study has compared these two techniques for PCT till now. However, long-axis in-plane technique eases needle visibility. The shallow horizontal skin incision (1–1.5 cm) can be given after successful guidewire insertion. A few authors confirmed guidewire position by direct visualization by bronchoscopy.[23] Next, a small dilator is used to create the initial stoma and then a single stage curved dilator is used to dilate the stoma over the guidewire. This step can also be visualized in real-time ultrasound. Then a proper sized tracheostomy tube fitted over a loading tube is inserted over the guidewire. Chacko et al.[29] reported that USG image-guided tracheal puncture has a very high success rate.

CONFIRMATION OF PLACEMENT

Bilateral equal air entry in all lung fields on auscultation can be used to confirm proper placement of the TT. Other methods, such as direct bronchoscopic visualization can also be used. Ultrasound can be used to document bilateral lung sliding. Another advantage of ultrasound is that it can be used to exclude pneumothorax immediately.

COMPLICATIONS

Safety profile of PCT is similar to surgical tracheostomy and it is a cost-effective alternative.[30] In terms of mortality, intraoperative and postoperative hemorrhage, both surgical and percutaneous tracheostomies are equivalent.[31] Kearney et al.[32] reported complications of 827 PCT procedures and found that mortality rate was 0.6%. Reported procedure related complications were inadvertent extubation, bleeding, false passage in the subcutaneous tissue, pneumothorax, guidewire dislodgement, subcutaneous emphysema, difficult tracheal tube placement, tracheal laceration and tracheoesophageal fistula. Bleeding is the commonest postoperative complication that occurred in 1.6% of patients. Others were airway obstruction after decannulation, premature extubation, stomal infection and excessive granulation tissue. Another large series by Kost[33] in 2005 reported that overall complication rate following PCT is 9.2%; multiple dilator technique is associated with 13.6% and single dilator technique is associated with 6.5% complications. Body mass index (BMI) more than 30 kg/m² and inexperienced operator were the risk factors of complication. However, we should note that in none of the previous series, ultrasound was used as an aid for the procedure. A recent randomized trial reported a significantly lower desaturation, minor and major bleeding, and multiple puncture attempts with real-time ultrasound-guided PCT.[34] Another noninferiority trial also demonstrated similar complication rates between ultrasound-guided and bronchoscopy-aided tracheostomy.[17] In this trial, procedure-related minor complications occurred in 33.3% patients in the ultrasound group and in 20.7% patients in the bronchoscopy group, which was statistically insignificant. Whether obesity increases complication rate of PCT is also a matter of debate now-a-days. A small prospective study reported similar complication rate between obese and nonobese patients.[27]

Whether ultrasound is a better alternative to bronchoscopy is a matter of debate in the present time. There is some evidence from a small single prospective randomized controlled trail that ultrasound-guided PCT may be a better alternative than bronchoscopy-guided technique. But, we feel further large randomized trials are required to confirm this finding.

ULTRASOUND-GUIDED CRICOTHYROTOMY

Emergency cricothyrotomy is a life-saving procedure required in "cannot intubate cannot ventilate situation" during general anesthesia or in acute upper airway obstruction.[35] Identification of the cricothyroid membrane based on anatomical landmarks is often inaccurate.[36] Ultrasound offers rapid assessment of the airway and has been found to be significantly helpful in the setting of emergency cricothyrotomy. Anatomical landmark-guided cricothyroid membrane puncture was first described as high cricothyrotomy in 1909 by Dr Chevalier Jackson, a laryngologist.[37]

CRICOTHYROTOMY TECHNIQUE

Various methods have been described in the literature for anatomical landmark-guided cricothyrotomy.[38,39] In standard technique, the larynx is immobilized and the cricothyroid membrane palpated, the skin is then incised vertically 3–5 cm overlying the membrane, and a horizontal incision of 1 cm is given over the cricothyroid membrane, with the tip of the scalpel facing caudad. Then, a tracheal hook is placed under the thyroid cartilage, traction is applied anteriorly, incision is enlarged by Trousseau dilator, following which tracheostomy tube[40] is inserted.

"Seldinger technique" for percutaneous cricothyrotomy has also been demonstrated to be significantly faster as compared to standard cricothyrotomy.[41] A number of commercial kits with all necessary equipment are available (e.g. Cook Melker Kit).

In this method, skin overlying the cricothyroid membrane is incised vertically by a stab knife. Then, cricothyroid membrane is punctured with an 18-gauge needle attached to a saline filled syringe through the incision. Needle placement in the trachea is confirmed by aspiration of air. A guidewire is inserted through the needle following which the latter is removed. An airway catheter with a 15 mm connector, along with a dilator is sled over the guidewire, followed by removal of the guidewire and the dilator. The Difficult Airway Society (DAS) 2015 guidelines recommend cricothyrotomy as the emergency surgical procedure either by scalpel technique or by cannula technique.[42]

ROLE OF ULTRASOUND IN CRICOTHYROTOMY

Conventional landmark-guided techniques of cricothyroid membrane identification are inaccurate particularly in patients with obesity and altered neck anatomy.[43,44] In an observational study, cricothyroid membrane (CTM) was accurately identified only in 39% of the obese pregnant patients by digital palpation compared to 71% in nonobese parturients. Though it is controversial whether ultrasound improves accuracy of CTM identification, there is some evidence that inadvertent injuries to the adjacent structures during needle puncture may be reduced with the use of ultrasound.[45,46] However, the time to identify CTM was significantly higher with ultrasound.[45] The DAS guideline 2015 has emphasized the use of ultrasound before airway management in elective settings when identification of the trachea and cricothyroid membrane is not possible with inspection and palpation alone. In emergency setting, ultrasound should be used to identify the key anatomical landmarks only, if available immediately, but its use should not delay emergency airway access. The guidelines also recommend training in using ultrasound for airway evaluation.[42] Nicholas et al. devised a method to identify the CTM in ultrasound image and other relevant structures in four cadaveric models and assessed its applicability in a cohort of 50 emergency department patients by two emergency physicians.[47] In a literature review, Kristensen et al.[48] identified two distinct techniques for performing ultrasound-guided cricothyrotomy.

THE TRANSVERSE TACA TECHNIQUE (THYROID CARTILAGE-AIRLINE-CRICOID CARTILAGE-AIRLINE)

In a transverse view of the anterior neck, the thyroid cartilage appears like a hyperechoic triangular structure. The transducer is next moved caudally to visualize a hyperechoic white line sometimes with parallel white lines underneath formed by reverberation artifacts. When further scanned caudally, the cricothyroid cartilage appears as "C" in black with a shiny lining.

THE LONGITUDINAL "STRING OF PEARL" TECHNIQUE

When a linear array transducer placed on the anterior neck, trachea is visualized as a black structure with a posterior shiny lining. In a midline longitudinal scan, trachea appears as a number of dark rings or "pearls" anterior to the white hyperechoic line of air-mucosal interface (string of pearls). The transducer is then sled cephalad until the cricoid cartilage (larger dark "pearl" compared to the tracheal rings) comes into view. Further cephalad, the distal part of the thyroid cartilage is noted. With a needle used as a pointer to create shadow, the midpoint of cricothyroid membrane is identified and skin overlying the membrane is marked with a pen.

Curtis et al.[49] described a novel technique for ultrasound-guided emergent cricothyrotomy. They used a linear array ultrasound probe longitudinally on the midline of the neck, incised the CTM with a number 20 scalpel blade. A curved-tip bougie was then inserted through the incision, followed by railroading a 6.0 mm internal diameter endotracheal tube over the bougie. This technique appears to be rapid and safe, with a median time to identification of the cricothyroid membrane of 3.6 seconds and median time to completion of the procedure of 26.2 seconds, with a failure rate of just 1 out of 21, where the incision placed between the cricoid cartilage and the first tracheal ring.

WHICH TECHNIQUE TO FOLLOW?

Compared to the longitudinal technique, the transverse technique appears to be faster and is advantageous in patients with short neck and severe neck flexion deformity.[48] The longitudinal technique can reveal additional information about the cricotracheal and tracheal interspaces. It can also identify overlying blood vessels. It is also useful while securing emergency tracheal airway access, especially in small children and in the presence of tumors over the cricothyroid membrane.[48] Therefore, it is worth learning both the techniques, as they supplement each other when used in tandem.[48]

In a randomized crossover trial, Kristensen and colleagues[50] compared the longitudinal and the transverse ultrasound-guided techniques for identification of the cricothyroid membrane. Forty two anesthetists were enrolled after a predefined teaching, followed by application of both techniques in a randomized, cross-over sequence to obese females with BMI 39.0–43.9 kg/m^2. The transverse technique was faster than the longitudinal technique (mean time to identify the cricothyroid membrane 24 s vs 37.6 s). The cricothyroid membrane was successfully identified by 90% of the anesthetists using either of the technique, and all anesthetists could successfully identify the cricothyroid membrane with at least one of the techniques. Therefore, the authors highlighted the importance of training and implementation of both the techniques for identification of the cricothyroid membrane in anticipated difficult airway scenario.

COMPLICATIONS

No study has yet reported any complication specific to ultrasound-guided technique of cricothyrotomy. Complications reported from cricothyrotomy can be classified as immediate (bleeding, pneumothorax, difficult cannulation), early (subglottic stenosis, acute respiratory failure, tracheoesophageal fistula, temporary or chronic dysfunction of vocal cords) and late (tracheal granuloma, nonhealing wounds, and scars). Subglottic stenosis has been reported with a frequency of 2.9–5%.[50]

Siddiqui et al.[46] studied the impact of ultrasound-guided cricothyrotomy using Portex device in human cadavers as compared to anatomical landmark-guided digital palpation method by group of trainees. The use of ultrasound significantly reduced laryngeal and tracheal injury (74% vs 25%) and significantly increased probability of proper placement in cadavers with difficult anatomy (digital palpation 8.3% vs ultrasound 46.7%).

In conclusion, it is needless to say that ultrasound scanning improves success rate of cricothyroid membrane identification. However, further human studies are required to know its value in cricothyrotomy in difficult airway setting.

REFERENCES

1. Delaney A, Bagshaw SM, Nalos M. Percutaneous dilatational tracheostomy versus surgical tracheostomy in critically ill patients: a systematic review and meta-analysis. Crit Care. 2006;10:R55.
2. Cox CE, Carson SS, Holmes GM, et al. Increase in tracheostomy for prolonged mechanical ventilation in North Carolina, 1993-2002. Crit Care Med. 2004;32(11):2219-26.
3. Needham DM, Bronskill SE, Calinawan JR, et al. Projected incidence of mechanical ventilation in Ontario to 2026: Preparing for the aging baby boomers. Crit Care Med. 2005;33(3):574-9.
4. Siempos II, Ntaidou TK, Filippidis FT, et al. Effect of early versus late or no tracheostomy on mortality and pneumonia of critically ill patients receiving mechanical ventilation: a systematic review and meta-analysis. Lancet Respir Med. 2015;3(2):150-8.
5. Szakmany T, Russell P, Wilkes AR, et al. Effect of early tracheostomy on resource utilization and clinical outcomes in critically ill patients: meta-analysis of randomized controlled trials. Br J Anaesth. 2015;114(3):396-405.
6. Puentes W, Jerath A, Djaiani G, et al. Early versus late tracheostomy in cardiovascular intensive care patients. Anaesthesiol Intensive Ther. 2016;48(2):89-94.
7. Devarajan J, Vydyanathan A, Xu M, et al. Early tracheostomy is associated with improved outcomes in patients who require prolonged mechanical ventilation after cardiac surgery. J Am Coll Surg. 2012;214(6):1008-16.e4.
8. Upadhyay A, Maurer J, Turner J, et al. Elective bedside tracheostomy in the intensive care unit. J Am Coll Surg. 1996;183(1):51-5.
9. Toursarkissian B, Zweng TN, Kearney PA, et al. Percutaneous dilational tracheostomy: report of 141 cases. Ann Thorac Surg. 1994;57(4):862-7.
10. Brass P, Hellmich M, Ladra A, et al. Percutaneous techniques versus surgical techniques for tracheostomy. Cochrane Database Syst Rev. 2016;7:CD008045.
11. Vargas M, Sutherasan Y, Antonelli M, et al. Tracheostomy procedures in the intensive care unit: an international survey. Crit Care. 2015;19:291.
12. Romero CM, Cornejo R, Tobar E, et al. Fiber optic bronchoscopy-assisted percutaneous tracheostomy: a decade of experience at a university hospital. Rev Bras Ter Intensiva. 2015;27(2):119-24.
13. Reynolds S, Zurba J, Duggan L. A single-centre case series assessing the Ambu® aScope™ 2 for percutaneous tracheostomies: A viable alternative to fibreoptic bronchoscopes. Can J Respir Ther. 2015;51(2):43-5.
14. Gobatto AL, Besen BA, Tierno PF, et al. Comparison between ultrasound- and bronchoscopy-guided percutaneous dilational tracheostomy in critically ill patients: a retrospective cohort study. J Crit Care. 2015;30(1):220.e13-7.
15. Gadkaree SK, Schwartz D, Gerold K, et al. Use of Bronchoscopy in Percutaneous Dilational Tracheostomy. JAMA Otolaryngol Head Neck Surg. 2016;142(2):143-9.

16. Pattnaik SK, Ray B, Sinha S. Griggs percutaneous tracheostomy without bronchoscopic guidance is a safe method: a case series of 300 patients in a tertiary care Intensive Care Unit. Indian J Crit Care Med. 2014;18(12):778-82.
17. Gobatto AL, Besen BA, Tierno PF, et al. Ultrasound-guided percutaneous dilational tracheostomy versus bronchoscopy-guided percutaneous dilational tracheostomy in critically ill patients (TRACHUS): a randomized noninferiority controlled trial. Intensive Care Med. 2016;42(3):342-51.
18. Rudas M, Seppelt I, Herkes R, et al. Traditional landmark versus ultrasound guided tracheal puncture during percutaneous dilatational tracheostomy in adult intensive care patients: a randomised controlled trial. Crit Care. 2014;18(5):514.
19. Ciaglia P, Firsching R, Syniec C. Elective percutaneous dilatational tracheostomy. A new simple bedside procedure; preliminary report. Chest. 1985;87(6):715-9.
20. Flint AC, Midde R, Rao VA, et al. Bedside ultrasound screening for pretracheal vascular structures may minimize the risks of percutaneous dilatational tracheostomy. Neurocrit Care. 2009;11(3): 372-6.
21. Hatfield A, Bodenham A. Portable ultrasonic scanning of the anterior neck before percutaneous dilatational tracheostomy. Anesthesia. 1999;54(7):660-3.
22. Mitra S, Kapoor D, Srivastava M, et al. Real-time ultrasound guided percutaneous dilatational tracheostomy in critically ill patients: A step towards safety! Indian J Crit Care Med. 2013;17(6): 367-9.
23. Dinh VA, Farshidpanah S, Lu S, et al. Real-time sonographically guided percutaneous dilatational tracheostomy using a long-axis approach compared to the landmark technique. J Ultrasound Med. 2014;33(8):1407-15.
24. Rajajee V, Fletcher JJ, Rochlen LR, et al. Real-time ultrasound-guided percutaneous dilatational tracheostomy: a feasibility study. Crit Care. 2011;15(1):R67.
25. Kollig E, Heydenreich U, Roetman B, et al. Ultrasound and bronchoscopic controlled percutaneous tracheostomy on trauma ICU. Injury. 2000;31(9):663-8.
26. McCormick B, Manara AR. Mortality from percutaneous dilatational tracheostomy. A report of three cases. Anesthesia. 2005;60:490-5.
27. Guinot PG, Zogheib E, Petiot S, et al. Ultrasound-guided percutaneous tracheostomy in critically ill obese patients. Crit Care. 2012;16(2):R40.
28. Alansari M, Alotair H, Al Aseri Z, et al. Use of ultrasound guidance to improve the safety of percutaneous dilatational tracheostomy: a literature review. Crit Care. 2015;19:229.
29. Chacko J, Nikahat J, Gagan B, et al. Real-time ultrasound-guided percutaneous dilatational tracheostomy. Intensive Care Med. 2012;38(5):920-1.
30. Hill BB, Zweng TN, Maley RH, et al. Percutaneous dilational tracheostomy: report of 356 cases. J Trauma. 1996;41(2):238-43.
31. Johnson-Obaseki S, Veljkovic A, Javidnia H. Complication rates of open surgical versus percutaneous tracheostomy in critically ill patients. Laryngoscope. 2016;126(11):2459-67.
32. Kearney PA, Griffen MM, Ochoa JB, et al. A single-center 8-year experience with percutaneous dilational tracheostomy. Ann Surg. 2000;231(5):701-9.
33. Kost KM. Endoscopic percutaneous dilatational tracheotomy: a prospective evaluation of 500 consecutive cases. Laryngoscope. 2005;115(10 Pt 2):1-30.
34. Ravi PR, Vijay MN. Real time ultrasound-guided percutaneous tracheostomy: Is it a better option than bronchoscopic guided percutaneous tracheostomy? Med J Armed Forces India. 2015;71(2): 158-64.
35. Henlin T, Michalek P, Tyll T, et al. Oxygenation, ventilation, and airway management in out-of-hospital cardiac arrest: a review. Biomed Res Int. 2014:376871.
36. Barbe N, Martin P, Pascal J, et al. Locating the cricothyroid membrane in learning phase: value of ultrasonography? Ann Fr Anesth Reanim. 2014;33:163-6.
37. Jackson C. Tracheotomy. Laryngoscope. 1909;19:285-90.

38. Davis DP, Bramwell KJ, Vilke GM, et al. Cricothyrotomy technique: standard versus the Rapid Four-Step Technique. J Emerg Med. 1999;17:17-21.
39. Schaumann N, Lorenz V, Schellongowski P, et al. Evaluation of Seldinger technique emergency cricothyroidotomy versus standard surgical cricothyroidotomy in 200 cadavers. Anesthesiology. 2005;102(1):7-11.
40. Bair AE. (2016). Emergency cricothyrotomy (cricothyroidotomy). [online]. Available from https://www.uptodate.com/contents/emergency-cricothyrotomy-cricothyroidotomy. [Accessed Nov. 2017].
41. Kundra P, Mishra SK, Ramesh A. Ultrasound of the airway. Indian J Anaesth. 2011;55:456-62.
42. Frerk C, Mitchell VS, McNarry AF, et al. Difficult Airway Society intubation guidelines working group. Difficult Airway Society 2015 guidelines for management of unanticipated difficult intubation in adults. Br J Anaesth. 2015;115:827-48.
43. You-Ten KE, Desai D, Postonogova T, Siddiqui N. Accuracy of conventional digital palpation and ultrasound of the cricothyroid membrane in obese women in labour. Anesthesia. 2015;70:1230-4.
44. Bair AE, Chima R. The inaccuracy of using landmark techniques for cricothyroid membrane identification: a comparison of three techniques. Acad Emerg Med. 2015;22:908-14.
45. Yıldız G, Göksu E, Şenfer A, et al. Comparison of ultrasonography and surface landmarks in detecting the localization for cricothyroidotomy. Am J Emerg Med. 2016;34:254-6.
46. Siddiqui N, Arzola C, Friedman Z, et al. Ultrasound improves cricothyrotomy success in cadavers with poorly defined neck anatomy: a randomized control trial. Anesthesiology. 2015;123:1033-41.
47. Nicholls SE, Sweeney TW, Ferre RM, et al. Bedside sonography by emergency physicians for the rapid identification of landmarks relevant to cricothyrotomy. Am J Emerg Med. 2008;26:852-6.
48. Kristensen MS, Teoh WH, Rudolph SS. Ultrasonographic identification of the cricothyroid membrane: best evidence, techniques, and clinical impact. Br J Anaesth. 2016;117:i39-48.
49. Curtis K, Ahern M, Dawson M, et al. Ultrasound-guided, Bougie-assisted cricothyroidotomy: a description of a novel technique in cadaveric models. Acad Emerg Med. 2012;19:876-9.
50. Kristensen MS, Teoh WH, Rudolph SS, et al. A randomised cross-over comparison of the transverse and longitudinal techniques for ultrasound-guided identification of the cricothyroid membrane in morbidly obese subjects. Anesthesia. 2016;71:675-83.

Chapter 5

Lung Ultrasound

Soumya Ray

OVERVIEW

Ultrasound of lung and pleura is an important element of critical care ultrasound. Lung pathologies are easily accessible through the chest wall. It can be performed quickly at the bedside and can give us important real time information. Goal directed ultrasound can help us institute appropriate management in complex critically ill-patients. In fact, it can be more useful than chest X-ray in diagnosing and managing multitude of pulmonary pathologies like pleural effusion, pneumothorax, pulmonary edema, and pulmonary infiltrates. Using lung ultrasound saves time and decreases the need for CT[1] thereby avoiding ionizing radiation and potentially dangerous transfers. Lung ultrasound is also used as a guide to weaning and fluid management in critically ill-patients. Luckily such a useful management tool is relatively easy to learn with a steep learning curve.[2]

HISTORY

Physics of Lung Ultrasound

Ultrasound images are formed by analysis of reflected sound waves transmitted from tissue interfaces. Normal air filled lungs are very poor conductors of sound. High acoustic impedance between pleura and air results in refection of most of the ultrasound waves with very little ultrasound waves travelling to the air-filled lungs. This makes it easy to appreciate pleural lines. Normal air filled lungs on the other hand are not visible on the ultrasound machine. Hence, we have to rely on pleural movements and artifacts to appreciate underlying normal lung. Pathological lungs (consolidated, fluid filled or collapsed), on the other hand, are devoid of air, transmit ultrasound waves and can be seen on the ultrasound machine.

EQUIPMENT FOR LUNG ULTRASOUND

Lung ultrasound can be performed with a wide variety of machines and probes. High frequency linear array probes (vascular probes) are excellent for visualization of pleura in thinly built persons. Their limited penetration makes it difficult to appreciate deeper structures, especially in patients with thick chest walls. A low frequency 3.5–5 MHz transducer allows better tissue penetration and is recommended for this purpose. In most ICUs, curvilinear probes are commonly used for lung ultrasound as well as abdominal imaging. A phased array probe can also be used for convenience when performing lung ultrasound along with

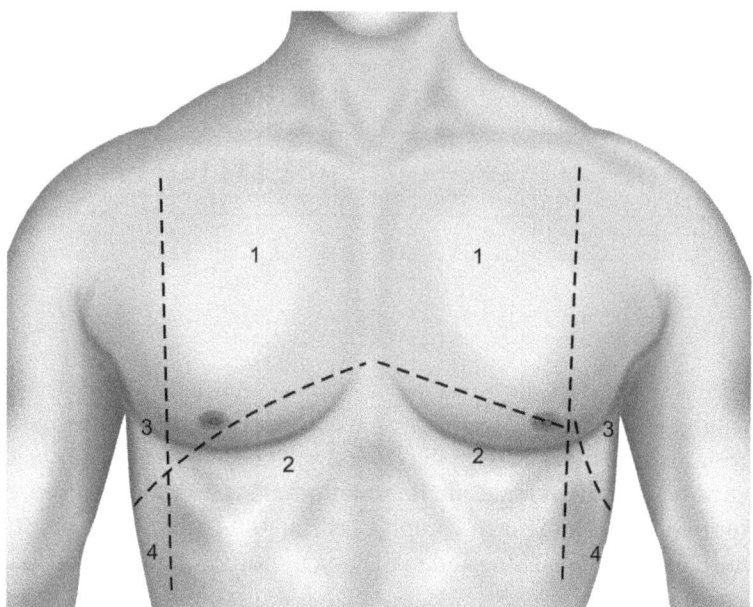

Fig. 1: Four practical lung zones to perform lung ultrasound in a ventilated intensive care unit patient.

echocardiography.[2] Lung ultrasound relies on visualization of artifacts. Artifact suppressing measures in modern ultrasound machines may need to be temporarily disabled (i.e. turning off harmonics, compound imaging, and speckle reduction) and focal zone may be adjusted to enhance imaging of artifacts like A lines, B lines. M mode and color Doppler can be utilized in addition to 2D imaging.

HOW TO PERFORM LUNG ULTRASOUND?

By convention, lung ultrasound is performed with the probe in a longitudinal orientation, perpendicular to the chest wall with the probe marker pointing cranially. Most critically ill patients remain in a supine position with limited access to posterior thorax. This can be accessed in stable patients by sitting them up, rolling to a lateral decubitus position or simply by sliding the probe as posteriorly as possible between the patient and the bed. However, in most of the patients, a bilateral assessment of the anterior and lateral chest wall, each divided into an upper and lower zone is a reasonable and validated method.[3] This may be labeled as zone 1–4 while performing the scan (Fig. 1).

Ultrasound of lung and pleura is almost always performed as a goal directed study in ICU. The diagnosis of any pathology begins with clinical suspicion. The pretest probability determines the significance of ultrasonographic findings and helps in answering the clinical question. This way ultrasound works as a valuable adjunct to our clinical skill.

Normal Lung

The transducer is positioned to focus on the intercostal space between adjacent ribs. The pleural line is easily visible as a hyperechoic horizontal line sliding between two rib shadows,

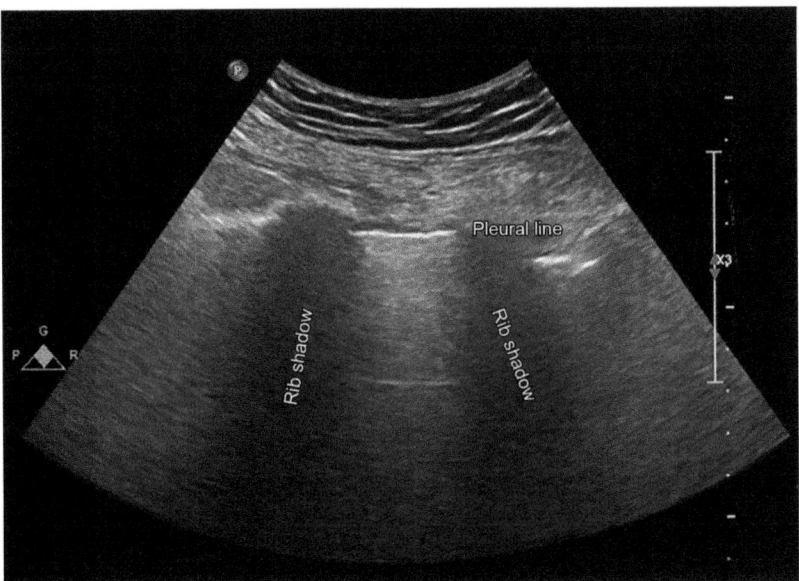

Fig. 2: Bat sign.

slightly deeper to the rib shadows. It is formed by ultrasound waves reflected from parietal form and visceral pleura in contact with each other in normal lung. The bright hyperechoic ribs with the intervening pleural line has been stated as the bat sign (Fig. 2). As the lung moves during respiratory cycle, the visceral pleura slides over the parietal pleura. This gives rise to lung sliding, in sync with respiratory cycle. Pleural movement can sometimes be appreciated in synchrony with cardiac cycle due to transmission of cardiac pulsations. This is termed as lung pulse. Lung pulse may be more easily seen when lung movements are minimal, e.g. endobronchial intubation. Both lung sliding and lung pulse are dynamic signs, which reflect normal lung with visceral and parietal pleura in contact. Both of these signs are lost in pneumothorax with air intervening between the visceral and parietal pleura.

Lung sliding can be confirmed on M mode by the seashore sign (Fig. 3). The top half of the images with horizontal lines reflect static chest wall structures and pleura. They represent the waves. The bottom half of the images representing the beach is formed by mobile lung tissue forming a coarse granular pattern.

Air in normal lung prevents transmission of ultrasound waves into the parenchyma. Hence, the ultrasound image of the lung is composed of artifacts arising from the pleural line. The scanning depth is set to examine deeper structures when looking for air artifacts.

A line is horizontal, equally spaced line recurring at intervals similar to distance between pleural line and the skin surface (Fig. 4). They are motionless and similar in appearance to pleural line. They are caused by reverberation of ultrasound waves repeatedly reflected between the pleura and the skin surface (Fig. 5). A lines in presence of lung sliding is strongly correlated with normal aeration pattern on CT scan.[4] A line dominant pattern indicates dry interlobular septa predicting a low pulmonary artery occlusion pressure and low risk of developing pulmonary edema with fluid therapy.[5]

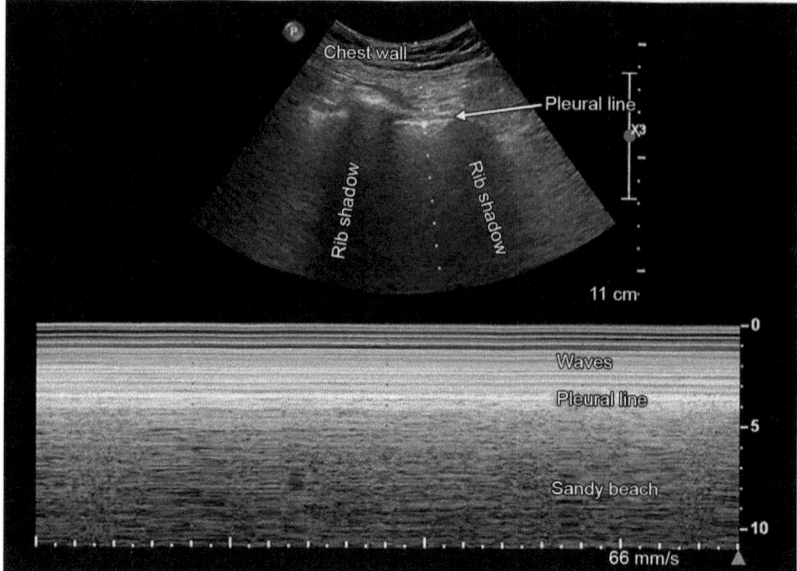

Fig. 3: Seashore sign.

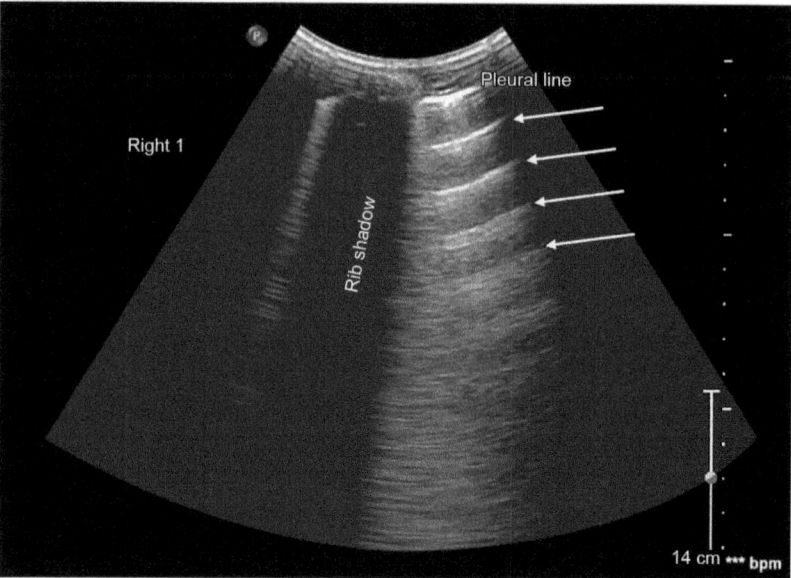

Fig. 4: A lines (reverberation artifacts from pleural line of diminishing echodensity are marked with arrows).

B lines are vertical hyperechoic lines originating from pleural surface and extending in a ray-like pattern from the transducer surface to the bottom of the screen (Fig. 6). They obliterate the A lines at the point of intersection. They move with the sliding of the pleural line with respiratory movements. However, static B lines can also be seen in the absence of lung sliding. They are also sometimes termed as lung rockets. They are a type of comet tail artifacts.

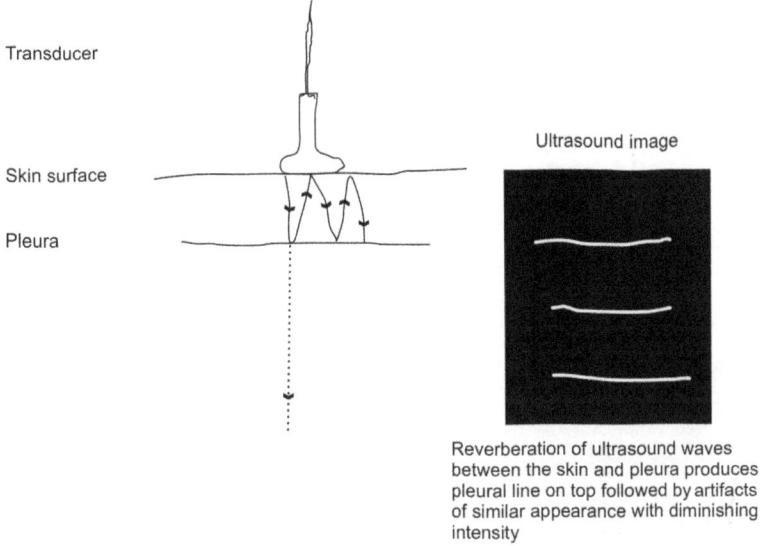

Fig. 5: Reverberation artifacts.

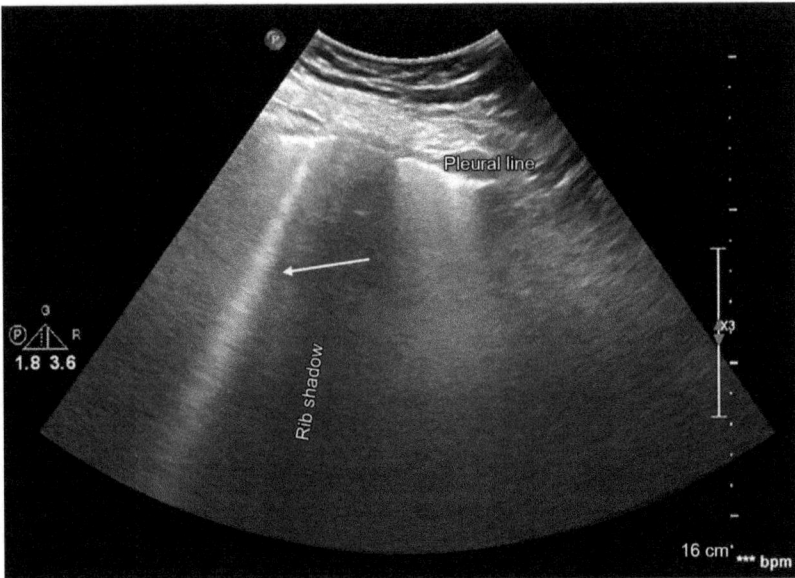

Fig. 6: Solitary B line (arrow denotes the solitary B line originating from pleural surface).

They are thought to be formed as ultrasound reverberates within small gas bubbles with little loss of energy producing an artifact which extends at full brightness to the maximum depth on screen. B lines are indicative of infiltration of pulmonary interstitium, e.g. pulmonary edema, pneumonitis, neoplasm and scarring. 1–2 B lines are frequently seen in last interstitial space in normal subjects (Fig. 6). More than two B lines in a single field is considered significant.

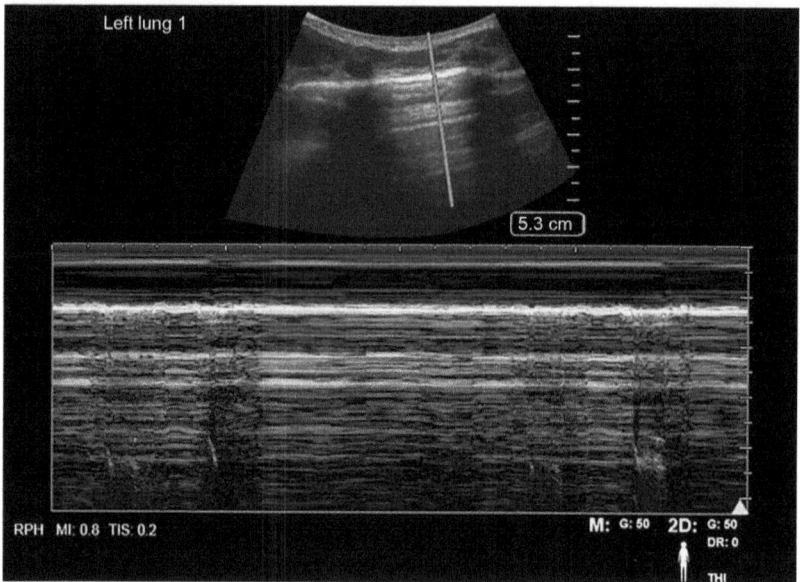

Fig. 7: Barcode sign. Apical pneumothorax on left side displaying the barcode sign on M mode.

B line pattern is strongly correlated with alveolar or interstitial pattern abnormalities on the CT scan.[4]

Several other comet trail artifacts can be confused with B lines. Z lines are ill-defined lines arising from pleural surface and fade quickly without obliterating A lines. They are commonly seen in normal lung and are of little significance.

Pneumothorax

Presence of air between parietal and visceral pleura results in inability to visualize the movements of the underlying visceral pleura and lung. This is appreciated as loss of lung sliding and lung pulse on the ultrasound image. This can be confirmed on M mode while keeping the probe still. This results in appearance of parallel horizontal lines extending throughout the image. The lines in the bottom of the image beyond the pleural line are reverberation artifacts. This is termed as barcode sign or stratosphere sign (compared to sea shore sign in normal lung) (Fig. 7). A line indicates presence of air below the pleura and could be seen in presence of pneumothorax. However, B lines are never seen in presence of pneumothorax.

Absence of lung sliding is a very sensitive but not very specific sign of pneumothorax. Absent lung sliding with bar code sign can also be seen in conditions with impaired lung movement like collapse, atelectasis, massive pneumonia, endobronchial intubation, apnea, acute severe asthma or pleural disease like pleurodesis and pleural effusion. More importantly presence of lung sliding has a high negative predictive value and virtually excludes pneumothorax at the site of transducer placement.[6]

Pleural air tends to move to the anterior thorax in supine ICU patients and can be easily detected by ultrasound. Anterior thorax should be the first site to be checked when performing ultrasound to rule out pneumothorax. Lung ultrasound has high accuracy comparable to CT scan and much higher than chest X-ray for detection of pneumothorax.[7] Lung ultrasound can be performed before and after a procedure like subclavian central line insertion, supraclavicular block, tracheostomy to quickly rule out pneumothorax post procedure.

In some patients, the precise point of separation of pleura by air can be seen. This transition point can be seen on ultrasound as a change from normal lung signs (lung sliding, B lines) to pneumothorax signs as described above. This is termed as the lung point. It is highly specific for diagnosis of pneumothorax, i.e. it rules in the diagnosis of pneumothorax.[8] Detection rate of lung point may depend on operator experience. The position of lung point (lateral chest vs posterior chest) is also indicative of size of pneumothorax.[8] It can also be useful in planning for a pleural drain to evacuate a pneumothorax.

Pleural Effusion

Pleural effusion can be seen in around two-third of patients in medical ICU.[9] Point-of-care ultrasound is more sensitive than physical examination and chest radiography to detect and characterize pleural fluid with a diagnostic accuracy of 93%.[4]

Pleural fluid appears as anechoic or hypoechoic space surrounded clear anatomic boundaries, i.e. diaphragm and subdiaphragmatic organs inferiorly, chest wall laterally and lungs (or heart on the left side) medially. Identification of diaphragm and subdiaphragmatic organs is critical to localize pleural effusions and safely perform thoracentesis (Figs. 8 and 9). Movement of the lungs with respiratory cycle can produce certain characteristic ultrasonographic signs.

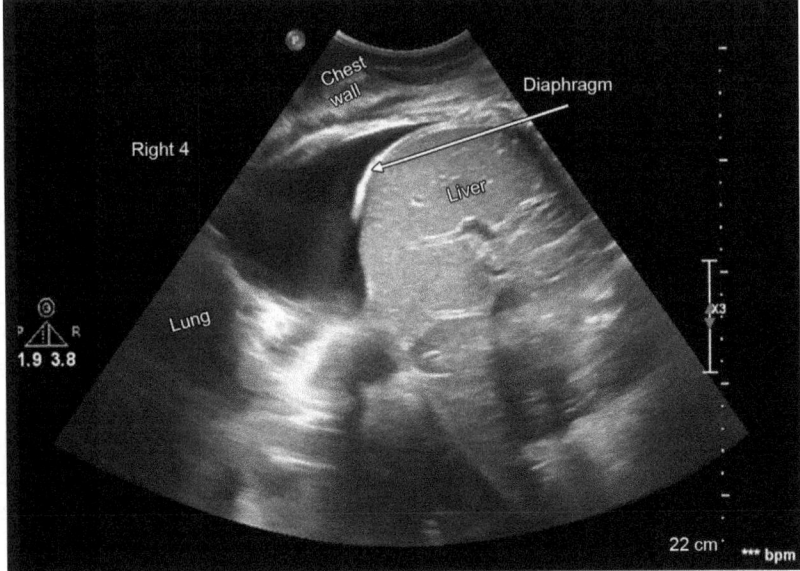

Fig. 8: Pleural fluid: surrounded by diaphragm, chest wall and right lung.

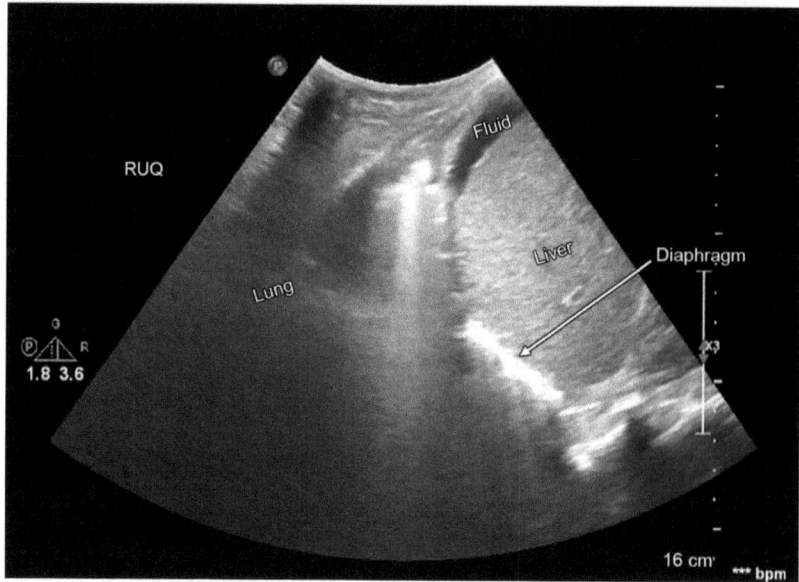

Fig. 9: Subdiaphragmatic fluid. Perihepatic fluid is seen inferior to the diaphragm (note small right basal consolidation with air bronchogram).

- *Flapping lung or jellyfish sign:* Movement of atelectatic lung in moderate to large effusions
- *Curtain sign:* Aerated lung partially obstructing the visualization of pleural fluid during inspiration
- *Sinusoid sign:* Undulating movement of visceral pleura with respiratory cycle on M mode
- *Plankton sign:* Movement of debris within pleural fluid
- *Hematocrit sign:* Settling of cellular debris within pleural fluid (Fig. 10).

Size of the pleural fluid is usually sufficiently described in a qualitative manner as mild, moderate and severe. Various quantitative formulae have been also derived to estimate volume of pleural fluid. One of such formula estimated the volume by multiplying the maximal distance between parietal and visceral pleura (in millimeter) at end-expiration on a transverse plane at posterior axillary line, by twenty.[10]

The next clinically important question is to identify the type of pleural fluid. Simple anechoic pleural effusions are usually transudative (Fig. 11). Transudative fluid can occasionally have a few swirling fibrin strands inside it. Exudative pleural fluid, on the other hand, are echogenic due to cellular debris and fibrin strands inside it. They may display the plankton sign and hematocrit sign as described above, along with undulating movement of strands with respiratory cycle. Heterogeneously echogenic effusions with septations (Fig. 12) may be seen in parapneumonic effusion and empyema. Loculated effusions may be seen in nondependent parts of chest. A complete assessment of pleural cavity is needed to pick up small, loculated effusions. Subacute hemothorax appears as homogeneously echogenic, whereas acute hemothorax can appear anechoic.

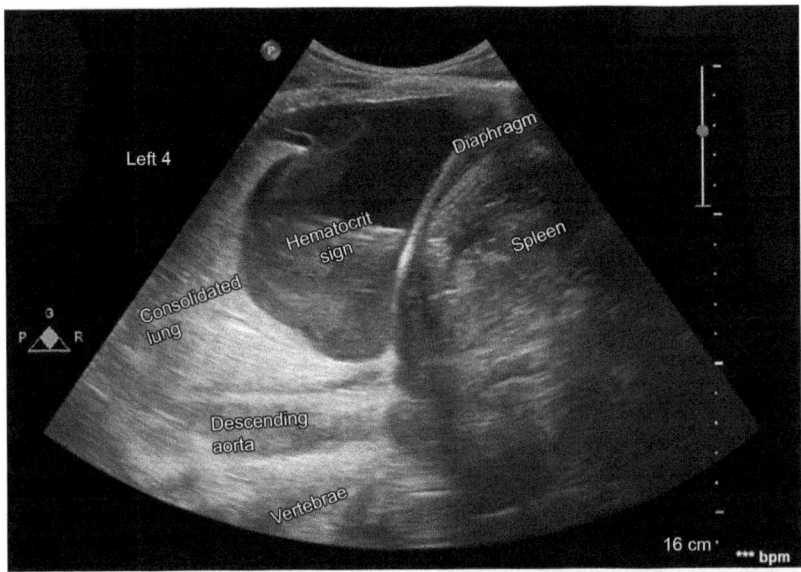

Fig. 10: Hematocrit sign: sedimentation of cellular debris within pleural fluid producing hematocrit sign. The descending aorta is visible through the pleural fluid window with the thoracic vertebrae (V line) lying posterior to it.

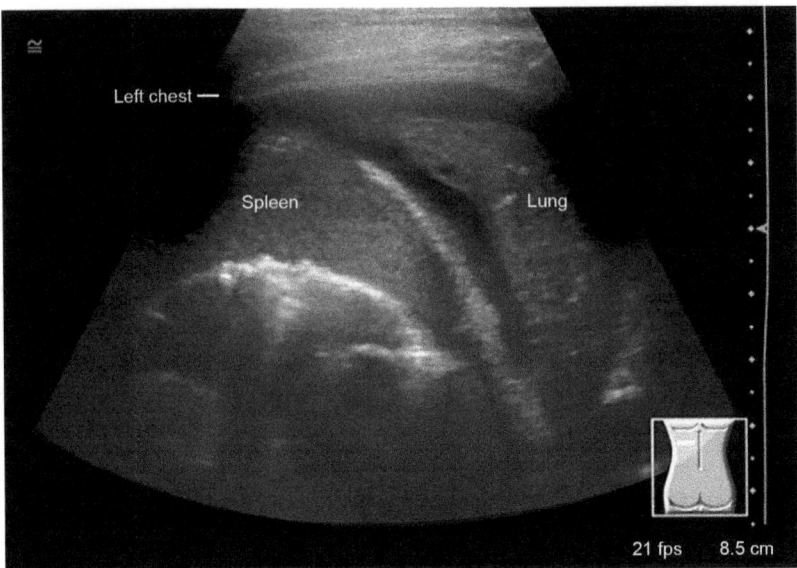

Fig. 11: Transudative pleural effusion. Clear anechoic pleural fluid is seen surrounding the collapsed lung.

Consolidation

Alveoli filled with fluid or inflammatory cells and devoid of air are good transmitter of ultrasound waves. Most of the collapse or consolidation of the lung reach the lung surface and

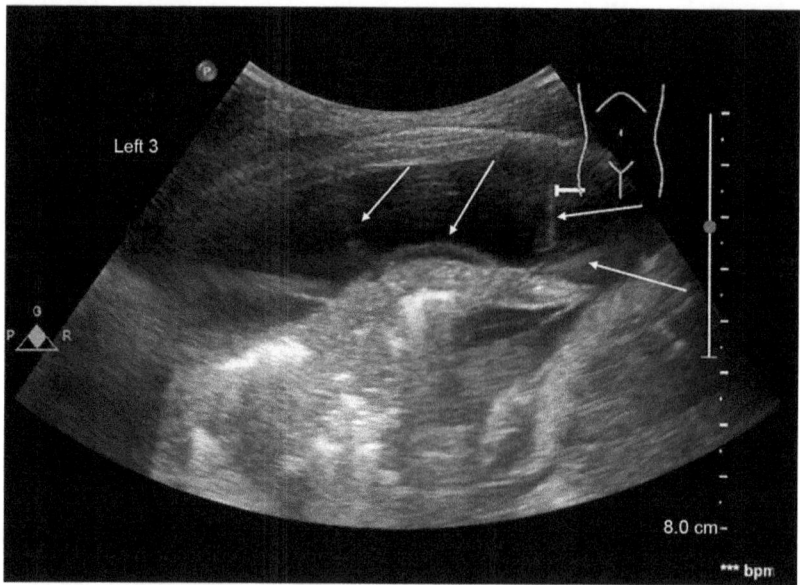

Fig. 12: Exudative pleural effusion. Consolidated lung with air bronchograms is seen surrounded by echogenic pleural fluid with multiple fibrin strands (marked by arrows) extending from lungs, suggestive of a parapneumonic effusion.

allowing the transmission of ultrasound beam. Deep seated consolidation not reaching the lung surface may not be seen on lung sonography.

Consolidated lung has a similar echotexture to solid tissue. It is often referred to as hepatization when echotexture of lung resembles that of liver (Fig. 13). Air bronchograms can be seen at times within the tissue-like lung. As air-tissue interface is highly refractive, sonographic air bronchograms appear as bright or hyperechoic, tubular artifacts (Fig. 14).

Sonographic air bronchograms can be seen to move centrifugally with respiratory effort in more than half of the patients with consolidation. This is termed as dynamic air bronchogram, indicative of patent proximal airway in pneumonia. This produces a sinusoidal pattern on M mode. Static air bronchograms, on the other hand, are more commonly seen in atelectasis, indicated of blocked proximal airway. Static air bronchograms produce straight lines on M mode.[11]

Interstitial Syndrome

Interstitial syndrome produces a B line pattern on lung ultrasound, as previously described. It is pathologically correlated with:
- Pulmonary edema (may be cardiogenic or noncardiogenic as in acute lung injury)
- Interstitial pneumonitis
- Pulmonary fibrosis.

B lines in pulmonary edema are correlated Kerley's B lines on chest radiography. As pulmonary edema becomes more severe, B lines become more frequent and involve the apical

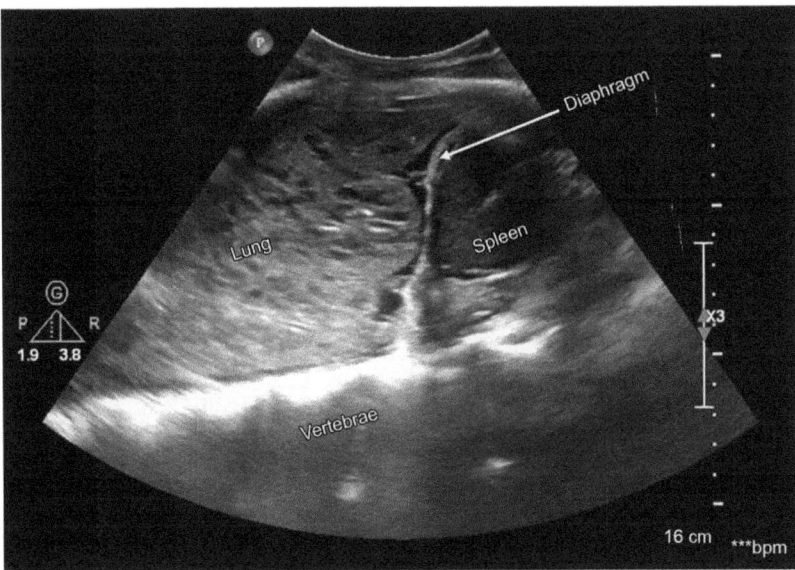

Fig. 13: Hepatization of lungs. Solid, liver like echotexture of consolidated lung [note the relatively flat diaphragm in this patient with underlying emphysema. The vertebral line (V line) is seen posterior to the lung].

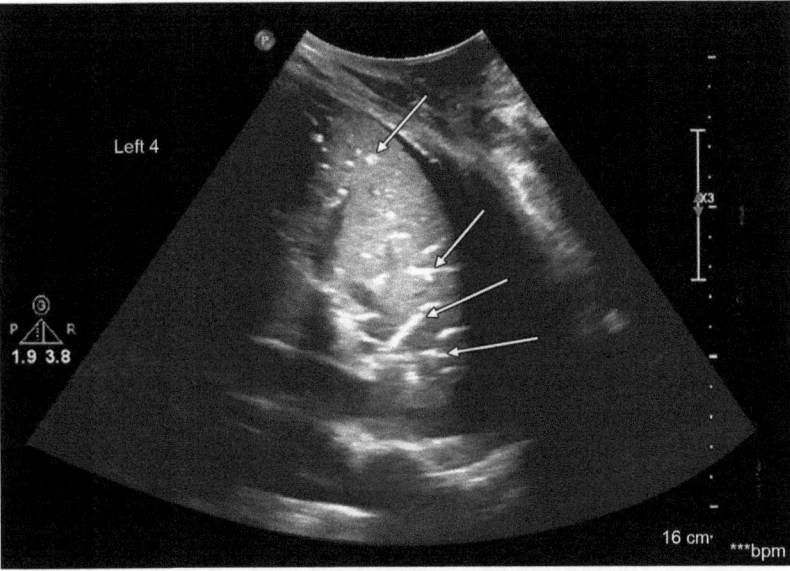

Fig. 14: Air bronchograms. Sonographic air bronchograms representing branching bronchi filled with air, marked with arrows. Bronchi cut in short axis appear as bright hyperechoic dots. The air may appear to move with inspiration in early stages of consolidation, i.e. dynamic air bronchogram (in contrast to lung collapse where they appear to be static).

as well as the basal lung regions. In early stages, B lines are about 7 mm apart. This corresponds to subpleural interlobular septa. It is described as B7 lines (Fig. 15). In more severe stages,

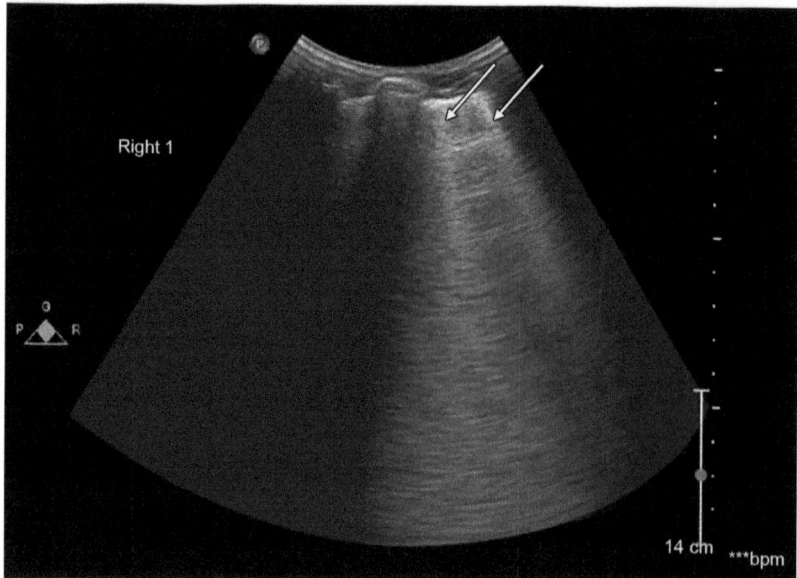

Fig. 15: B7 lines. Couple of B lines (m) are seen traversing the A line, extending deep into the frame.

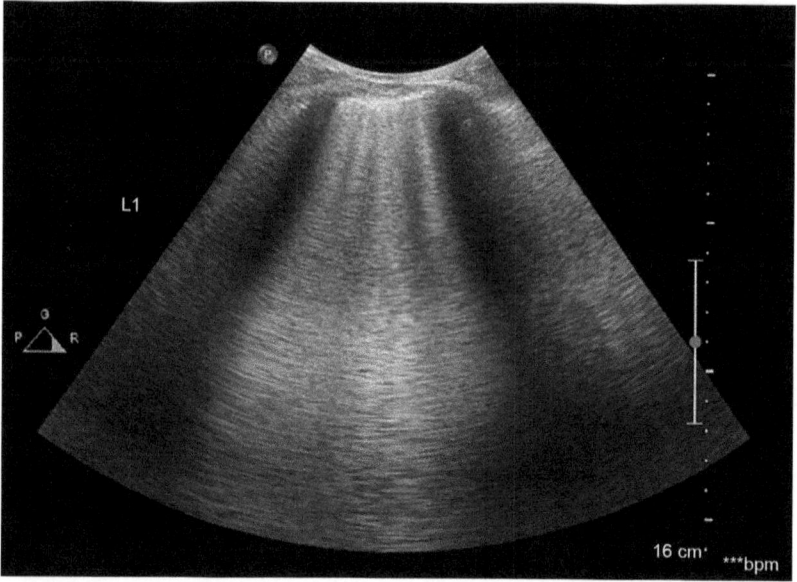

Fig. 16: B3 lines. Confluent B lines in this patient with hypoxic respiratory failure from acute pulmonary edema.

they come closer together, about 3 mm apart. This is described as B3 lines (Fig. 16) and it corresponds to ground glass appearance on CT scan. In very severe cases, they produce a hyperechoic, confluent pattern described as white lung (Fig. 17). The number of B lines correlates with the amount of extravascular lung water.[12]

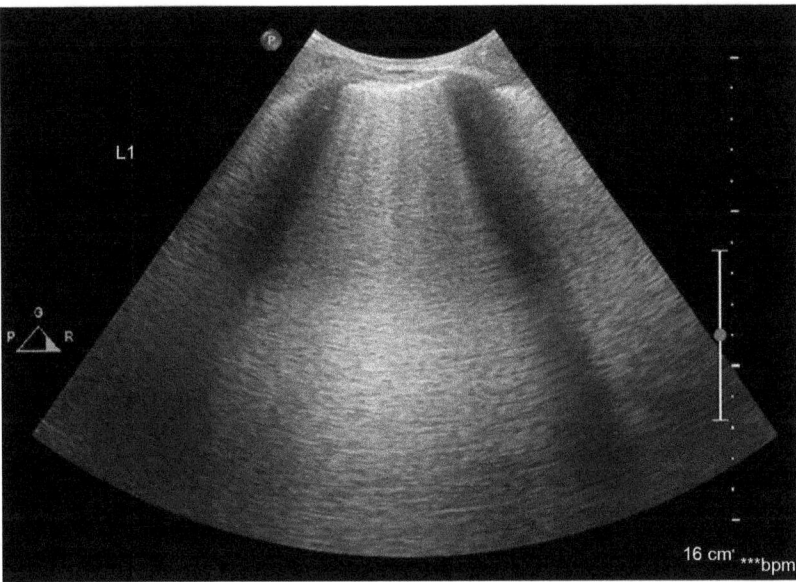

Fig. 17: White lung (note the confluent B lines producing a "white lung" appearance).

Lung ultrasonography can be useful in differentiating cardiogenic pulmonary edema seen in fluid overload from noncardiogenic pulmonary edema seen in acute lung injury. Cardiogenic pulmonary edema is commonly associated with a smooth pleura with normal lung sliding, absence of consolidation and presence of pleura effusion. They tend to be distributed in a homogeneous, symmetrical pattern. Noncardiogenic pulmonary edema, on the other hand, produces irregular thickened pleura with patchy areas of consolidation. Lung sliding is reduced and pleural effusion is less common. Lung pulse may be seen next to consolidated lung on the left side.[13]

LUNG ULTRASOUND IN INTENSIVE CARE UNIT

Use of lung and pleural ultrasound to look for specific pathologies has been described above. We now look at its integrated use in a few common intensive care scenarios.

- Trauma:
  - Ultrasound has the potential to be quite useful in the trauma setting allowing real time assessment of multiple chest pathologies. This can be challenging as well due to presence of subcutaneous air, skin loss or presence of bandages, and an uncooperative patient affecting image acquisition.
  - Ultrasound examination of trauma can begin with assessment of anterior nondependent chest to look for lung sliding. Presence of lung sliding with an A line pattern rules out pneumothorax. Importantly, absence of above signs is not specific enough to rule in a diagnosis of pneumothorax. The exception is lung point; if seen, it is 100% specific and rules in the diagnosis of pneumothorax.

- We can then proceed to examine the basal lung regions to look for free fluid. This can be combined as a part of extended focused assessment with sonography in trauma (eFAST) scan. Presence of any fluid is thought to be related to hemothorax. Acute hemothorax can appear anechoic, which develops clots, fibrin strands or debris with passage of time.
- Lung contusion can be detected earlier than ultrasound compared to chest radiography.[14] Early stages produce an interstitial pattern with multiple B lines, which later on produces features of consolidation.
- Acute respiratory distress syndrome:
  - Acute respiratory distress syndrome (ARDS) is characterized by nonuniform lung injury leading to noncardiogenic pulmonary edema. There is diffuse lung consolidation with multifocal patchy involvement visible on CT scan. The dependent posterior region is worse affected. The nonuniform nature may not be appreciated on chest X-ray making the distinction from cardiogenic pulmonary edema difficult. In contrast, lung ultrasound as a noninvasive, bedside technique allows us to make that distinction with good sensitivity compared to CT scan, without exposing the patient to potentially harmful radiation.
  - Acute respiratory distress syndrome produces a B line pattern on lung ultrasound. The number of B lines tend to be more as we move posteriorly with ultrasound probe. They can be confluent to produce a white lung appearance in posterior lung fields. Lung changes tend to be symmetrical, especially in extrapulmonary cases of ARDS. The changes tend to be patchy with characteristic spared areas, especially in anterior lung fields. The pleural line tends to be thickened and irregular with areas of small, subpleural consolidation (Fig. 18). Small areas of consolidation may be seen in lung bases. Pleural effusion is seen infrequently and tend to be small, if present.

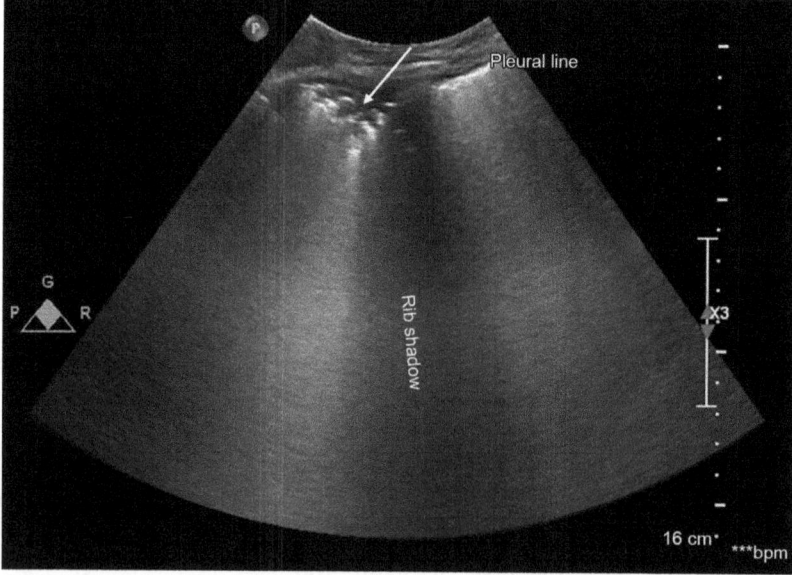

Fig. 18: Subpleural consolidation. Note the small area of subpleural consolidation with spared area in the next interspace).

- Distinction between cardiogenic and noncardiogenic has been described before in the section on interstitial syndrome.
- Optimizing mechanical ventilation:
  - Recruitment maneuvers are commonly used to open up collapsed alveoli and keep them open as part of an open lung approach ventilation. Appropriately used, it can improve oxygenation and minimize harmful effects of mechanical ventilation.
  - Traditionally, CT scan and pressure volume loops can be used to assess the effectiveness of recruitment maneuvers. Lung ultrasound can also be used to look for improvement in lung aeration and thereby optimize lung recruitment by adjusting positive end-expiratory pressure (PEEP).
  - Four distinct ultrasound patterns have been described in ARDS, representing increasing amount of extravascular lung water, i.e. normal A profile, two grades of B profile with increasing number of B lines and collapse/consolidation of C profile. Each lung is subdivided into six regions and scored on the basis of improvement in ultrasonographic appearance of lung postrecruitment. The calculated lung aeration score has been shown to correlate with improved recruitment by traditional pressure volume loop analysis.[15] An improvement in lung aeration score of more than 8 was associated with a PEEP-induced lung recruitment of greater than 600 mL in the same study.
  - A simplified approach might be to look for lung pulse representing collapsed, potentially recruitable lung. The loss of lung pulse with normal pleural sliding post recruitment suggests successful re-expansion of lung.
  - Ultrasound can also be used to look for complications of mechanical ventilation including pneumothorax, right ventricular dysfunction from high PEEP. It also helps us in appropriately managing these complications.
- Assessing weaning:
  - Ultrasound can be a useful tool to look for potential causes of difficulty in weaning and address them appropriately.
  - Diaphragmatic dysfunction is common in critically ill-patients on mechanical ventilation. It can be assessed by examining diaphragmatic excursion using M mode from anterior axillary line. Vertical excursion distance of less than 10 mm or paradoxical diaphragmatic movement during spontaneous breathing trial has been shown to suggest weaning failure with similar predictability as rapid shallow breathing index[16] (Fig. 19).
  - Pleural space could be examined and volume of effusion can be estimated using the formula used by Balik et al. as described above.[10] Complicated effusions or large effusions could be drained under ultrasound guidance to facilitate weaning.
  - Significant changes in lung aeration during spontaneous breathing trial has been shown to be associated with post-extubation respiratory distress. Lung ultrasound may also be used to dynamically assess the effects of reducing positive pressure ventilation on cardiac function (increasing preload on right heart and afterload on left heart) in patients with compromised cardiovascular reserve. Patients dependent on PEEP would be prone to develop pulmonary edema with a B profile with reduction of PEEP.

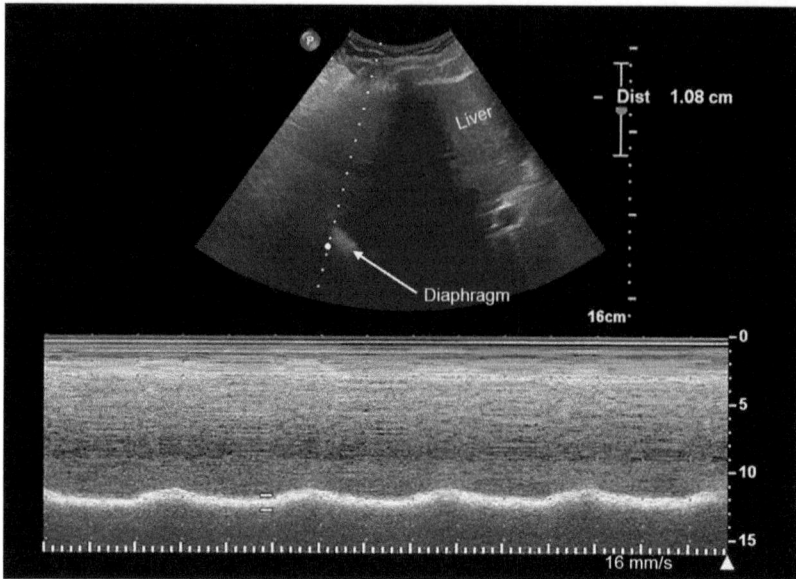

Fig. 19: Diaphragmatic excursion on M mode (note that the sweep speed has been reduced to the minimum to improve the display on M mode).

- Rapid evaluation of acute respiratory failure:
    - Lichtenstein initially described the utility of lung ultrasound in making a rapid diagnosis of acute respiratory failure in adult ICU patients using a simplified protocol, described as bedside lung ultrasound in emergency (BLUE) protocol.[1] It consists of lung ultrasound looking for pleural sliding and pattern of underlying lung—A profile indicating aerated lung, B profile indicating alveolar-interstitial syndrome and C profile indicating collapse/consolidation. The last step is to look for pleural effusion.
    - In presence of pleural sliding, bilateral anterior B profile indicates pulmonary edema. A profile in presence of deep venous thrombosis indicates pulmonary embolism. A profile in absence of deep venous thrombosis needs assessment for posterolateral alveolar/pleural syndrome (PLAPS). Presence of PLAPS suggests pneumonia whereas absence of PLAPS suggests chronic obstructive pulmonary disease or asthma.
    - In the absence of pleural sliding, B profile indicates pneumonia (remember B lines indicates ultrasound beam passing to underlying lungs, excluding pneumothorax). A profile in presence of lung point conformed presence of pneumothorax, whereas A profile in absence of lung point needed further investigation.
    - Use this diagnostic algorithm that is shown to provide an accurate diagnosis in 90.5% cases of acute respiratory failure.

LIMITATIONS OF LUNG ULTRASOUND

Like other imaging techniques, bedside ultrasound is operator dependent and needs focused supervised training to ensure correct interpretation of findings. Captured images should be ideally stored in a retrievable system and findings should be clearly documented for the purpose of comparison as well as medicolegal reasons. This can be difficult in a busy ICU. Modern ultrasound machines may result in deterioration of lung artifacts due to extensive image processing. Machine settings should be adjusted appropriately when performing lung ultrasound. Lung ultrasound can be technically challenging in patients with thick chest walls, chest wall dressing and subcutaneous emphysema.

CONCLUSION

Lung and pleural ultrasound is rapidly becoming a very useful tool in the hand of a qualified intensivist. It can be performed easily and quickly at bedside providing instantons diagnosis. It avoids radiation exposure and can be repeated as many times as needed. It is a highly sensitive and usually specific diagnostic or therapeutic modality available in the armamentarium of the current intensivist.

REFERENCES

1. Lichtenstein DA, Mezière GA. Relevance of lung ultrasound in the diagnosis of acute respiratory failure: the BLUE protocol. Chest. 2008;134(1):117-25.
2. Volpicelli G, Elbarbary M, Blaivas M, et al. International evidence-based recommendations for point-of-care lung ultrasound. Intensive Care Med. 2012;38(4):577-91.
3. Volpicelli G, Mussa A, Garofalo G, et al. Bedside lung ultrasound in the assessment of alveolar-interstitial syndrome. Am J Emerg Med. 2006;24(6):689-96.
4. Lichtenstein D, Goldstein I, Mourgeon E, et al. Comparative diagnostic performances of auscultation, chest radiography, and lung ultrasonography in acute respiratory distress syndrome. Anesthesiology. 2004;100(1):9-15.
5. Lichtenstein DA, Mezière GA, Lagoueyte JF, et al. A-lines and B-lines: lung ultrasound as a bedside tool for predicting pulmonary artery occlusion pressure in the critically ill. Chest. 2009;136(4):1014-20.
6. Lichtenstein DA, Menu Y. A bedside ultrasound sign ruling out pneumothorax in the critically ill. Lung sliding. Chest. 1995;108(5):1345-8.
7. Soldati G, Testa A, Sher S, et al. Occult traumatic pneumothorax: diagnostic accuracy of lung ultrasonography in the emergency department. Chest. 2008;133(1):204-11.
8. Lichtenstein D, Mezière G, Biderman P, et al. The "lung point": an ultrasound sign specific to pneumothorax. Intensive Care Med. 2000;26(10):1434-40.
9. Mattison LE, Coppage L, Alderman DF, et al. Pleural effusions in the medical ICU: prevalence, causes, and clinical implications. Chest. 1997;111(4):1018-23.
10. Balik M, Plasil P, Waldauf P, et al. Ultrasound estimation of volume of pleural fluid in mechanically ventilated patients. Intensive Care Med. 2006;32(2):318-21.
11. Lichtenstein D, Mezière G, Seitz J. The dynamic air bronchogram. A lung ultrasound sign of alveolar consolidation ruling out atelectasis. Chest. 2009;135(6):1421-5.
12. Lichtenstein DA. Lung and interstitial syndrome. In: Lichtenstein DA (Ed.). Whole Body Ultrasonography in the Critically Ill. Berlin: Springer; 2010. pp. 15-161.
13. Copetti R, Soldati G, Copetti P. Chest sonography: a useful tool to differentiate acute cardiogenic pulmonary edema from acute respiratory distress syndrome. Cardiovasc Ultrasound. 2008;6:16.
14. Soldati G, Testa A, Silva FR, et al. Chest ultrasonography in lung contusion. Chest. 2006;130(2):533-8.
15. Bouhemad B, Brisson H, Le-Guen M, et al. Bedside ultrasound assessment of positive end-expiratory pressure-induced lung recruitment. Am J Respir Crit Care Med. 2011;183(3):341-7.
16. Kim WY, Suh HJ, Hong SB, et al. Diaphragm dysfunction assessed by ultrasonography: influence on weaning from mechanical ventilation. Crit Care Med. 2011;39(12):2627-30.

Chapter 6

Chest Ultrasound in Acute Respiratory Distress Syndrome

Shyam Madabhushi, Puneet Khanna

INTRODUCTION

Until 1967, it was believed that ultrasound as a technique was not applicable to the lung, as "air is the enemy of ultrasound." This narrative has changed because of the realization that in aerated tissues, such as the lung, the loss of normal aeration accompanies most pathological entities, making the hitherto "invisible" lung, visible.

Current radiological methods like CT and MRI are capable of identifying differences in these conditions. However, these techniques are not amenable to deployment by the bedside. Issues with shifting unstable patients to the radiology suite remain, and the associated radiation exposure and cost are also a concern. In this regard, lung ultrasound is a repeatable, inexpensive, and radiation-free investigation that can readily be performed at the bedside. The main barrier to the use of ultrasound is the operator-dependent and subjective nature of the modality.

Acute respiratory distress syndrome (ARDS) is characterized by acute inflammatory lung injury, increased vascular permeability, and leads to an increase in lung weight, poor gas exchange, reduced compliance, and dysfunction of the right ventricle. However, in the acute care or critical care setting, diseases like acute cardiogenic pulmonary edema and pneumonia need to be distinguished from ARDS. This would enable early implementation of the low tidal volume protocol and avoids ventilator-induced lung injury.

LUNG MORPHOLOGY IN ACUTE RESPIRATORY DISTRESS SYNDROME

In ARDS, the distribution of lung injury is inhomogeneous and nongravity dependent. Affected areas of the lung demonstrate B-line patterns, which, when confluent, appear as areas of "white lung". The affected areas also demonstrate pleural line abnormalities (thickening and irregularity of the pleural margin, the latter also being known as the "pleural shredding" sign; subpleural consolidations). The intervening, relatively normal lung appears dark and has A-lines. These areas of normal lung are more common anteriorly than posteriorly. This is because the posterior areas of the lung become more homogenous and compact. This is owing to the supine position of the critically ill-patient and the effect of gravity. In addition, consolidations are seen in ARDS, mainly in the basal and posterior areas. Consolidated areas appear hyperechoic, and are accompanied by static or dynamic air bronchograms.

The B-line pattern described above is not specific for ARDS. B-lines are generated due to subpleural interlobular septal thickening, and can be seen in cardiogenic pulmonary edema, pneumonia, pulmonary fibrosis, and pulmonary masses. Therefore, pattern recognition (distribution of B-lines, density, homogeneity, the sonographic image pattern of the contralateral lung, etc.) is important.

DIFFERENTIATING ACUTE RESPIRATORY DISTRESS SYNDROME FROM OTHER CAUSES OF RESPIRATORY FAILURE

Lichtenstein et al. in their study used ultrasound for the diagnosis of acute respiratory failure using a protocol. Ultrasound patterns were used to recognize diagnoses, and these were compared to the final diagnosis arrived at by the ICU team.[1] In their study, the following parameters were used to grade the image seen in a scan window.
- A-lines or B-lines and their density:
  - A scan window with three or more B-lines was called B+
- Lung sliding: Present or absent
- Alveolar consolidation and/or pleural effusion:
  - A term called posterolateral alveolar and/or pleural syndrome (PLAPS)—a combination of pleural effusion, consolidation and pleural line abnormality.

Based on these parameters, common lung conditions leading to acute respiratory failure were diagnosed, and the results may be summarized according to the Table 1.

A further study[2] by Sekiguchi and colleagues used a combination of lung and cardiac critical care ultrasound (CCUS). They proposed an algorithm based on their findings (Flowchart 1).

In summary, a combination of lung ultrasound, clinical features, and critical care echocardiography can be used to identify patients with ARDS from other causes of respiratory failure.

LUNG ULTRASOUND AND EXTRAVASCULAR LUNG WATER

Acute respiratory distress syndrome is, at least initially, a nonstarling pulmonary edema, and the extravascular lung water (EVLW) is increased. The degree of increase in EVLW has been shown to influence outcome in ARDS. EVLW can be evaluated by transpulmonary thermodilution and CT densitometry. Lichtenstein et al. were the first to report on the use of ultrasound as a technique to quantify EVLW depending on the appearance of B-lines.[3] Subsequently, scoring systems quantifying the number and thickness of B-lines were shown to correlate with lung weight or density by CT analysis. For example, Baldi et al.[4] in their pilot study compared lung ultrasound to CT—determined lung weight and density. In their study, patients were scanned both with ultrasonography and CT. B-line profile score in each scan window was noted. The lung ultrasound was derived by scanning 28 areas, and a score of less than or equal to 5 was considered normal. They concluded that lung ultrasound correlates well with CT-determined lung density, which is a reflection of EVLW.

PROGNOSIS OF ACUTE RESPIRATORY DISTRESS SYNDROME DETERMINED BY ULTRASOUND

Factors determining prognosis in ARDS, germane to the use of lung ultrasound, are:
- Higher EVLW

Table 1: Accuracy of the ultrasound profiles.

Disease	Ultrasound sign used	Sensitivity, %	Specificity, %	Positive predictive value, %	Negative predictive value, %
Cardiogenic pulmonary edema	Diffuse bilateral anterior B + lines associated with lung sliding (B profile)	97 (62/64)	95 (187/196)	87 (62/71)	99 (187/189)
COPD or asthma	Predominant anterior A lines without PLAPS and with lungs sliding (normal profile), or with absent lung sliding without lung point	89 (74/83)	97 (172/177)	93 (74/79)	95 (172/181)
Pulmonary embolism	Predominant anterior bilateral A lines plus venous thrombosis	81 (17/21)	99 (238/239)	94 (17/18)	98 (23/242)
Pneumothorax	Absent anterior lung sliding, absent anterior B lines and present lung point	88 (8/9)	100 (251/251)	100 (8/8)	99 (251/252)
Pneumonia	Diffuse bilateral anterior B + lines associated with abolished lung sliding (B' profile)	11 (9/83)	100 (177/177)	100 (9/9)	70 (177/251)
	Predominant anterior B + lines on one side, predominant anterior A lines on the other (A/B profile)	14.5 (12/83)	100 (177/177)	100 (12/12)	71.5 (177/248)
	Anterior alveolar consolidation (C profile)	21.5 (18/83)	99 (75/177)	90 (18/20)	73 (175/240)
	A profile plus PLAPS	42 (35/83)	96 (170/177)	83 (35/42)	78 (170/218)
	A profile plus PLAPS, B', A/B or C profile	89 (74/83)	94 (167/177)	88 (74/84)	95 (167/176)

*Data in parentheses indicate no. of patients (total).
Abbreviations: COPD: Chronic obstructive pulmonary disease; PLAPS: Posterolateral alveolar and/or pleural syndrome. (Adapted from Ref. No 1)

- Pulmonary vascular dysfunction, which may manifest in the extreme as right ventricular dysfunction (RVD)
- Severity of hypoxemia, expressed as the PaO_2/FiO_2 ratio.

Lung ultrasound to determine prognosis of ARDS is beset by the fact that lung ultrasound is a qualitative examination, with repeatability and operator dependence being built in. However, attempts have been made to introduce some objectivity by scoring the B-line density, and the lung ultrasound score is one such scoring system. EVLW has been demonstrated to be an independent predictor of mortality by Craig et al.[5] In this study, EVLW was

Flowchart 1: Weighted scoring system to distinguish ARDS from other etiologies of hypoxemia

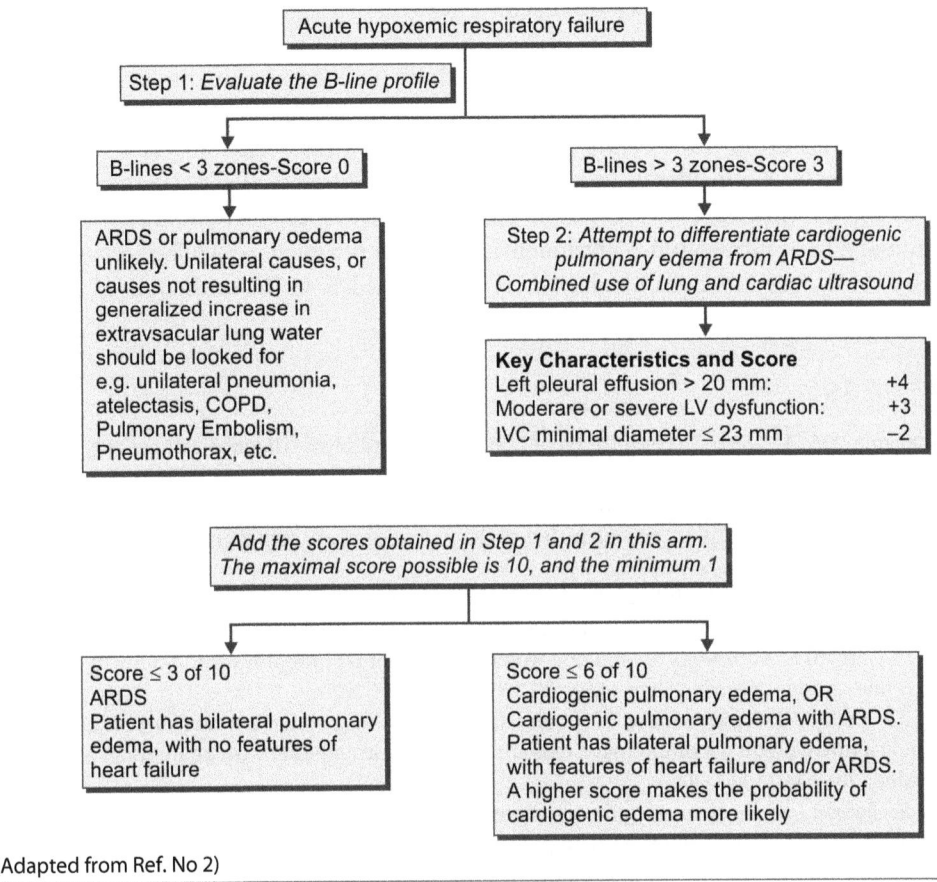

(Adapted from Ref. No 2)

measured using the PiCCO (pulse contour cardiac output) system, and it was found that the EVLW indexed to predicted body weight was significantly increased in nonsurvivors.

Measuring the EVLW using the PiCCO, however, requires an invasive approach, and is expensive. To compare the utility of the lung ultrasound score in the measurement of EVLW, Zhao and colleagues[6] compared the lung ultrasound score with the PiCCO system, and the receiver operating characteristic (ROC) curve for lung ultrasound was 0.846, while it was 0.918 for PiCCO. Therefore, it may be concluded that EVLW can be predicted using the lung ultrasound.

The status of the right ventricle is important in ARDS. Lhéritier et al.[7] showed that ARDS patients with RVD (assessed by transthoracic (TTE) or transesophageal echocardiography (TEE) received prone mechanical ventilation and vasoactive therapy more frequently, and required higher dose of inhaled nitric oxide (NO) as a rescue therapy, than those without RVD. Assessment of RVD requires echocardiography, which is also a part of critical care ultrasonography. Measures of RVD include the ratio of right ventricular end-diastolic area and left ventricular end-diastolic area (RVEDA: LVEDA), tricuspid annular plane systolic excursion (TAPSE), Doppler tissue imaging, RV strain, etc.

Studies evaluating the relationship of PaO_2/FiO_2 ratio and lung ultrasound are few. As an example, Bilotta et al.[8] performed a study of the lung ultrasound score on 45 patients in a neuro critical care unit. Lung ultrasound was calculated from eight lung fields, four on each side. Any field with up to three B-lines was considered normal, while any field with more than three B-lines was considered to be significant for increased EVLW. Upon comparing the lung ultrasound with the PaO_2/FiO_2 ratio, a significant correlation was noticed.

In summary, Lung ultrasound is invaluable in ARDS for diagnosis and differentiation from other causes of hypoxemia, for assessing the severity of ARDS, in the evaluation of the right ventricle, and for assessing prognosis.

REFERENCES

1. Lichtenstein DA, Mezière GA. Relevance of Lung ultrasound in the diagnosis of acute respiratory failure. The BLUE Protocol. Chest. 2008;134:117-25.
2. Sekiguchi H, Schenck LA, Horie R, et al. Critical care ultrasonography differentiates ARDS, pulmonary edema, and other causes in the early course of acute hypoxemic respiratory failure. Chest. 2015;148(4):912-8.
3. Lichtenstein D, Mézière M, Philippe B, et al. The comet-tail artifact–an ultrasound sign of alveolar-interstitial syndrome. Am J Respir Crit Care Med. 1997;156:1640-6.
4. Baldi G, Gargani L, Abramo A, et al. Lung water assessment by lung ultrasonography in intensive care: A pilot study. Intensive Care Med. 2013;39(1):74-84.
5. Craig TR, Duffy MJ, Shyamsundar M, et al. Extravascular lung water indexed to predicted body weight is a novel predictor of intensive care unit mortality in patients with acute lung injury. Crit Care Med. 2010;38(1):114-20.
6. Zhao Z, Jiang L, Xi X, et al. Prognostic value of extravascular lung water assessed with lung ultrasound score by chest sonography in patients with acute respiratory distress syndrome. BMC Pulm Med. 2015;15(1):98.
7. Lhéritier G, Legras A, Caille A, et al. Prevalence and prognostic value of acute cor pulmonale and patent foramen ovale in ventilated patients with early acute respiratory distress syndrome: A multicenter study. Intensive Care Med. 2013;39(10):1734-42.
8. Bilotta F, Giudici LD, Zeppa IO, et al. Ultrasound imaging and use of B-lines for functional lung evaluation in neurocritical care: a prospective, observational study. Eur J Anaesthesiol. 2013;30(8):464-8.

Chapter 7

Ultrasonography in Cardiac Evaluation

Sarvesh Pal Singh, Sonali Bansal

INTRODUCTION

Echocardiography has become an important investigation in the intensive care unit (ICU). In the present era, when nearly all ICUs are "closed" ICUs, intensivists favor doing echocardiogram by themselves. Echocardiography requires skill to perform and knowledge to interpret the findings. Therefore, it is imperative for an intensive care physician to be certified in echocardiography. Transthoracic echocardiography has been shown to contribute positively in 97.4% patients in an ICU. In 24.5% of cases, the information was decisive, in 37.3% supplemental and in 35.6% supportive.[1]

INDICATIONS FOR PERFORMING ECHOCARDIOGRAPHY IN THE CRITICALLY ILL PATIENT[2]

- *Category I:* Conditions for which there is evidence and/or general agreement that a given procedure is useful and effective:
  - The hemodynamically unstable patient
  - Suspected aortic dissection
  - Serious blunt or penetrating chest trauma (suspected pericardial effusion or tamponade)
  - Mechanically ventilated multiple trauma or chest trauma patient
  - Suspected preexisting valvular or myocardial disease in the trauma patient
  - Widening of the mediastinum, suspected aortic injury
  - Potential catheter, guidewire, pacer electrode or pericardiocentesis needle injury with or without signs of tamponade.
- *Category IIA:* Conditions for which there is conflicting evidence or divergence of opinion, but the weight of evidence or opinion is in favor of usefulness:
  - Evaluation of hemodynamics in multiple-trauma or chest trauma patients with pulmonary artery catheter monitoring and data disparate with clinical situation
  - Follow-up study on victims of serious trauma.
- *Category III:* Conditions for which there is evidence and/or general agreement that the procedure is not useful and in some cases may be harmful:
  - The hemodynamically stable patient not expected to have cardiac disease
  - Re-evaluation follow-up studies on hemodynamically stable patients
  - Suspected myocardial contusion in the hemodynamically stable patient with a normal electrocardiography who has no abnormal physical findings.

ADVANTAGES AND DISADVANTAGES OF ECHOCARDIOGRAPHY IN THE INTENSIVE CARE UNIT (TABLE 1)

Table 1: Advantages and disadvantages of echocardiography in the intensive care unit.

S. N.	Advantages	Disadvantages
1.	Noninvasive	Noncontinuous
2.	Real time monitoring	Difficult to get views in postoperative and trauma patients
3.	Requires less time	Learning curve
4.	Offline analysis possible	

GENERAL POINTS

Imaging

Images in transthoracic echocardiography are inverted as opposed to images in transesophageal echocardiography which are upright. Also, the orientation of anterior and posterior is opposite.

Probe Selection

The frequency of probe is expressed in hertz (Hz). The depth (amplitude) at which the probe can interrogate a specific structure and the frequency of probe are inversely related. Therefore, higher frequency probes delineate deeper structures poorly. The transthoracic echocardiography probes are curvilinear probes available for adults, children and neonates. For example, on a Philips iE33 machine, there is a S5 probe (adult), S8 probe (pediatric) and S12 probe (neonatal).

Probe Placement

All transthoracic echocardiography probes have a marker on one end of the probe (Fig. 1).

The position of probe is described with respect to the marker. Conventionally, there are four places (windows) where an echocardiography probe can be placed to form different views. These are (Fig. 2):

- Apical — Left fourth to fifth intercostal space
- Parasternal — Left second to fourth intercostal space
- Subcostal — Just below xiphisternum
- Suprasternal — In the suprasternal notch

Different views are formed by adjusting the site, angulations and rotation of the probe.

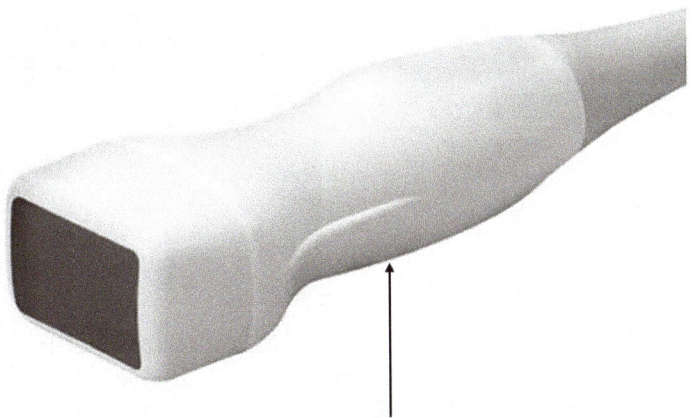

Fig. 1: Transthoracic echocardiography probe showing marker on one side (S5 probe for iE 33 Philips Machine).

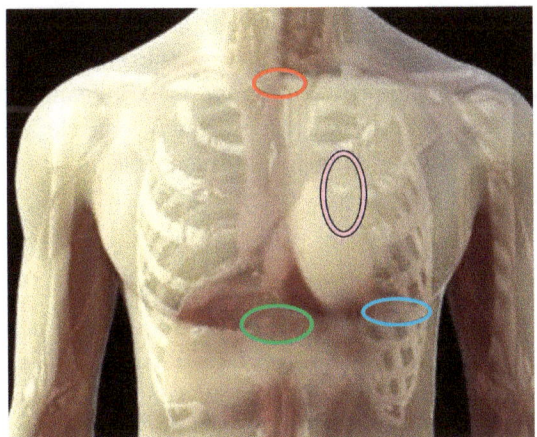

Fig. 2: Four conventional windows for transthoracic echocardiography: left parasternal (pink), apical (blue), subcostal (green) and suprasternal (red).

COMMON TRANSTHORACIC ECHOCARDIOGRAPHY VIEWS FOR INTENSIVE CARE UNIT

- Parasternal long-axis view (PLAX)
- Parasternal short-axis view (PSAX)
- Apical four-chamber view
- Subcostal four-chamber view
- Subcostal inferior vena cava view.

Parasternal Long-axis View

To achieve PLAX, the echo probe is placed in the left second intercostal space with the marker facing the right shoulder of the patient. Then, the probe is gently swept down (from second intercostal space towards the third or fourth intercostal space) to achieve the long-axis view.

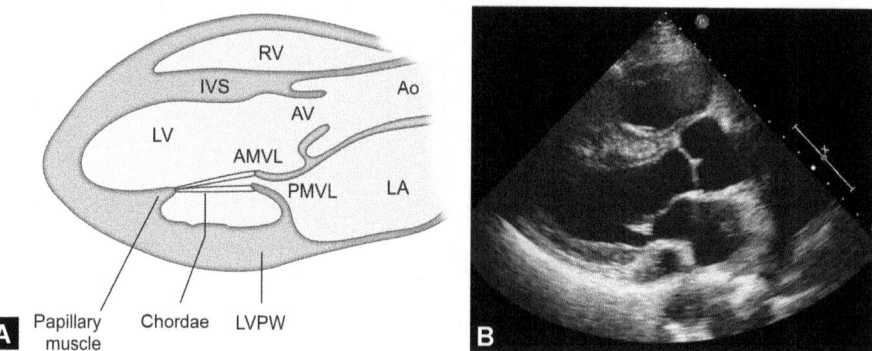

Figs. 3A and B: Parasternal long-axis view showing left atrium, left ventricle, interventricular septum, right ventricle, aortic valve, and aortic root, ascending aorta, mitral valve, papillary muscle, chordae and apex.

The structures seen in the PLAX view are left atrium, left ventricle, interventricular septum, right ventricle, aortic valve, and aortic root, ascending aorta, mitral valve, papillary muscle, chordae and apex (Figs. 3A and B). This view is useful to see aortic and mitral valve pathologies, left ventricular function, left ventricular posterior wall and apical aneurysm. The dimensions of right ventricle, interventricular septum, left ventricle, and thickness of posterior left ventricular wall in both systole and diastole are also measured in this view. The size of left atrium and ascending aorta can also be measured in this view.

Parasternal Short-axis View

To achieve a PSAX view, the probe is rotated clockwise (at the same place where PLAX was achieved) to the left so that the marker on the probe is directed towards the left shoulder of the patient. The view thus obtained is a basal PSAX view. PSAX can be obtained at three levels—basal, mitral level and mid ventricle level. To achieve the other two views, the probe is titled downwards towards the apex of left ventricle.

Basal PSAX view: The structures seen in this view are right atrium, right ventricle, main pulmonary artery, bifurcation of pulmonary artery into left and right branches, left atrium, tricuspid valve, pulmonary valve and aortic valve. The view is important to assess the morphology of aortic valve (bicuspid or tricuspid, stenosis), pulmonary artery dimensions, thrombus in main pulmonary artery and branch PAs (pulmonary embolism or chronic thromboembolic pulmonary hypertension), left atrial thrombus or tumors and tricuspid regurgitation or stenosis. This view is also valuable to calculate the right ventricle function by measuring fractional shortening of the right ventricular cavity. The position of the pulmonary artery catheter is seen in this view (Figs. 4A to C).

PSAX at mitral valve level: The structures visible in this view are free wall of right ventricle, basal interventricular septum, left ventricle (anterior, lateral and inferior walls) and mitral valve (anterior and posterior leaflets). This view may be used for evaluating contraction of left ventricle, movement of mitral leaflets, area of mitral valve (stenosis), regional wall motion abnormality (RWMA) of left ventricle and ventricular septal defect. Pericardial effusion may also be evaluated in this view (Figs. 5A and B).

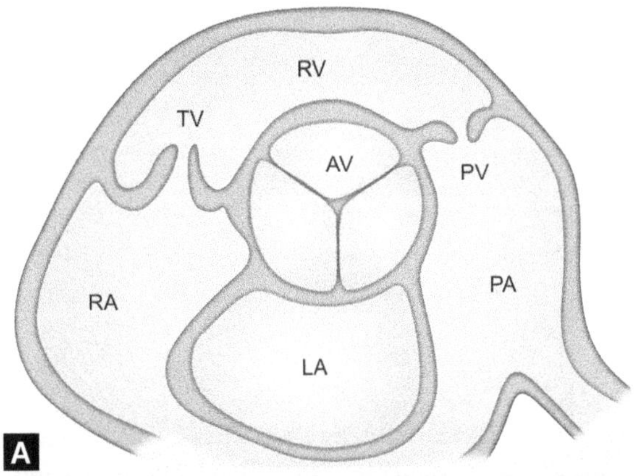

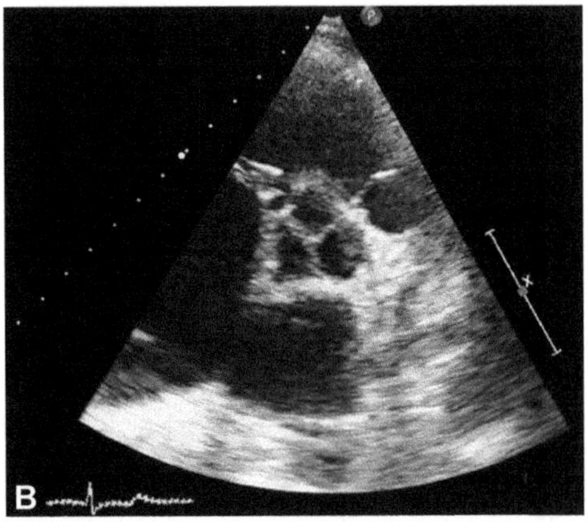

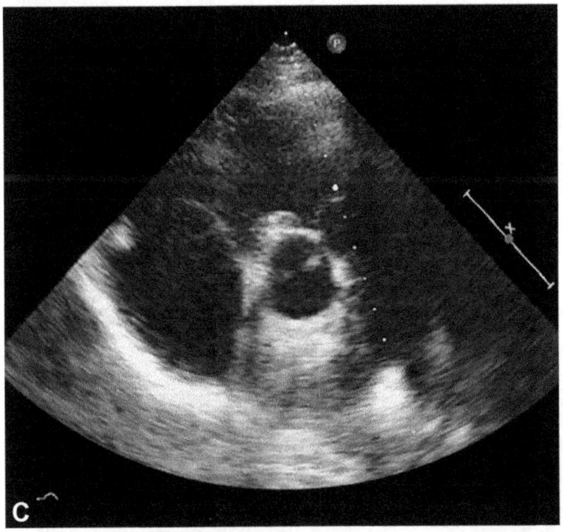

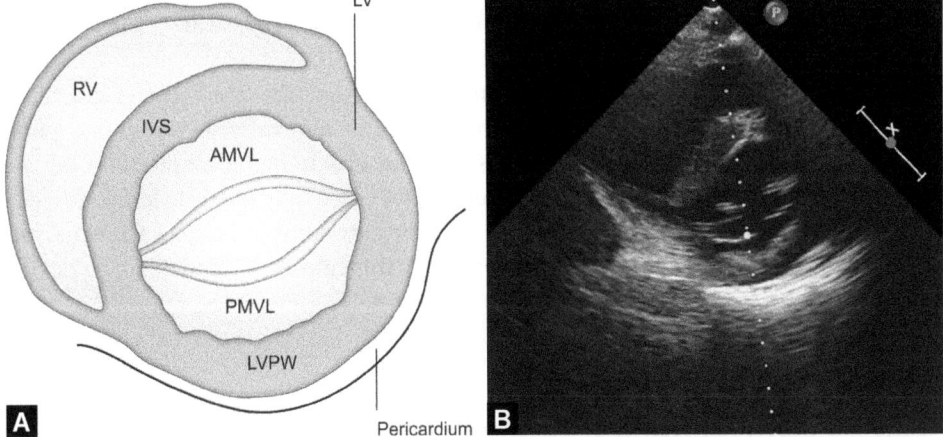

Figs. 4A to C: Parasternal sternal short-axis view at basal level. (A) Right atrium, right ventricle, main pulmonary artery, bifurcation of pulmonary artery into left and right branches, left atrium, tricuspid valve, pulmonary valve and aortic valve; (B) Bifurcation of MPA into branch pulmonary arteries is not seen. This bifurcation is depicted in Fig. 4C.

Figs. 5A and B: Parasternal sternal short-axis view at mitral valve level showing free wall of right ventricle, basal interventricular septum, left ventricle (anterior, lateral and inferior walls) and mitral valve (anterior and posterior leaflets).

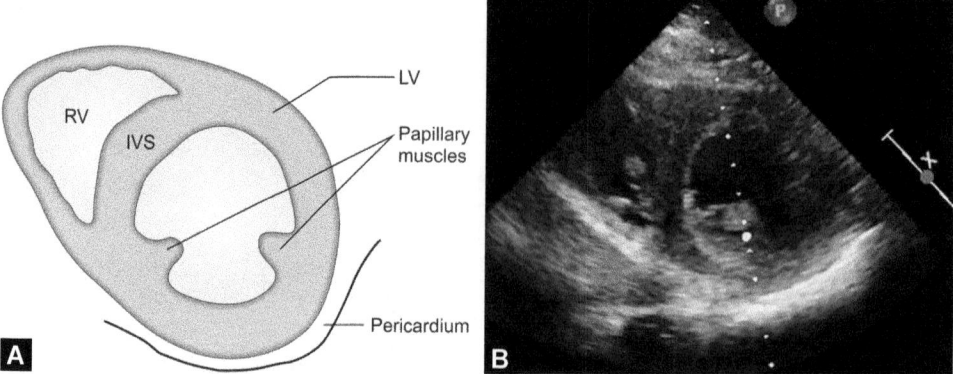

Figs. 6A and B: Parasternal sternal short-axis view at apical level showing free wall of right ventricle, muscular interventricular septum, left ventricle (anterior and inferior walls) and papillary muscles (anterior and posterior).

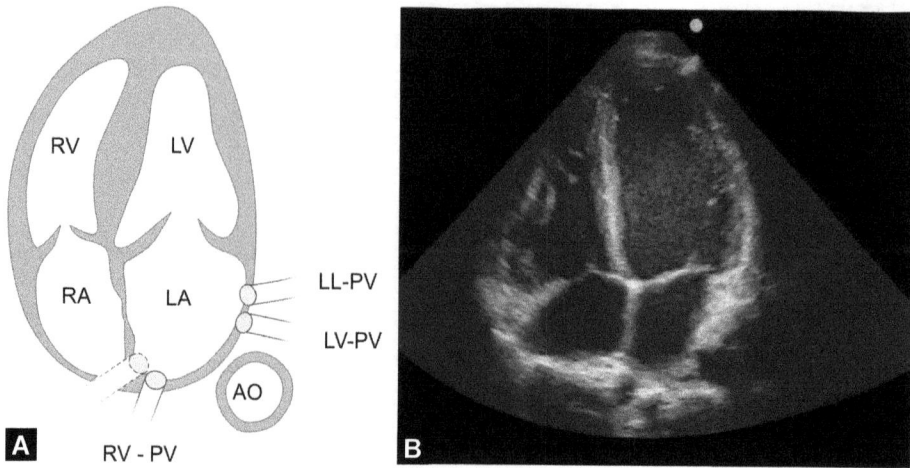

Figs. 7A and B: Apical four-chamber view showing right atrium, right ventricle, interatrial septum, interventricular septum, left atrium, left ventricle, tricuspid and mitral valves, and three pulmonary veins.

PSAX at mid ventricle level: The structures visible in this view are free wall of right ventricle, muscular interventricular septum, left ventricle (anterior and inferior walls) and papillary muscles (anterior and posterior). This view is used for evaluating contraction of left ventricle, RWMA of left ventricle and ventricular septal defect. In a hypovolemic left ventricle, with normal ejection fraction (EF), both the papillary muscles come close to each other during systole and are known as "kissing papillary muscles". This finding is specific for hypovolemia (Figs. 6A and B).

Apical Four-chamber View

This is the most common view used to visualize the heart. The probe is placed just lateral to the apex in the left fourth to fifth intercostal space and adjusted to get the view shown in Figures 7A and B.

The probes should be shifted left or right gently to orient the interventricular septum parallel to the screen to get a true four-chamber view. This view shows right atrium, right ventricle, interatrial septum, interventricular septum, left atrium, left ventricle, tricuspid and mitral valves, and three pulmonary veins. The view is useful for demonstrating septal defects, volume status of heart, right ventricular function (tricuspid annular plane systolic excursion), tricuspid and mitral regurgitation, position of central venous line (should not be seen in right atrium), progress and evaluation of balloon mitral valvotomy, tumors of heart and pulmonary vein stenosis. This view also delineates the pericardial effusion, if any.

A slight modification of the four-chamber view is the apical five-chamber view. The probe is placed just lateral to the apex in the left fifth or sixth intercostal space with anterior angulation. The difference from four-chamber view is that the aortic valve is seen along the long axis of left ventricle and pulmonary veins are no longer visible. This view is good for alignment of Doppler across the left ventricle outflow tract (LVOT) and aortic valve. The gradient across the aortic valve is measured in this view (e.g. aortic stenosis).

Subcostal Four-chamber View

The disadvantage of the apical and parasternal windows is that they are not very useful in postcardiac surgery patients in the immediate postoperative period. The probability of achieving a subcostal view is more in these patients. This view may also be useful in patients with chronic obstructive pulmonary disease or malposition of heart. The view is slightly tilted in comparison to the apical four-chamber view but shows the same structures as visible in the apical four-chamber view except for pulmonary veins. To obtain the view, the probe is placed in the subcostal area with the marker facing left and is tilted upwards to look at the heart (Figs. 8A and B).

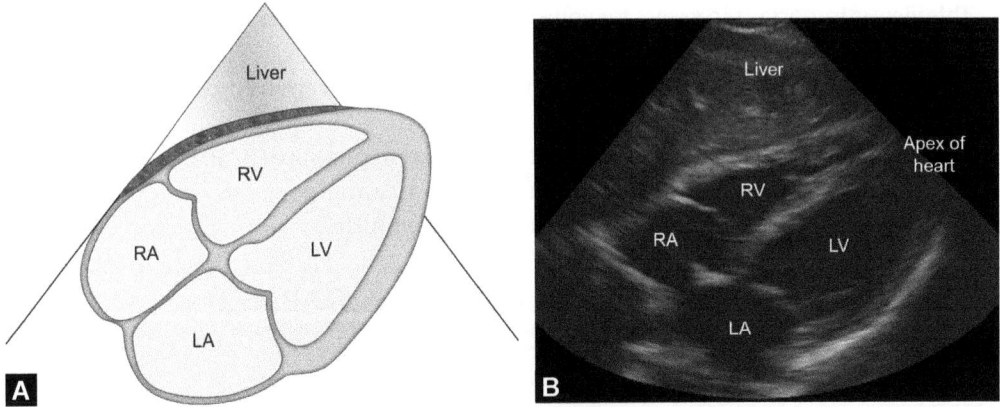

Figs. 8A and B: Subcostal four chamber view showing right atrium, right ventricle, interatrial septum, interventricular septum, left atrium, left ventricle, tricuspid and mitral valves. Also visible is liver in the superior part of the screen.

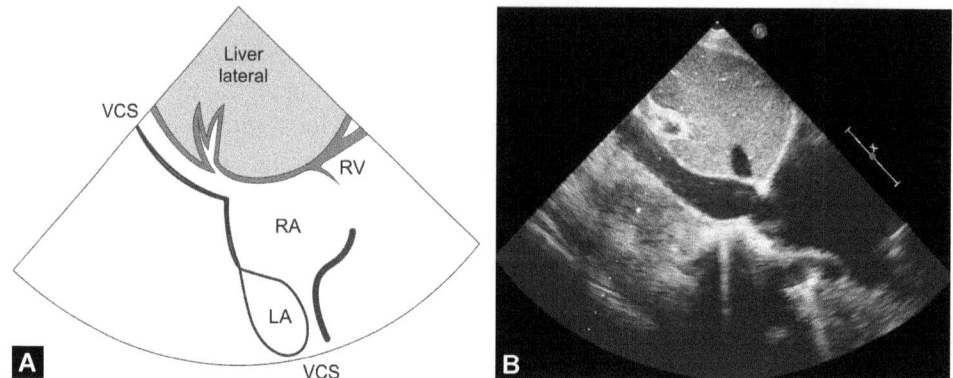

Figs. 9A and B: Subcostal IVC view showing IVC, right atrium and right ventricle along with liver and hepatic veins.

Table 2: Ventricular function assessment on basis of individual segment motion and thickening.

		Wall motion	% Radius change		Wall thickening
1.	Normal	Inward	>30%	+++	30–50%
2.	Mild	Inward	10–30%	++	30–50%
3.	Severe	Inward	<10%	+	<30%
4.	Akinesis	None	None	0	<10%
5.	Dyskinesis	Outward	None	0	None/thinning

Subcostal Inferior Vena Cava View

The probe is rotated counter clockwise, from subcostal four-chamber view, positioning it parallel to the superior vena cava and inferior vena cava (SVC-IVC) axis, as shown in Figures 9A and B.

This view is important to assess for volume status, pericardial tamponade and normal inferior vena cava drainage into right atrium.

In adults following correlation is present between central venous pressure, IVC diameter and respiration.[3]

- Size ≤ 2.1 cm; collapses >50% during sniff = Right atrial pressure (RAP) 0–5 mm Hg
- Size > 2.1 cm; collapses >50% during sniff = RAP 5–10 mm Hg
- Size > 2.1; collapses <50% during sniff = RAP 10–20 mm Hg.

ASSESSMENT OF VENTRICULAR FUNCTION (TABLE 2)

- Left ventricle
  - Systolic function and RWMA
  - Cardiac output measurement
  - Diastolic function.
- Right ventricle.

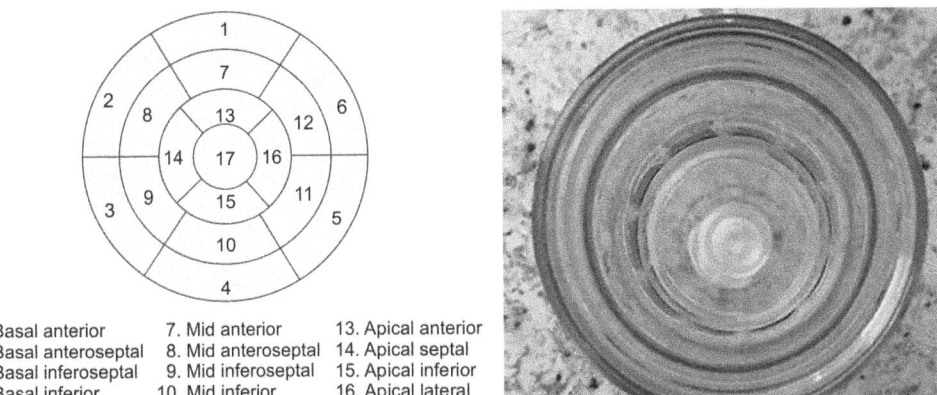

1. Basal anterior
2. Basal anteroseptal
3. Basal inferoseptal
4. Basal inferior
5. Basal inferolateral
6. Basal anterolateral
7. Mid anterior
8. Mid anteroseptal
9. Mid inferoseptal
10. Mid inferior
11. Mid inferolateral
12. Mid anterolateral
13. Apical anterior
14. Apical septal
15. Apical inferior
16. Apical lateral
17. Apex

Figs. 10A and B: (A) The 17-segment model of left ventricle; (B) The view when someone looks inside a tumbler from the top. Similarly when looking inside the ventricle from the inflow (mitral valve) different segments at different levels may be seen and distinguished as shown in Figure 10A.

Left Ventricle

The 17-segment model of left ventricle divides left ventricle into 17 segments and each segment is evaluated individually in terms of motion, wall-thickening and radius change. Any segment which is contracting less than normal is hypokinetic, not contracting at all is akinetic and the segment contracting out of synchrony with other segments is called dyskinetic. Any deviation from normal contractile pattern is called as regional wall motion abnormality (Figs. 10A and B).

The 17-segment model can be understood by comparing the left ventricle (LV) with a tumbler. If seen from top, this looks like Figure 2. If we cut segments at three different levels, it replicates the 17-segment model.

Wall motion scoring index (WMSI) = Sum of scores/Number of visualized segments
1—Normal
1-1.9—Smaller infarct
>2—Predictive of complications.

ASSESSMENT OF SYSTOLIC FUNCTION OF LEFT VENTRICLE

Fractional Shortening

The decrease in the end diastolic diameter of the left ventricle at the end of systole (LVes), expressed as a fraction of the left ventricle end-diastolic diameter (LVed). It is calculated using M mode across left ventricle in either PLAX view or PSAX view at mid ventricular level and expressed as percentage.

Fractional Shortening (Fs) = (LVDed−LVDes)/LVDed × 100
25–40%—normal
20–25%—mild dysfunction

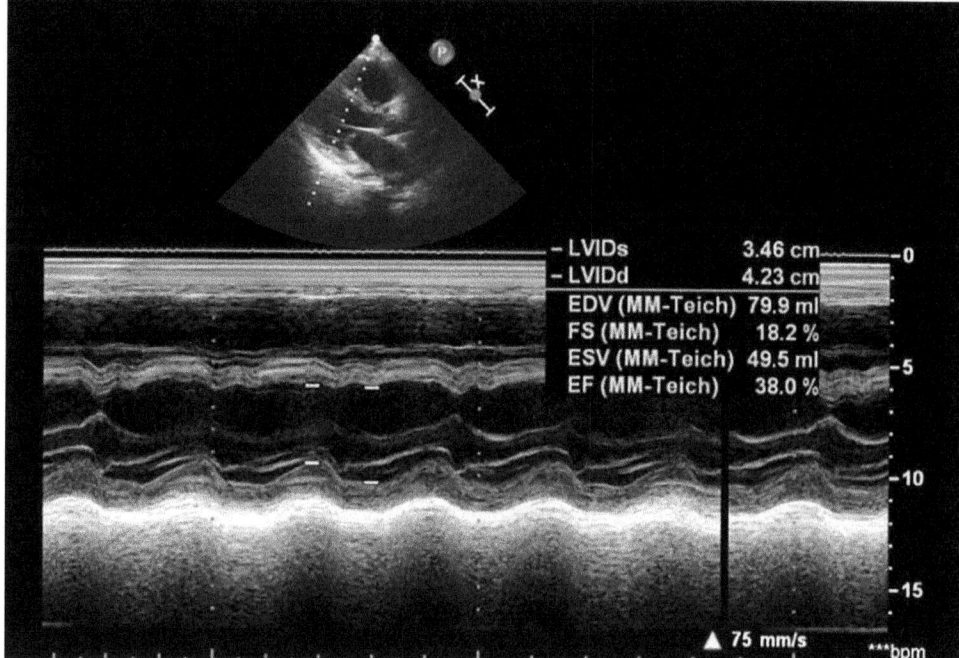

Fig. 11: Fractional shortening and ejection fraction in PLAX view on 2D transthoracic echocardiography.

15-20%—moderate dysfunction
<15%—severe dysfunction.

Quinones Method

It measures the EF, most commonly used parameter for assessment of left ventricular function. The EF is also calculated by applying M mode across the left ventricle in either PLAX view or PSAX view at mid ventricular level and expressed as percentage. All machines display both Fs and EF together, once LVed and LVes dimensions are measured. End diastolic and end systolic volumes are also displayed (Fig. 11).

EF % = $(LVDed^2 - LVDes^2) / LVDed^2 \times 100$
>55% EF—normal
40-50%—mild dysfunction
30-40%—moderate dysfunction
20-30%—moderately severe dysfunction
<20%—severe dysfunction.

Modified Simpson's Method of Discs

An apical four chamber or two chamber view is selected and the endocardial border of the left ventricle is traced during diastole and systole. The trace begins and ends at the mitral

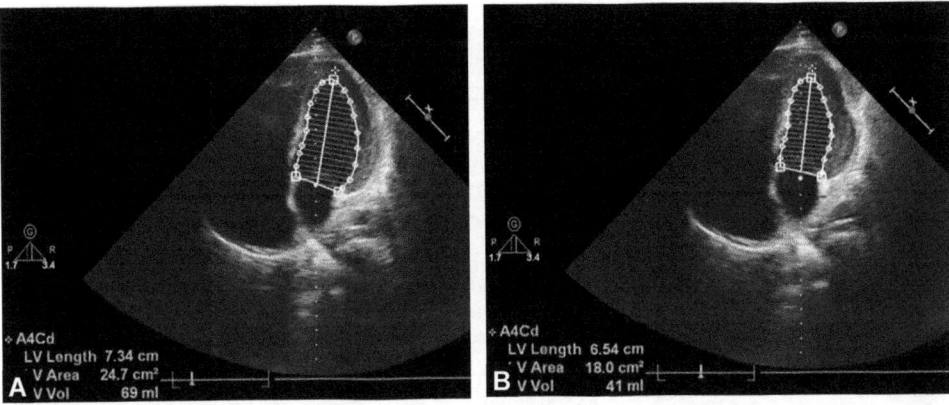

Figs. 12A and B: (A) End-diastolic volume by Simpson's method; (B) End-systolic volume by Simpson's method.

valve annulus. The software automatically converts the ventricle cavity into 20 discs and calculates the total volume of discs at both end diastole (EDV) and end systole (ESV). The stroke volume is calculated as the difference of EDV and ESV (Figs. 12A and B). EF (%) is calculated as:

EF (%) = (SV/EDV) × 100
EF = (69–41)/69 × 100 = 40.5%.

Cardiac Output Measurement

Cardiac output is the amount of blood pumped by heart into the systemic circulation per minute. Stroke volume is the amount of blood ejected by heart in one cardiac cycle (per heart beat). The amount of blood ejected in one cardiac cycle travels some distance in the aorta known as stroke distance. The product of area of the orifice through which the blood is ejected (left ventricle outflow) and stroke distance equals the stroke volume.

In left ventricle, the orifice is LVOT or the aortic valve. The cross-section area of LVOT is measured as CSA = $\pi d^2/4$, where d = diameter of LVOT. For aortic valve the area can be measured by using planimetry (tracing the borders of open aortic valve) in the basal PSAX view.

Stroke distance is measured as velocity time integral (VTI) with either pulse wave Doppler (for LVOT) or continuous wave Doppler (for aortic valve). An apical five-chamber view is ideal to align for Doppler measurement across LVOT or aortic valve. The normal value of VTI for LVOT is more than 20 cm. A value less than 18 cm implies normal stroke volume (Fig. 13).

VTI = 21.8 cm. If aortic annulus is 2.0 cm then area = (3.14 × 2 × 2)/4 = 3.14 cm^2
SV = 21.8 × 3.14 = 68.5 mL.

Tissue Doppler Imaging

Tissue Doppler imaging (TDI) mode measures the velocities of myocardial tissues at any given point. For example, tricuspid annulus, interventricular septum. To measure the tissue velocity, TDI mode is selected and pulse wave Doppler is used to measure the velocity at the defined location. The velocity in systole is designated as s'". A S' velocity greater than 7.5 cm/s at the basal segments of left ventricle implies an LVEF greater than 50%[4] (Fig. 14).

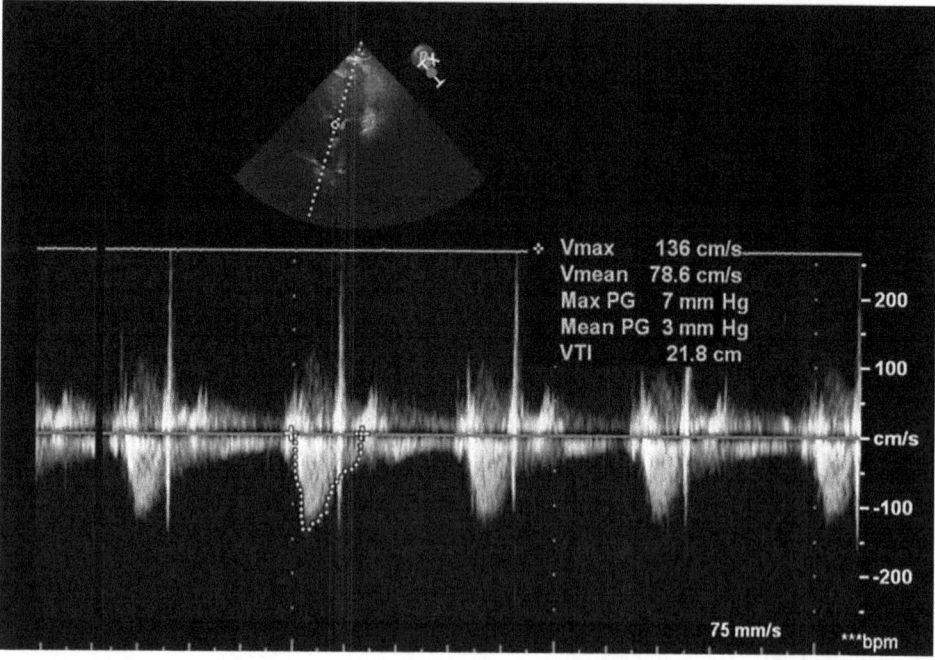

Fig. 13: Velocity time integral (VTI) across aortic valve to calculate cardiac output in apical five-chamber view.

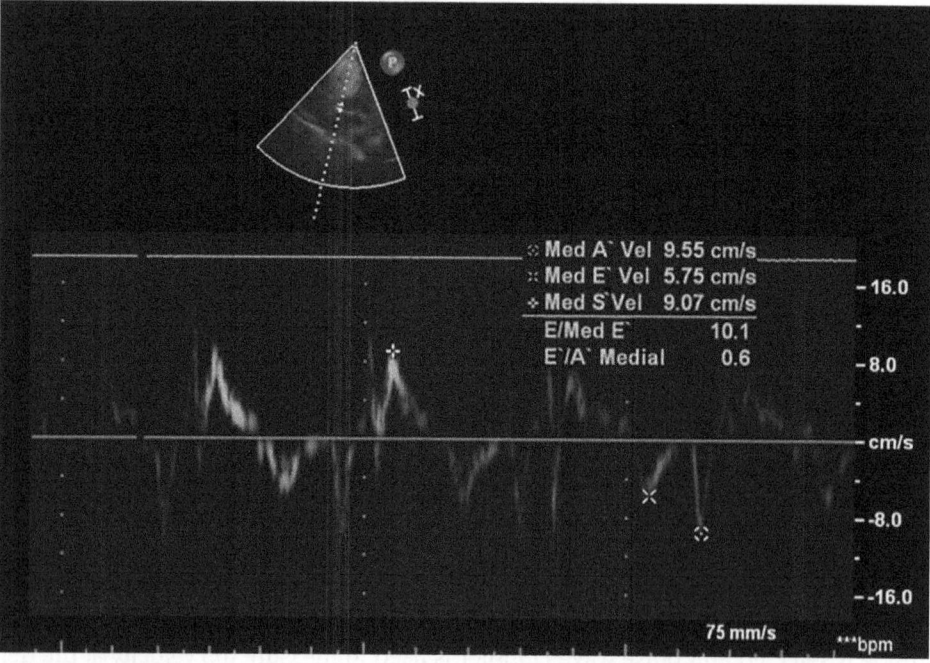

Fig. 14: Tissue Doppler imaging (TDI) showing early filling (E'), atrial (A') and systolic (S') velocities.

ASSESSMENT OF DIASTOLIC FUNCTION OF LEFT VENTRICLE

Diastolic dysfunction is the inability of the ventricle to fill to its normal end diastolic volume without an inappropriate increase in end diastolic pressure.

In a normal diastole, there are four phases of filling:
- *Isovolumetric relaxation:* Duration between aortic valve closure and mitral valve opening.
- *Rapid early filling (E wave):* Mitral valve opening leading to flow into the LV and gradual increase in EDV and end-diastolic pressure.
- *Diastasis:* Period of "no" flow when LA and LV pressures are equal.
- *Atrial contraction (A wave):* increase in LAP due to atrial contraction causes filling of LV towards the end of diastole. It contributes about 15–20% of normal LV EDV.

Transmitral Pulse Wave Doppler

When pulse wave Doppler is placed across the mitral valve E and A waves are seen as shown below. E wave represents early diastolic filling and A wave atrial contribution to the left ventricular preload. There is a period of diastasis in between E and A waves.

With the envelope, thus obtained, isovolumic relaxation time (IVRT), E and A wave velocities, and E/A ratio can be measured. The normal E wave velocity is 0.8 ± 0.2 m/s and A wave velocity is 0.5 ± 0.2 m/s.

On the basis of E/A ratio, the phases of diastolic dysfunction are classified as:

Normal:	E/A = 1–2
Impaired relaxation:	E/A < 1.0
Pseudonormal filling:	E/A = 1–1.5
Restrictive filling:	E/A > 2.0

To differentiate between normal and pseudonormal patterns, pulmonary vein pulse wave Doppler is done.

The echocardiograms in Figures 15A and B show transmitral Doppler of two different patients. E/A ratio in both echocardiograms is normal. However, the E wave velocity in Figure 15A is only 45.9 cm/s which is less than normal. In Figure 15B the E wave velocity is normal 89.4 cm/s.

Pulmonary Vein Pulsed-wave Doppler

Similarly, when pulsed-wave Doppler is placed across pulmonary vein, the following envelope is seen. Pulmonary venous systolic and diastolic velocities and duration are calculated. Normal "s" wave velocity is 28–82 cm/s; "d" wave velocity is 27–72 cm/s and "a" (atrial reversal) wave is 14–25 cm/s (Fig. 16).

Further characterization of diastolic dysfunction is done as follows:

	TM Doppler	*Pulsed-wave Doppler*
Normal:	E/A = 1–2	PVs > PVd
Impaired relaxation:	E/A < 1.0	PVs > PVd
Pseudonormal filling:	E/A = 1–1.5	PVs < PVd
Restrictive filling:	E/A > 2.0	PVs << PVd

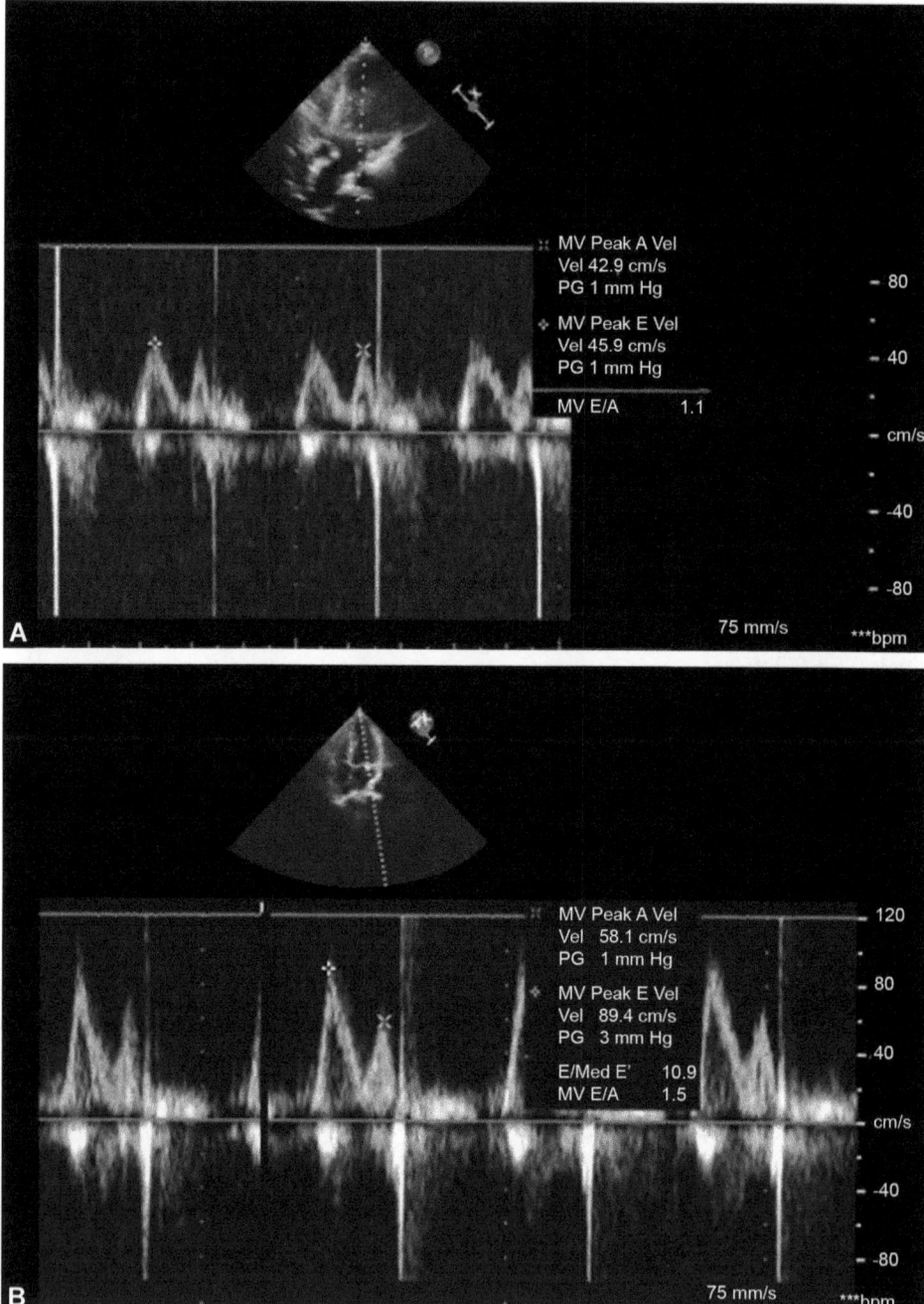

Figs. 15A and B: (A) Transmitral pulse wave Doppler showing normal E/A ratio and A wave velocity but decreased E wave velocity; (B) Transmitral pulse wave Doppler showing normal E/A ratio with normal E and A wave velocity.

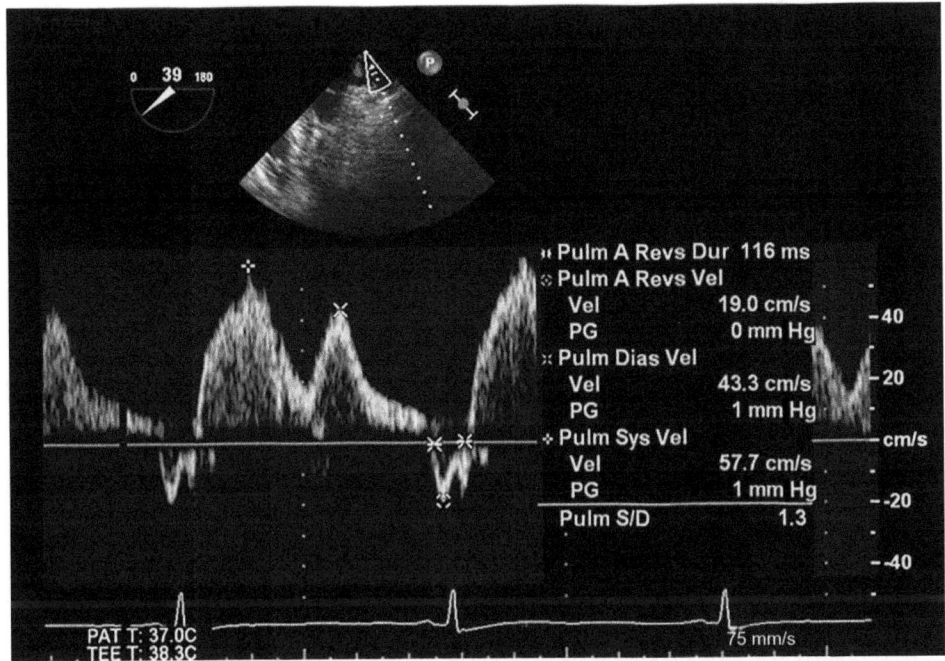

Fig. 16: A normal pulmonary vein pulsed-wave Doppler S, D and A wave velocities and S/D ratio in normal limits.

ASSESSMENT OF RIGHT VENTRICULAR FUNCTION

Dilatation of Right Ventricle

Normally, RV area or LV area is less than 0.6 and RV length is less than 0.6 LV length. The apex is formed by left ventricle. The dilatation of RV can be graded in the following stages:
- *Mild:* RV area /LV area = 0.6
- *Moderate:* RV area /LV area = 1.0 (Fig. 17)
- *Severe:* RV area /LV area >1.0.

Causes of RV dilatation and dysfunction:
- Pulmonary artery hypertension
- Coronary artery disease
- Tricuspid or pulmonary regurgitation
- Atrial or ventricular septal defect
- Dilated cardiomyopathy (Idiopathic, chemotherapy induced, pregnancy induced, etc.)

Tricuspid Annular Plane Systolic Excursion

The simplest method to evaluate right ventricular function is tricuspid annular plane systolic excursion (TAPSE). In the M mode, the cursor of the echo probe is placed at the lateral annulus of the tricuspid valve and the excursion of the annulus is measured in the systole. The TAPSE has been correlated with EF of right ventricle, a normal right ventricular ejection fraction (RVEF) being taken as 40%. A simple equation for prediction of right ventricle

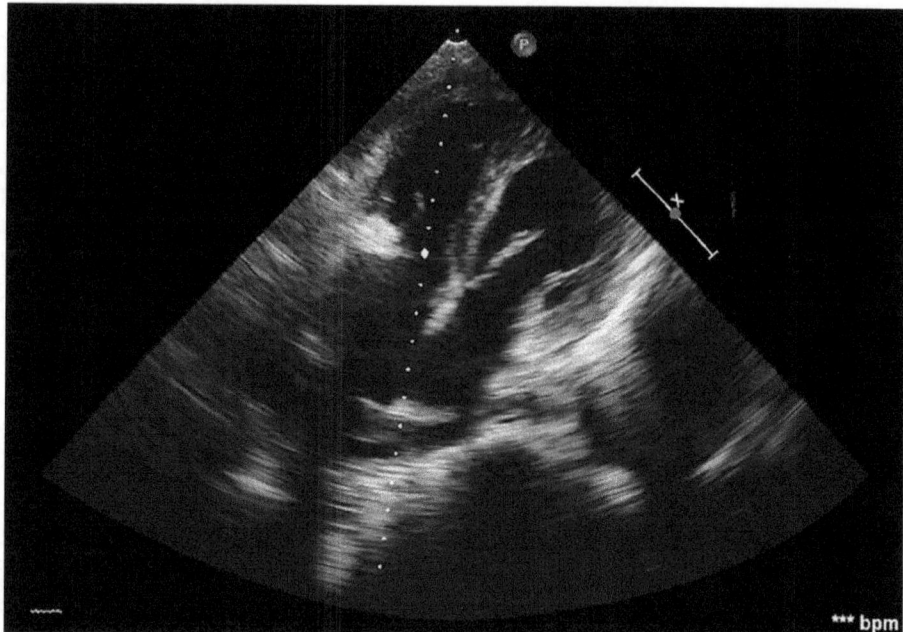

Fig. 17: Echocardiogram showing a dilated right ventricle (RV) with an RV and left ventricle ratio of 1—moderate RV dilatation.

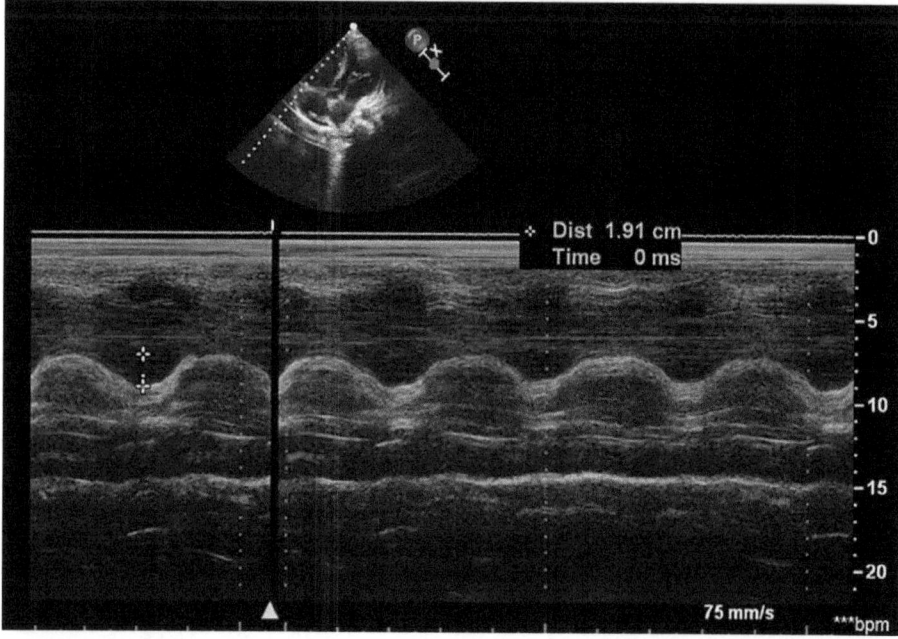

Fig. 18: M mode echocardiography showing normal tricuspid annular plane systolic excursion (19.1 mm).

ejection fraction is RVEF = 2 × TAPSE. Normal TAPSE is 20–25 mm. The cut off value of 19.7 mm predicts RV dysfunction with a sensitivity and specificity of 88.9% and 84.6%, respectively[5] (Fig. 18 and Table 3).

Table 3: Relationship between TAPSE and RVEF.	
TAPSE (mm)	RVEF (%)
20	40
15	30
10	20
5	10

Abbreviations: TAPSE: Tricuspid annular plane systolic excursion; RVEF: Right ventricular ejection fraction.

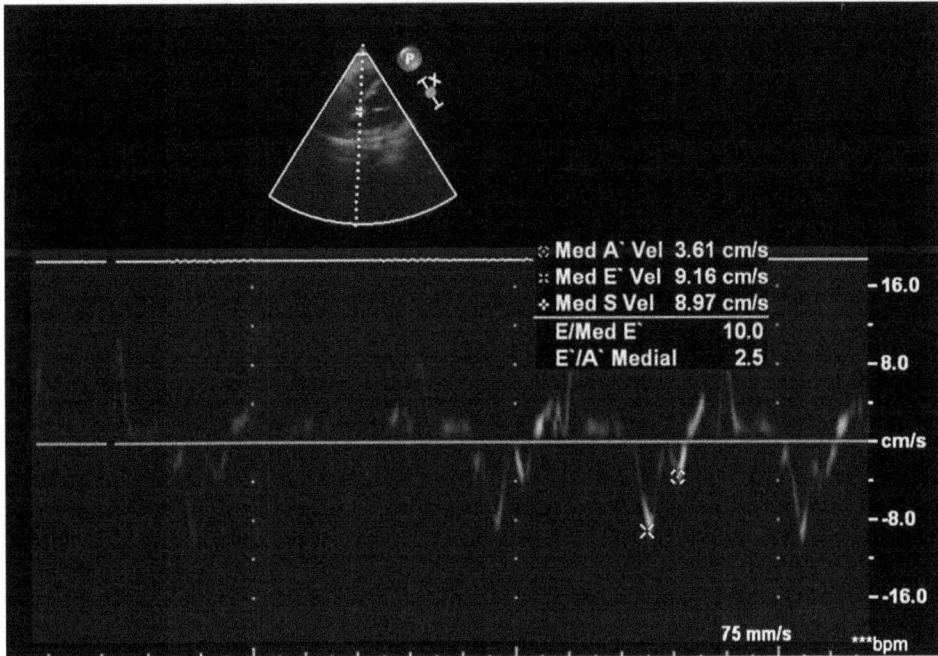

Fig. 19: Tissue Doppler imaging showing a near normal e' velocity (9.16 cm/s) and a normal E/e' ratio of 10.

Tissue Doppler Imaging: E/e' Ratio

With TDI a lateral e' less than 10 cm/s and septal less than 7 cm/s is suggestive of diastolic dysfunction.[6] The ratio of E/e' less than 8 is normal and a ratio more than 15 is associated with LV diastolic dysfunction in nonventilated patients. A ratio of more than 12 is sufficient to identify diastolic dysfunction in patients on mechanical ventilation[7] (Fig. 19).

Hypovolemia

Following abnormalities are seen in patients with hypovolemia:

Echocardiogram showing kissing walls: In hypovolemia, normal ventricles when underfilled try to compensate by increasing the contractility to maintain stroke volume. Due to less

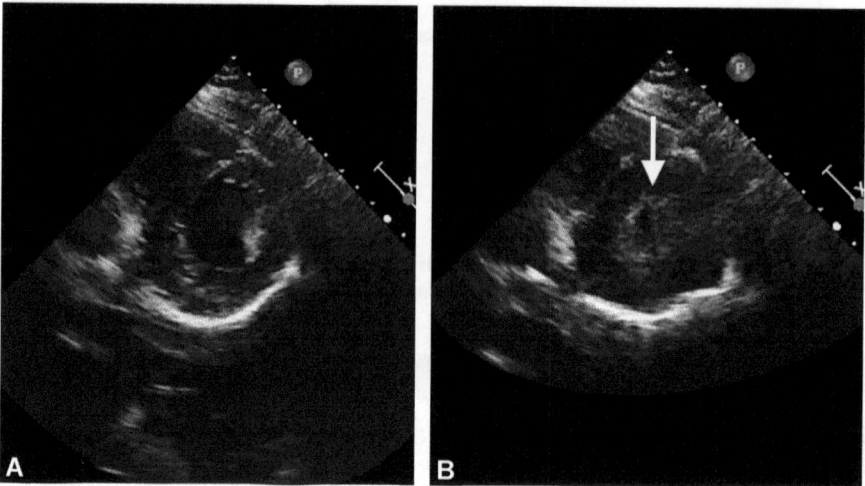

Figs. 20A and B: (A) Hypovolemic ventricle in diastole; (B) Hypovolemic ventricle in systole-cavity (arrow) is hardly seen "kissing papillary muscles".

volume in the ventricle in diastole and a hyper contractile state both papillary muscles come close to each other in systole typically known as "kissing papillary muscles". The sign is typically seen in PSAX view at the papillary muscle level (Figs. 20A and B).

LVED area: An LVEDA of less than 10 cm² or a LVEDA index (LVEDA/BSA) of less than 5.5 cm²/m² indicates significant hypovolemia.

IVC size and collapsibility: The IVC size and collapsibility is assessed in the subcostal IVC view.
- Size ≤ 2.1 cm; collapses >50% during sniff = RAP 0–5 mm Hg; benefit from fluid administration
- Size > 2.1 cm; collapses >50% during sniff = 5–10 mm Hg; benefit from fluid administration
- Size > 2.1; collapses <50% during sniff = 10–20 mm Hg; will not benefit from fluid administration (Fig. 21).

Fluid response: An increase of more than 15% in stroke volume or cardiac output is inferred as fluid responsiveness. The surrogate measure of SV on echocardiography is VTI and therefore can be used as a guide.

LVOT VTI variation with respiration: During respiration, a variation of more than or equal to 12% in VTI is suggestive of fluid responsiveness and may be equivalent to a rise in stroke volume by 15%. Instead of tracing the curve to calculate VTI, the maximum and minimum velocity may also be measured. Any change of more than or equal to 12% predicts fluid responsiveness.

LVOT VTI variation with passive leg raise: An increase in VTI by more than 12% after raising both legs by more than 45° is a predictor of fluid requirement.[8-10]

Dynamic LVOT obstruction: It may occur in hypovolemic and hyperdynamic ventricles, as happens in poorly filled ventricles on high inotropic support.

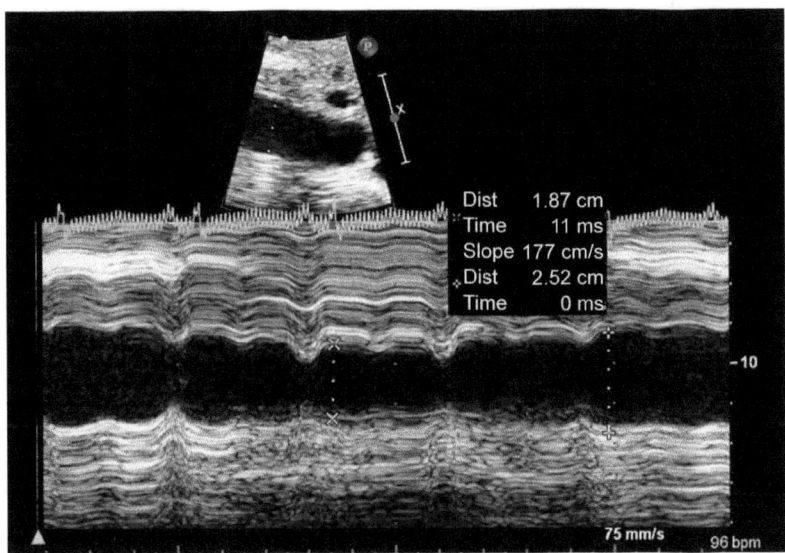

Fig. 21: Inferior vena cava size is 2.52 cm and collapsibility [(2.52–1.87)/2.52] is 26%. This patient is unlikely to benefit from fluid administration and inotropes may be required.

Pericardial Effusion and Tamponade

Sinus tachycardia raised jugular venous pressure, pulsus paradoxus and pericardial rub points to the clinical diagnosis of cardiac tamponade. However, echocardiography aids in diagnosis. Diastolic collapse of RA and RV, more specifically of LA, collection of pericardial fluid around compressed cardiac chambers, IVC plethora are the common signs of cardiac tamponade. When the intrapericardial pressure exceeds the diastolic pressure of right ventricle, there is impaired filling of chambers resulting in tamponade.

Following finding should be looked for to diagnose as tamponade:
- Pericardial effusion
- Diastolic collapse of right ventricle
- Right atrial collapse
- Dilated IVC with no collapsibility
- Increase in TV inflow velocity >25% and mitral inflow velocity >15% (Figs. 22 and 23).

Acute Pulmonary Embolism

Echocardiography is not highly sensitive and specific in diagnosing pulmonary embolism, but act as a surrogate in appropriate clinical scenarios. The echocardiographic findings are suggestive of pulmonary hypertension and right ventricle failure.

Dilatation of right ventricle diagnosed with a RV/LV area ratio more than 0.6 is an early and significant finding. The ratio may approach a value of 1.0 with more severe dilatation and failure. The hypokinesia of RV free wall with normokinetic apex (McConnell's sign) may be seen. Occasionally, thrombus may be visualized in right ventricle or in pulmonary arteries (Figs. 24 and 25).

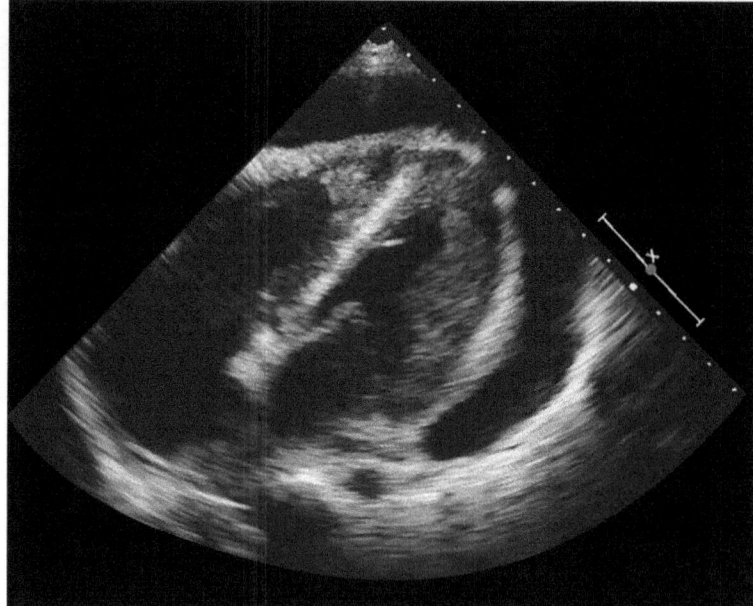

Fig. 22: Echocardiogram showing pericardial effusion only with "NO" collapse of RV; therefore no tamponade.

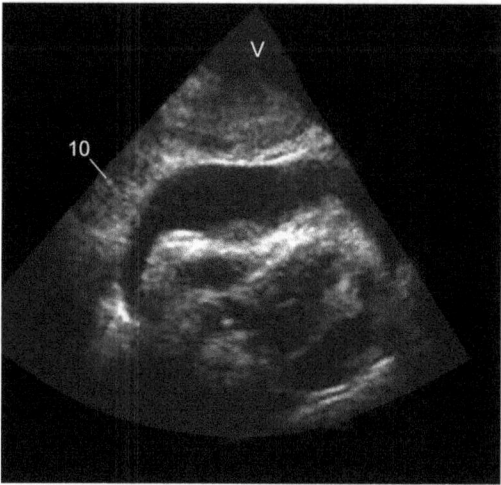

Fig. 23: Echocardiogram showing diastolic collapse of right ventricle in pericardial tamponade.

Pulmonary Hypertension

Pulmonary hypertension is defined as mean pulmonary artery pressure (mPAP) more than 25 mm Hg at rest. Pulmonary hypertension is graded as:
- Mild   mPAP 25–40 mm Hg
- Moderate   mPAP 40–55 mm Hg
- Severe   mPAP >55 mm Hg

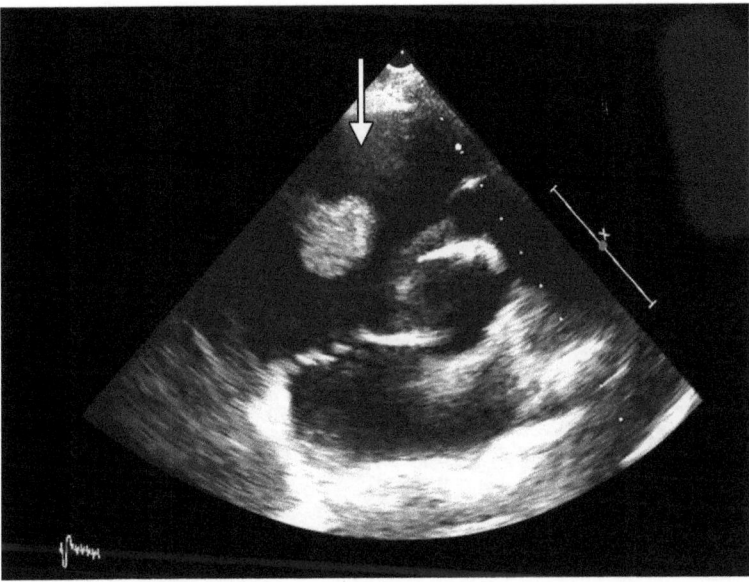

Fig. 24: Parasternal short-axis view showing thrombus in right ventricle.

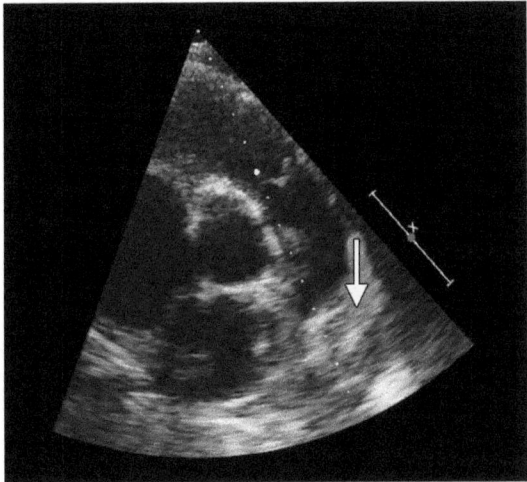

Fig. 25: Parasternal short-axis view showing thrombus in main pulmonary artery.

The following echocardiographic findings may be seen in pulmonary hypertension:
- Dilated right ventricle and right atrium
- Right ventricle hypertrophy (wall thickness >7 mm)
- Interventricular septum convex towards left ventricle
- Dilated main pulmonary artery
- Impaired right ventricle function as measured by TAPSE.

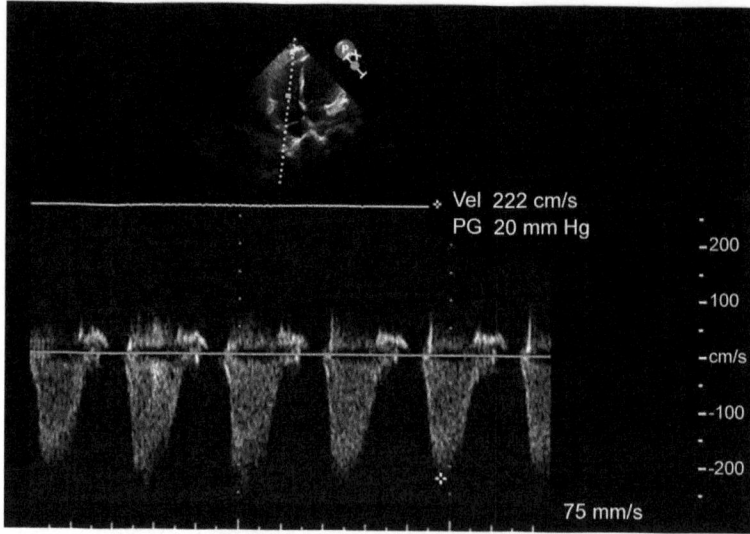

Fig. 26: Pulmonary artery systolic pressure (PASP) estimation by TR jet.

- *Tricuspid regurgitation jet with high velocity:* Continuous wave Doppler is used to measure tricuspid regurgitant jet velocity. Pulmonary artery systolic pressure (PASP) is calculated as PASP (in mm Hg) = Gradient across TR ($4V^2$) + RAP (Fig. 26).
  PASP (in mm Hg) = Gradient across TR ($4V^2$) + RAP; $4 \times (2.22)^2$ + RAP = 20 mm Hg + RAP
  RAP = IVC size and collapsibility
- Size ≤ 2.1 cm; collapses >50% during sniff = RAP 0–5 mm Hg.
- Size > 2.1 cm; collapses >50% during sniff = 5–10 mm Hg.
- Size > 2.1 cm; collapses <50% during sniff = 10–20 mm Hg.

Cardiac Abnormalities in Severe Sepsis

The following abnormalities have been found to occur in sepsis:[10]
- Left ventricular dilatation
- Left ventricular contraction impairment:
  - Global
  - Segmental
- Left ventricular diastolic dysfunction
- Right ventricle systolic/diastolic dysfunction
- Ventricular outflow obstruction
- Valvular lesions:
  - Functional
  - Endocarditis.

REFERENCES

1. Jensen MB, Sloth E, Larsen KM, et al. Transthoracic echocardiography for cardiopulmonary monitoring in intensive care. Eur J Anaesthesiol. 2004;21:700-7.
2. American College of Cardiology. (2003). Cheitlin MD, Armstrong WF, Aurigemma GP, et al. ACC/AHA/ASE 2003 guidelines update for the clinical application of echocardiography: a report of the American College of Cardiology/American Heart Association Task Force on Practice Guidelines 2003. [online]. American College of Cardiology website. Available from www.acc.org/clinical/guidelines/echo/index.pdf. [Accessed November 2017].
3. Porter TR, Shillcutt SK, Adams MS, et al. Guidelines for the Use of Echocardiography as a Monitor for Therapeutic Intervention in Adults: A Report from the American Society of Echocardiography. J Am Soc Echocardiography. 2015;28:40-56.
4. Alam M, Wardell J, Andersson E, et al. Effects of first myocardial infarction on left ventricular systolic and diastolic function with the use of mitral annular velocity determined by pulsed wave Doppler tissue imaging. J Am Soc Echocardiogr. 2000;13:343-52.
5. Sato T, Tsujino I, Oyama-Manabe N, et al. Simple prediction of right ventricular ejection fraction using tricuspid annular plane systolic excursion in pulmonary hypertension. Int J Cardiovasc Imaging. 2013;29:1799-1805.
6. Flachskampf FA, Biering-Sørensen T, Solomon SD, et al. Cardiac imaging to evaluate left ventricular diastolic function. JACC Cardiovasc Imaging. 2015;8:1071-93.
7. Vignon P, AitHssain A, François B, et al. Echocardiographic assessment of pulmonary artery occlusion pressure in ventilated patients: a transoesophageal study. Crit Care. 2008;12:R18.
8. Maizel J, Airapetian N, Lorne E, et al. Diagnosis of central hypovolemia by using passive leg raising. Intensive Care Med. 2007;33:1133-8.
9. Monnet X, Rienzo M, Osman D, et al. Passive leg raising predicts fluid responsiveness in the critically ill. Crit Care Med. 2006;34:1402-7.
10. Mc Lean AS. Echocardiography in shock management. Crit Care. 2016;20:275

Chapter 8

Volume Status Assessment and Predicting Volume Responsiveness

Tanvir Samra

INTRODUCTION

The primary objective of administration of fluids in hypotensive patients is to increase the stroke volume/cardiac output. Previous studies have demonstrated that only 50% of fluid bolus in critically ill-hypotensive patients lead to an increase in stroke volume (SV), i.e. only 50% are "fluid responsive" (FR).[1] In order to avoid complications of fluid overload it is thus of paramount importance to identify "fluid responsive" patients. Various invasive and non-invasive techniques have been described for assessment of fluid status but only some are capable of estimating the fluid responsiveness.

Volume Status Assessment

A patient may be euvolemic, hypovolemic or hypervolemic and thus an accurate assessment of the volume status is necessary for clinical decision making and further management. Traditionally volume status assessment has been done based on the results of the clinical examination; tachycardia, hypotension, loss of skin turgor signify hypovolemia. Sunken eyes and fontanels are seen in dehydrated infants. Hypervolemia leads to jugular vein distension, peripheral edema, rales, and congestive heart failure (appearance of third heart sound). But critically ill-patients have numerous other systemic abnormalities and thus a reliable estimate of volume status cannot be made on the basis of physical examination. Also subtle changes are missed and only extreme abnormalities of body water are recognized.

STATIC VERSUS DYNAMIC PARAMETERS

Fluid loading increases stroke volume only if the ventricles are on the steep portion of the Frank-Starling curve (condition of ventricular preload dependence). "Fluid responsiveness" is defined as a more than 15% increase in cardiac index in response to a volume expansion.[2] Ventricles operating on flat portion of the curve are not responsive (condition of ventricular preload independence). Static parameters are markers of ventricular preload but dynamic parameters are markers of fluid responsiveness (Figs. 1 and 2).[3]

STATIC INDICES OF VENTRICULAR PRELOAD

Central venous pressure (CVP), pulmonary artery occlusion pressure (PAOP), left ventricle end-diastolic volume (LVEDV) index, and right ventricle end-diastolic volume (RVEDV) index

Volume Status Assessment and Predicting Volume Responsiveness 139

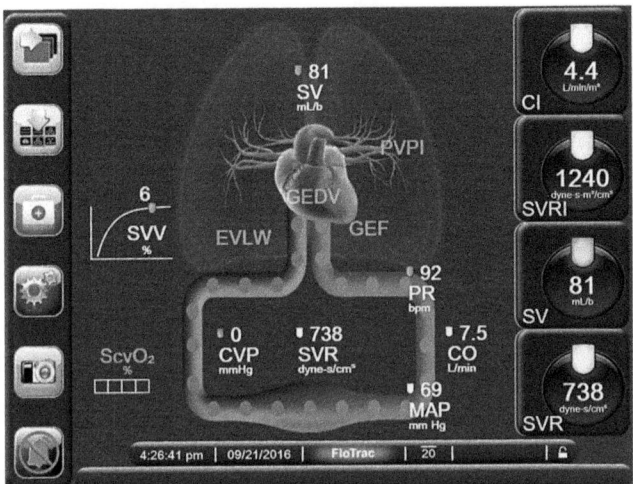

Fig. 1: Use of Vigileo™ monitor (Vigileo™; FloTrac; Edwards™; Lifesciences, Irvine, CA, USA) for depiction of static and dynamic parameters to enable prediction of fluid responsiveness.
Abbreviations: SV: Stroke volume; SVV: Stroke volume variation; CVP: Central venous pressure; SVR: Systemic vascular resistance; SVRI: Systemic vascular resistance index; EVLW: Extravascular lung water; GEF: Global ejection fraction; MAP: Mean arterial pressure; CO: Cardiac output; CI: Cardiac Index; ScvO$_2$: Central venous oxygen saturation.

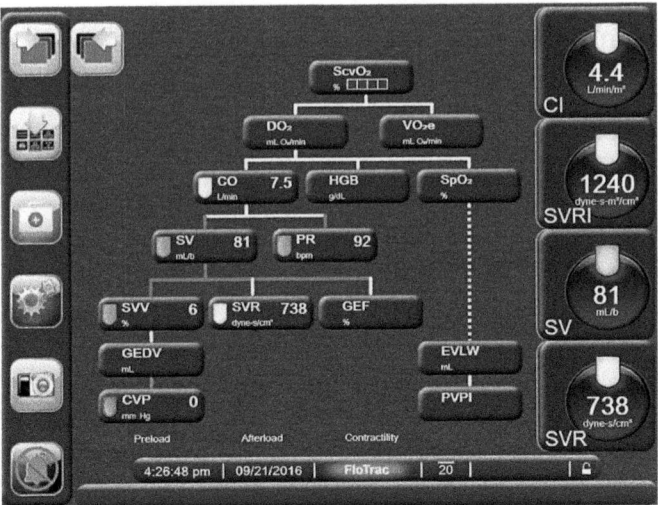

Fig. 2: Fluid optimization based on static and dynamic parameters of hemodynamic monitoring.
Abbreviations: DO$_2$: Oxygen delivery; VO$_2$: Oxygen consumption; HGB: Hemoglobin; PVPI: Pulmonary vascular permeability index; CVP: Central venous pressure; SVR: Systemic vascular resistance; SVRI: Systemic vascular resistance index; EVLW: Extravascular lung water; GEF: Global ejection fraction; MAP: Mean arterial pressure; CO: Cardiac output; CI: Cardiac index; ScvO$_2$: Central venous oxygen saturation.

Central Venous Pressure

It is the pressure within the superior vena cava (SVC) which is located in the thorax. In the past it was one of the most commonly used techniques for estimating volume status in critically ill-patients. But it has gone out of favor as preload or venous return is determined not just by the pressure in SVC but by the combination of the following variables (Figs. 1 and 2):

Preload = (MCFP − RAP) ÷ SVR[4]

(MCFP: Mean circulating filling pressure; RAP: Right atrial pressure; SVR: Systemic vascular resistance).

It cannot be used in intensive care unit (ICU), operating room (OR) or emergency department (ED) to predict hemodynamic response to a fluid challenge.[5]

Pulmonary Artery Occlusion Pressure

Values derived from CVP and PAC are dependent on myocardial compliance. In critically ill patients pulmonary artery wedge pressure (PAWP) correlates poorly with LVEDV. Use of PAC has phased out due to high rates of complications and misinterpretation of data.[6]

Left Ventricle End-diastolic Volume Index and Right Ventricle End-diastolic Volume Index

It can be determined using a fast-response thermistor PAC or by cardiac scintigraphy. A response to fluid is likely with an RVEDV index of less than 90 mL/m^2 and unlikely with an index of more than 138 mL/m^2, respectively.[7]

Left Ventricle End-diastolic Area

This can be determined using echocardiography. A hyperdynamic left ventricle (LV) with an end-diastolic area in the parasternal short axis view of less than 10 cm^2 or papillary apposition (kissing ventricles) indicates hypovolemia.[8] Marik et al.[9] demonstrated failure of LVEDA to predict fluid responsiveness in patients on positive pressure ventilation.

Global End-diastolic Volume

Transpulmonary thermodilution using a commercially available device is used to assess the largest volume of blood contained within the four heart chamber [intrathoracic blood volume comprises global end-diastolic volume (GEDV) and pulmonary blood volume]. GEDV has been validated as an indicator of cardiac preload and is indexed to body surface area, yielding global end-diastolic volume index (GEDVI). The reference range for GEDVI is 680–800 mL/m^2.[10]

DYNAMIC PARAMETERS: PREDICTING VOLUME RESPONSIVENESS

Dynamic parameters are derived when the circulating blood volume is provoked by either heart lung interaction or change in posture which leads to change in the loading condition of the heart.

Physiological Basis of Heart-lung Interaction

Inspiration increases intrathoracic pressure which has the following consequences:[11]
- Causes increased transmural pressure across the right ventricle (RV) wall, plethora within the IVC and compression of the SVC. The RV stroke volume immediately falls.
- Compresses pulmonary vasculature and forces blood into the LV causing an initial increase in LV stroke volume.

After the pulmonary transition time, the LV receives less blood and its stroke volume falls. This effect is exaggerated in states of low circulating volume and attenuated in the overloaded system or when either ventricle is failing.[12]

Indicators of fluid responsiveness in patients on mechanical ventilation:
- **Stroke volume variation:**

  Stroke volume variation (SVV) (%) = $(SV_{max} - SV_{min})/SV_{mean}$.

  It is assessed continuously by any beat-to-beat cardiac output monitor. Normal SVV values are less than 10–15% and the goal is to achieve a SVV of less than 13% with fluid resuscitation.[9]

  Pulse pressure variation:

  Pulse pressure variation (PPV) (%) = $(PP_{max} - PP_{min})/PP_{mean}$.

  Predicts preload nonresponders in patients with a PPV <13%.[12]

- **Systolic pressure variation and Δdown:** Systolic pressure variation (SPV) is less specific in predicting fluid responsiveness.[13]

  SPV = $SP_{max} - SP_{min}$ (values of systolic pressure over a single respiratory cycle)

  It is divided into two components (1) Δup and (2) Δdown and thus a reference systolic pressure reading is to be recorded during an end-expiratory pause.

  1. Δdown: Reference systolic pressure – Minimal value of systolic pressure. Venous return decrease due to increased intrathoracic pressure.
  2. Δup: Maximum SBP – Reference SBP during apnea.

  It is due to increase in left ventricular systolic volume from the blood expelled from lungs and decreased afterload. Improved left ventricular compliance due to temporary decrease in right heart chambers volume also contributes for Δup.

  Marik et al.[9] in a meta-analysis have reported pooled correlation coefficients between the baseline PPV, SVV, SPV, and the change in cardiac index as 0.78, 0.72, and 0.72, respectively. The area under the receiver operating characteristic curves were 0.94, 0.84, and 0.86, respectively, compared with 0.55 for the CVP, 0.56 for the GEDVI and 0.64 for the LVEDVI. Mean threshold values were 12.5 ± 1.6% for PPV and 11.6 ± 1.9% for SVV. The sensitivity, specificity, and diagnostic odds ratio were 0.89, 0.88, and 59.86 for PPV and 0.82, 0.86, and 27.34 for SVV, respectively.

- **Pleth variability index:** Tsuchiya et al.[14] have reported that preanesthesia pleth variability index (PVI) can predict a decrease in mean arterial pressure (MAP) during anesthesia induction. By classifying PVI >15 as positive, a MAP decrease more than 25 mm Hg could be predicted, with sensitivity, specificity, positive predictive, and negative predictive values of 0.79, 0.71, 0.73, and 0.77, respectively. The PVI threshold value of 17% discriminates between fluid responders and nonresponders with a sensitivity of 95% (95% confidence interval, 74–100%) and a specificity of 91% (95% confidence interval, 70–99%).[15]

- **Aortic blood velocity (ΔV_{peak}) and velocity time integral variation:**
  ΔV_{peak} = Maximal − Minimal peak velocity of aortic blood flow/mean of the two values.
  Assumption made is that aortic annulus diameter remains constant over the respiratory cycle. Aortic blood flow is measured by a pulsed-wave Doppler echocardiography at the level of the aortic valve. A ΔV_{peak} threshold value of 12% discriminates between responders and nonresponders.[14] Similarly, ΔV_{peak} is readily measured in the descending aorta using transesophageal Doppler.
- **Inferior vena cava collapsibility/distensibility index:** Curvilinear or phased array transducers (echo probes) are used to obtain a two-dimensional (2D) image of the IVC entering the right atrium. Transducer is placed just inferior to the xiphoid process with the index mark pointing directly toward the patient's head to obtain the long axis of the IVC. The isthmus is the narrowest part of the IVC and is most responsive to respiration. It is to be ensured that the IVC visualization is not lost during movements of respiration. We then place an M-mode line through the IVC 1 cm caudal from its junction with the hepatic vein, and obtain an M-mode tracing. This placement ensures that we do not measure the intrathoracic IVC during any part of the respiratory cycle. Freeze the M-mode image using callipers and measure the maximum and minimum diameter of the IVC tracing.

  Literature supports the use of IVC ultrasonography (USG) to optimize the administration of fluids in hemodynamically unstable critically ill-patients.[15,16] Preoperative detection of hypovolemia and intraoperative assessment of fluid responsiveness based on results of IVC USG are gaining popularity. But we need to be cautious due to the following limitations:
  - Inferior vena cava USG is an operator dependent modality
  - Inferior vena cava diameter varies with ethnicity and is variable
  - Calculations of IVC may not be taken at the isthmus.

  Interpretation of inferior vena cava diameter:
  Inferior vena cava is normally 1.5–2.5 cm in diameter (measured 3 cm from right atrium). IVC less than 1.5 cm suggests volume depletion and an IVC more than 2.5 cm suggests volume overload.[15]

  Inferior vena cava collapsibility index/caval index:
  In spontaneously breathing patients IVC diameter decreases on each inspiration.[16]
  Caval index = (IVC_{max} diameter − IVC_{min} diameter)/(IVC_{max} diameter) × 100
  It is to be noted that the IVC collapses nonuniformly and there is ambiguity in the exact site to be selected for measurement of IVC collapsibility. Thus, it is recommended that all measurements should be done just caudal to confluence of hepatic veins (~3 cm from right atrium). This is an index of volume status (hypovolemia, hypervolemia) and right atrial pressure. It gives a noninvasive estimate of the CVP.

  The 2010 American Society of Echocardiography (ASE) guidelines use IVC diameter and IVC collapsibility index to estimate RAP.[17]
- RAP is normal (0–5 mm Hg), if IVC diameter less than 2.1 cm and collapse more than 50%
- RAP is high (> 15 mm Hg), if IVC diameter more than 2.1 cm with less than 50% collapse

- RAP is intermediate (5–10 mm Hg), if:
  - IVC diameter less than 2.1 cm and less than 50% collapse or
  - IVC diameter more than 2.1 cm with more than 50% collapse correspond to an intermediate.

If the IVC is only visible with a partial or full inhalation then it is advised to record a long clip with the full deep breath and then ask the patient to exhale or inhale slightly until the IVC becomes visible.

Sniff test: If the IVC does not collapse more than 50% with normal respiration we direct the patient to perform "sniff test". With elevated venous pressure, the IVC will collapse less than 50%. To ensure visualization of the IVC it is advised to instruct the patient to perform a slow and powerful sniff.

Pulse wave (PW) Doppler is used for assessment of the pattern of blood flow through the IVC. If view of the IVC on 2D echo does not produce a sufficient Doppler shift for adequate blood flow assessment then we use a hepatic vein that drains into the IVC for the assessment of IVC blood flow.

Volume status by ΔIVC and distensibility index: Used to assess fluid responsiveness in mechanically ventilated patient. Values are not reliable, if tidal volume is less than 7 mL/kg ideal body weight, nonsinus rhythm and right ventricular dysfunction.

Mechanism: Positive pressure ventilation increases intrathoracic pressure, decreases venous return to the right atrium and distends the IVC due to resistance to right atrial filling. Thus IVC diameter increases in the inspiratory phase and decreases in the expiratory phase.

ΔIVC: Maximum-minimum diameters on the M-mode tracing/mean of the two. A 12% or more variation identified FR patients.

Distensibility index (DI) = (maxIVC − minIVC)/minIVC

Distensibility index less than 18% indicates patient is not volume responsive (unlikely to benefit from fluid bolus).[18]

- *Superior vena cava collapsibility index:* Diameter can be measured using TOE.

SVC collapsibility index = Maximum diameter on expiration − Minimum diameter on inspiration/Maximum diameter on expiration.

Increase in pleural pressure during positive pressure ventilation may completely collapse the vessel in hypovolemic adults. An SVC collapsibility index more than 36% has been shown to predict fluid responsiveness.[19]

PASSIVE LEG RAISING: PREDICTION OF FLUID RESPONSIVENESS IN SPONTANEOUSLY BREATHING ADULTS

Physiological Response to Change in Posture

Postural change (head up to head down) alters the loading conditions of the heart by transferring blood between the leg veins and the central circulation. Passive leg raising (PLR) is used in spontaneous breathing patients to assess fluid responsiveness but requires cardiac output measurement using transthoracic echo (measuring velocity time integral at the aortic valve as an index of aortic flow). Tilting the patient from a 45° semirecumbent head up position to a 45° leg up position, transfers up to 300 mL of blood into the central circulation. Stroke volume/VTI (velocity time integral) across either outflow tract is measured before and 1 minute after the maneuver; an increase of 10% suggests FR.[18]

CURRENT ROLE OF ULTRASOUND FOR VOLUME STATUS ASSESSMENT AND PREDICTION OF FLUID RESPONSIVENESS

Ultrasonography (USG) is an easy, noninvasive and repeatable point of care modality which has gained popularity for assessment of volume and to predict fluid responsiveness. Various different sites of placement of the probe and probe positioning have been mentioned namely:
- *Subcostal longitudinal*: Most commonly used
- *Subcostal transverse*: The RUSH Examination—Rapid Ultrasound for Shock and Hypotension describes this view
- *Midaxillary long axis*: EFAST—extended focused assessment with sonography for trauma describes this view.

Bedside ultrasound has been incorporated into the practice of emergency medicine, trauma and perioperative care to facilitate critical decision making and numerous protocols have been formulated.

ULTRASOUND PROTOCOLS

Focused Assessment with Sonography for Trauma (FAST)

It is a point-of-care ultrasound examination performed during primary survey of a trauma patient. It is done to identify intra-abdominal free fluid which is assumed to be hemoperitoneum in the context of trauma. This necessitates immediate transfer to operation theater in case of hemodynamic instability or CT for further evaluation in hemodynamically stable patients. It has replaced diagnostic peritoneal lavage and has a sensitivity of approximately 90% (range 75–100%) and a specificity of approximately 95% (range 88–100%) for detecting intraperitoneal fluid.[20] FAST also enables visualization of blood around the heart (pericardial effusion).

The patient is positioned supine but Trendelenburg position increases the sensitivity of the abdominal FAST examination. A 3.5–5 MHz convex transducer is used to scan four classic areas for free fluid; the perihepatic space (also called Morison's pouch or the hepatorenal recess), perisplenic space, pericardium, and the pelvis.

Scanning the Pericardial Space

Probe is placed in the subxiphoid space with the probe marker to the right at an angle of 5–10° with the skin. Utilize the liver as an acoustic window and adjust the depth to allow visualization of the posterior bright white pericardium. In traumatic pericardial effusion a black stripe of fluid is seen separating the hyperechoic pericardium from the gray myocardium.

Scanning the Hepatorenal Space and Right Pleural Space

Probe is placed in the anterior axillary line at the inferior portion of the thorax with the probe marker pointing cephalad in a coronal plane. Scan cephalad and caudad in this or the midaxillary line to obtain a clear interface of the liver and kidney (Fig. 3A). Probe is rotated obliquely to visualize the inferior pole of the right kidney. In intraperitoneal fluid collections a black hypoechoic or anechoic stripe is seen in the hepatorenal interspace (Fig. 3B). Slide the probe cephalad one rib space for evaluation of the area above the diaphragm. If this area appears black or anechoic, there should be concern for pleural fluid proximal to the diaphragm. If a rib shadow precludes the ability to evaluate this area, rotate the probe obliquely toward the back.

Scanning the Splenorenal Space

Probe is placed in the mid or posterior axillary line at the inferior portion of the thorax with the probe marker pointed toward the head in the coronal plane. For the left pleural space, the probe is placed in the posterior axillary line between the fourth and eighth intercostal spaces. Rotate the probe toward the back, if rib shadows prevent full evaluation.

Scanning the Rectovesicular Space

This space should be evaluated in both the longitudinal and transverse plane. Probe is placed above the pubic bone with the probe marker pointing to the patient's right side. A full bladder acts as an acoustic window to the space behind the bladder and facilitates evaluation of free fluid (an area of hypoechogenicity) in the anterior vesicouterine space and the posterior rectouterine space. Rotate the probe 90° clockwise so that the probe marker points toward the head for evaluation in the longitudinal plane.

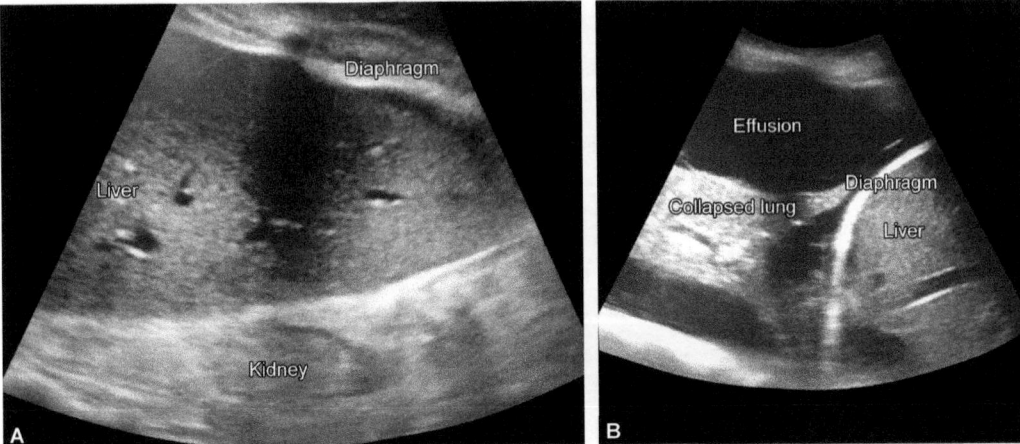

Figs. 3A and B: Normal hepatorenal space (no fluid collection) visualized during Focused Assessment with Sonography for Trauma (FAST) examination; (B) Pleural effusion in right lung.

Extended Focused Assessment with Sonography for Trauma (eFAST) Examination

It is an extension of the FAST examination and includes bilateral anterior thoracic sonography to assess for pneumothorax and hemothorax.[21]

Probe placement for scanning of the anterior lung: In the second or third intercostal space in the midclavicular line in a sagittal orientation.

Normal lung findings (Fig. 4) include visible sliding at the level of the pleura in B-mode and comet tails (vertical reverberation artifacts arising from the pleural line). Lung sliding and the presence of comet tails are evidence of movement of the visceral on the parietal pleura. In M-mode, the lung sliding pattern is called the seashore sign. In some cases lung sliding may be absent and the heartbeat may then be visualized as pulsations of the expanded lung corresponding to the heart rate. This sign is called "lung pulse" and is equivalent to lung sliding.[22]

Pneumothorax is detected with the absence of normal "lung-sliding" and "comet-tail" artefact. Stratosphere sign/bar code sign is seen on M-mode. The lung point is the transition between collapsed and normally expanded lung.[23]

Limitations of sonography are that both false negatives (due to obesity and subcutaneous emphysema) and false positives (fluid-filled bowel adjacent to the liver, spleen or kidneys/preexisting ascites/epicardial fat pad mimicing hemopericardium/seminal vesicles mistaking for pelvic free fluid) are common.

Rapid Ultrasound for Shock and Hypotension (RUSH) Exam

It is a resuscitative ultrasound protocol published in 2010 to accurately diagnose the cause of hemodynamic instability in a patient and is performed using a three step approach.[24-26]

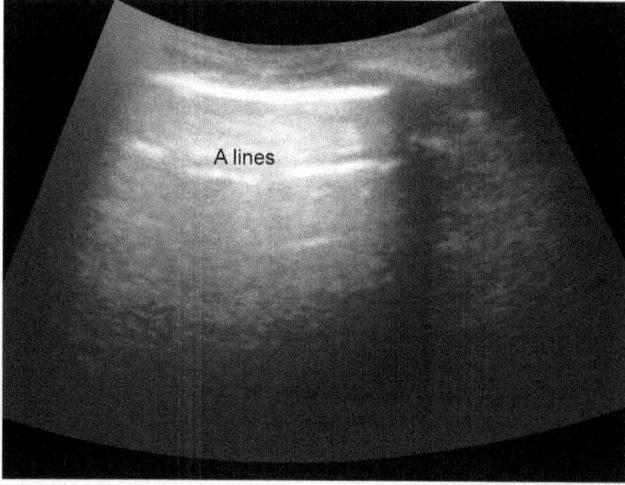

Fig. 4: A lines on lung ultrasonography (USG).

- **Step 1:** *The pump*
  Phased array transducer (3.5–5 MHz) is used to visualize the pericardial sac, assess the contractility of the LV and compare the size of LV and RV.
  Focused echocardiography will be done involving all four traditional views: Parasternal long/short axis, subxiphoid four chamber view and apical four chamber view (Figs. 5A to D).
- **Step 2:** *The tank*
  Linear array transducer (7.5–10 MHz) is used for determination of volume status of the patient by placement over the IVC as described above (Fig. 6). Change in size of the internal jugular vein (IJV) with respiration can be used.
- **Step 3:** *The pipes*
  The arterial vasculature (abdominal and thoracic aorta) to be evaluated for aneurysm and dissection and the venous system to be evaluated for deep venous thrombosis (DVT).

CONCLUSION

For optimum administration of fluids and better clinical outcome it is of paramount important that dynamic predictors of fluid responsiveness should be continuously monitored.

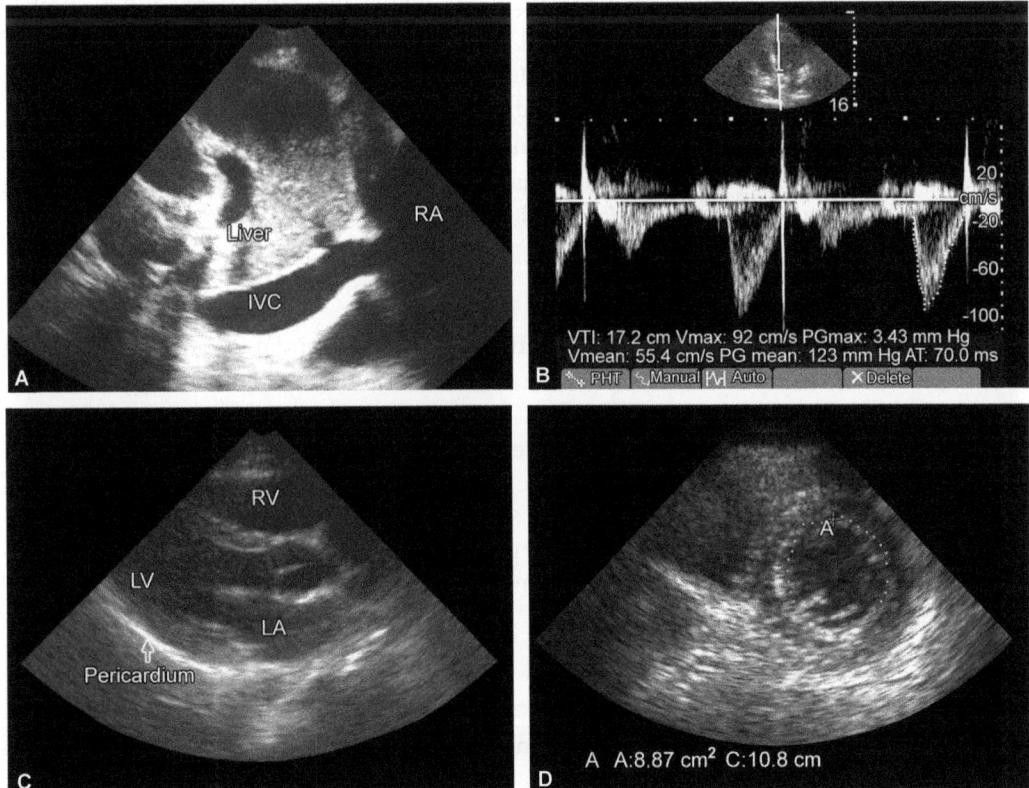

Figs. 5A to D: (A) Subcostal transthoracic echocardiography (TTE) with inferior vena cava (IVC) draining into right atrium (RA); (B) Calculation of velocity time integral (VTI) on transthoracic echocardiography (TTE); (C) Parasternal long axis view [Rapid Ultrasound for Shock and Hypotension (RUSH) exam]; (D) Parasternal short axis view [Rapid Ultrasound for Shock and Hypotension (RUSH) exam] at level of papillary muscle.

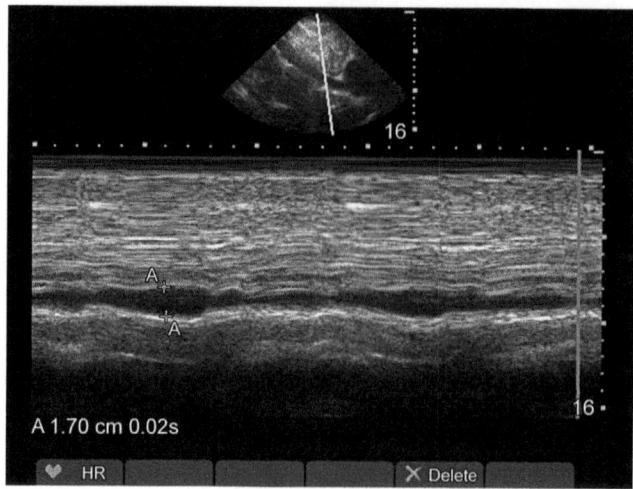

Fig. 6: Calculation of inferior vena cava (IVC) collapsibility index [(subcostal transthoracic echocardiography (TTE)].

But it is also to be noted that there is no single parameter that consistently and accurately predicts fluid responsiveness. The combined use of two or more parameters with clinical assessment and experience should enable the clinician to make fluid-related decisions for their patients. USG is a simple, bedside, reliable, and repeatable diagnostic modality which enables assessment of volume status of an individual and responsiveness to fluid administration and should be extensively used. Focused TTE should be encouraged and short-term training courses should be organized for the clinicians.

SUMMARY

Assessment of intravascular (IV) volume is challenging but essential for goal directed fluid administration. Euvolemia is achieved when IV volume allows adequate filling of cardiac chamber which facilitates optimum cardiac output and oxygen delivery. "Fluid responsiveness" is defined as a more than 15% increase in cardiac index in response to a volume expansion. Static and dynamic parameters have been derived from invasive [invasive arterial and central lines, pulmonary artery catheters (PACs), FloTrac®] and noninvasive monitoring devices [inferior vena cava (IVC) ultrasound, transthoracic echocardiography (TTE), transesophageal echocardiography (TOE)] to evaluate volume status and predict response to IV fluid administration.

REFERENCES

1. Michard F, Teboul JL. Predicting fluid responsiveness in ICU patients: a critical analysis of the evidence. Chest. 2002;121:2000-8.
2. Pinsky M, Teboul JL. Assessment of indices of preload and volume responsiveness. Curr Opin Crit Care. 2005;11:235-9.
3. Eyre L, Breen A. Optimal volaemic status and predicting fluid responsiveness. Anaesth Crit Care Pain. 2010;10:59-62.

4. Gelman S. Venous function and central venous pressure: a physiologic story. Anesthesiology. 2008;108:735-48.
5. Marik PE, Baram M, Vahid B. Does central venous pressure predict fluid responsiveness? A systematic review of the literature and the tale of seven mares. Chest. 2008;134:172-8.
6. Robin E, Costecalde M, Lebuffe G, et al. Clinical relevance of data from the pulmonary artery catheter. Critical Care. 2006;10(Suppl 3):S3.
7. Diebel LN, Wilson RF, Tagett MG, et al. End-diastolic volume: a better indicator of preload in the critically ill. Arch Surg. 1992;127:817-21; discussion 821-2.
8. Tousignant CP, Walsh F, Mazer CD. The use of transesophageal echocardiography for preload assessment in critically ill patients. Anesthesia Analg. 2000;90:351-5.
9. Marik PE, Cavallazzi R, Vasu T, et al. Dynamic changes in arterial waveform derived variables and fluid responsiveness in mechanically ventilated patients: a systematic review of literature. Crit Care Med. 2009;37:2642-7.
10. Kapoor PM, Bhardwaj V, Sharma A, et al. Global end-diastolic volume an emerging preload marker vis-a-vis other markers—Have we reached our goal? Ann Card Anaesth. 2016;19:699-704.
11. Michard F. Changes in arterial pressure during mechanical ventilation. Anesthesiology. 2005;103: 419-28.
12. Michard F, Boussat S, Chemla D, et al. Relation between respiratory changes in arterial pulse pressure and fluid responsiveness in septic patients with acute circulatory failure. Am J Respir Crit Care Med. 2000;162:134-8.
13. Tavernier B, Makhotine O, Lebuffe G, et al. Systolic pressure variation as a guide to fluid therapy in patients with sepsis-induced hypotension. Anesthesiology. 1998;89:1313-21.
14. Tsuchiya M, Yamada T, Asada A. Pleth variability index predicts hypotension during anesthesia induction. Acta Anaesthesiol Scand. 2010;54:596-602.
15. Loupec T, Nanadoumgar H, Frasca D, et al. Pleth variability index predicts fluid responsiveness in critically ill patients. Crit Care Med. 2011;39:294-9.
16. Feissel M, Michard F, Mangin I, et al. Respiratory changes in aortic blood velocity as an indicator of fluid responsiveness in ventilated patients with septic shock. Chest. 2001;119:867-73.
17. Jardin F, Vieillard-Baron A. Ultrasonographic examination of the venae cavae. Intensive Care Med. 2006;32:203-6.
18. Barbier C, Loubières Y, Schmit C, et al. Respiratory changes in inferior vena cava diameter are helpful in predicting fluid responsiveness in ventilated septic patients. Intensive Care Med. 2004;30:1740-6.
19. Jue J, Chung W, Schiller NB. Does inferior vena cava size predict right atrial pressures in patients receiving mechanical ventilation? J Am Soc Echocardiogr. 1992;5:613-9.
20. Vieillard-Baron A, Chergui K, Rabiller A, et al. Superior vena caval collapsibility as a gauge of volume status in ventilated septic patients. Intensive Care Med. 2004;30:1734-9.
21. Cavallaro F, Sandroni C, Marano C, et al. Diagnostic accuracy of passive leg raising for prediction of fluid responsiveness in adults: systematic review and meta-analysis of clinical studies. Intensive Care Med. 2010;36:1475-83.
22. Brenchley J, Walker A, Sloan JP, et al. Evaluation of focussed assessment with sonography in trauma (FAST) by UK emergency physicians. Emerg Med J. 2006;23:446-8.
23. Hamada SR, Delhaye N, Kerever S, et al. Integrating eFAST in the initial management of stable trauma patients: the end of plain film radiography. Ann Intensive Care. 2016;6:62.
24. Lichtenstein DA, Lascols N, Prin S, et al. The "lung pulse": an early ultrasound sign of complete atelectasis. Intensive Care Med. 2003;29:2187-92.
25. Lichtenstein D, Mezière G, Biderman P, et al. The "lung point": an ultrasound sign specific to pneumothorax. Intensive Care Med. 2000;26:1434-40.
26. Perera P, Mailhot T, Riley D, et al. The RUSH exam: Rapid ultrasound in shock in the evaluation of critically ill patient. Emerg Med Clin North Am. 2010;28:29-56.

Chapter 9

Ultrasound in Cardiopulmonary Resuscitation

Rakesh Garg

INTRODUCTION

Cardiac arrest situation may be observed in different areas of hospital including emergency unit, medical or surgical wards and critical care units. The resuscitation of such patients follows a definite protocol as recommended by various resuscitation councils. There are specific steps that need to be followed in time bound manner as time is essence during resuscitation for a positive outcome. During the initial resuscitation, present guidelines suggest assessment of etiological factors for cardiac arrest clinically usually and management accordingly. The various imaging tools are not feasible during initial resuscitation as they would interrupt the cardiac compression affecting the outcome. Through the usefulness of ultrasound during resuscitation in cardiac arrest situation has not been proven but the evidence is emerging for the usefulness of ultrasound during resuscitation in patients of cardiac arrest. Ultrasound guided evaluation of patient for underlying etiologies of cardiac arrest and effectiveness of resuscitation measures may assess using ultrasonography in a real time. With the emergence of point of care tool of ultrasonography, resuscitation protocols need modifications. The main drawback of other imaging tools is interruption of cardiac compression for imaging process and at times shifting to the imaging suite. This has been mitigated by ultrasonography as it may be used as bedside without shifting patient to imaging suite. Ultrasound appears to be a useful adjunct in patients' real time assessment and management of patient at bedside.

NEED OF ULTRASOUND IN RESUSCITATION

The management of the cardiac arrest victim depends on good quality cardiopulmonary resuscitation and the assessment of the underlying etiological cause and its correction. The chorus of the steps of cardiopulmonary resuscitation is well-documented in various resuscitation council guidelines. Identifying the underlying cause timely and without interruption in cardiac massage remains a big challenge. Accurate determination of these cause and timely therapeutic management would improve the patient outcome and return of spontaneous circulation. The therapeutic interventions and for affirming the quality of cardiopulmonary resuscitation, various tools have been described. However, these may have limitations including its availability, feasibility, interruptions in chest compressions. Ultrasound appears to have important role in providing diagnosis of underlying etiology, performing real time ultrasound aided interventions and real time assessment of the quality of resuscitation and response of the interventions. Ultrasound is useful to assess patients with cardiac arrest in real time.

This assessment is considered better and with better accuracy as compared to conventional physical examination based assessment of the patient. It can be performed at bedside as point of care management and is repeatable to assess for response to intervention as well.

The management of ventricular arrhythmias like ventricular fibrillation and pulseless ventricular tachycardia using defibrillation has brought an improvement in the outcome of the patient. However certain other etiologies leading to pulseless electrical activity (PEA) and asystole still need to be improved further by use of technological advancement in medicine. The cardiac arrest rhythms including PEA and asystole require correction of the underlying etiology while providing resuscitation including cardiac compression and ventilation. Ultrasound aids in diagnosing various reversible causes of cardiac arrest. Ultrasound is useful in assessing and managing etiologies like cardiac tamponade, hemorrhage leading to hypovolemia, simple or tension pneumothorax, and coronary or pulmonary embolus. Ultrasound assesses these causes faster and thus earlier management with faster return of spontaneous circulation.

Focused ultrasound by a trained noncardiologists and nonultrasonologist clinician has been found to be useful tool for management patients with cardiac arrest. It has been used for assessing and management of reversible causes even by nonultrasonologist who has some training in focused use of ultrasound. A protocolized approach and focused scanning is essential for optimal integration of ultrasound in resuscitation sequence in cardiac arrest victim. The focused training and protocolized approach would avoid unnecessary interruptions of cardiac compression and appropriate use of ultrasound during resuscitation. Utility of ultrasound as real time tool for management of cardiac arrest victim is beyond doubt. However, concrete evidence and protocolized guideline for integration of ultrasound in resuscitation protocol have yet to be emerged.

ULTRASOUND-GUIDED MANAGEMENT

The usual factors which need to be considered for management in cardiac arrest victims include hypothermia, hypoxia, hypovolemia, tension pneumothorax, hypoglycemia, electrolyte and metabolic imbalances, coronary thrombosis, pulmonary embolus, cardiac tamponade and drug over dosages and toxicity. These conditions not only require appropriate assessment but also require invasive intervention to correct the cardiac activity. The invasiveness of the intervention could be harmful, if either not diagnosed appropriately or not performed properly. The diagnosis is usually done based on clinical presentation and examination. The failure to identify the underlying etiology remains a major concern in such scenarios as clinical features and examination may be unreliable. This may lead to deferred or delayed interventions to correct the underling etiology. Also, the interventions are routinely done blindly or require shifting the patient to fluoroscopy suite or an imaging facility. Blind procedures have their own limitations, higher failure and complication rate. Shifting of patients during this phase may not be feasible in patient with cardiac arrest. On the other hand, patient would not have good outcome until and unless these are managed timely. So in view of these facts, it requires timely and accurately diagnosis and ultrasound guided intervention for positive outcome. Ultrasound has found its important role in diagnostic clarity for identifications of reversible cause of cardiac arrest and thus timely and effective clinical decision-making. So it appears ultrasound would found its important role in identifying and correcting reversible causes during resuscitation (Fig. 1). Also, ultrasound would aid in identifying true asystole. In addition, many interventions are required like assessment of pulse, airway management, etc. during the resuscitation steps. Majority of these are clinically based on history, presentation and examination. Ultrasound has separately been found to be useful in many of these clinical scenarios for assessment and real time management. The utility in ultrasound in management of underlying etiology is being described in following sections.

Pericardial Effusion/Cardiac Tamponade

Various traumatic conditions may lead to collection of blood in the pericardial space. This collection is suspected based on the mechanism of injury and assessed based on clinical signs like distended neck veins and muffled heart sounds on auscultation. During acute event and in busy emergency department, it would be difficult to identify the heart or respiratory

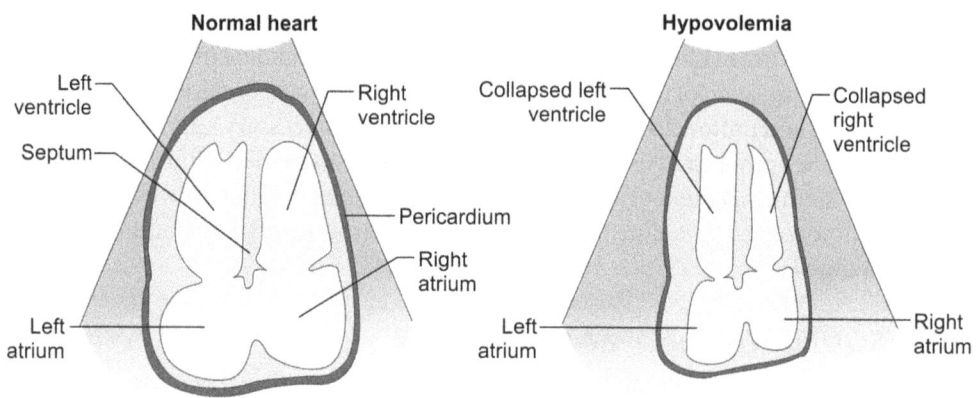
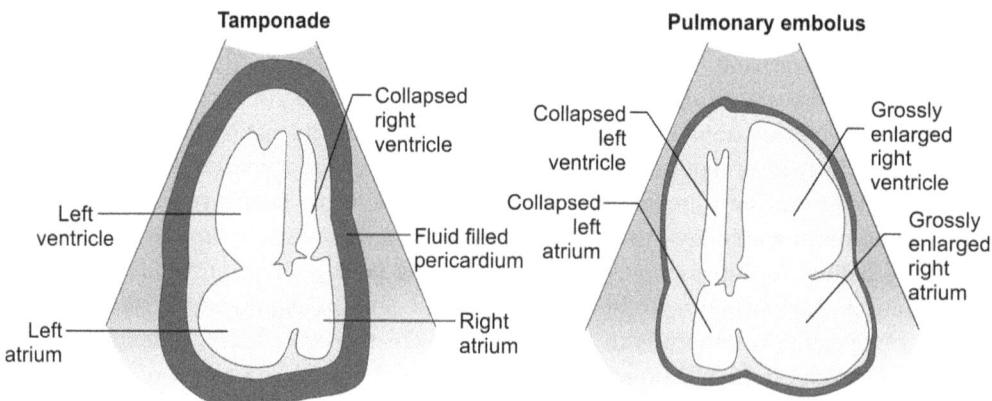

Fig. 1: Ultrasound scans in cardiac tamponade, pulmonary embolus, and hypovolemia (apical four-chamber view).
Source: (Adapted from—Hernandez C, Klaus Shuler, Hannan H, Sonyika C, Likourezos A, Marshall J. CAUSE: Cardiac arrest ultra-sound exam-A better approach to managing patients in primary non-arrhythmogenic cardiac arrest. Resuscitation. 2008;76:198-206).

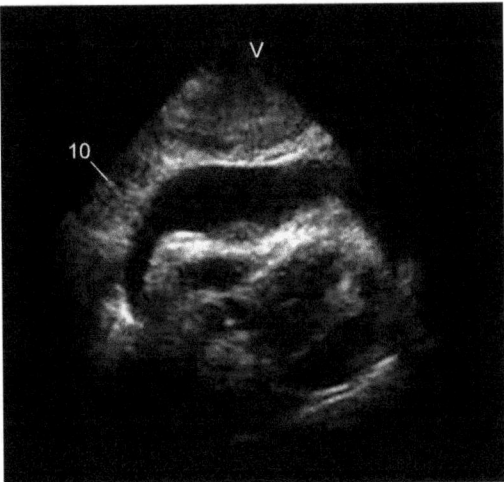

Courtesy by: Sarvesh Pal Singh, Sonali Bansal

Fig. 2: Echocardiogram showing diastolic collapse of right ventricle in pericardial tamponade.

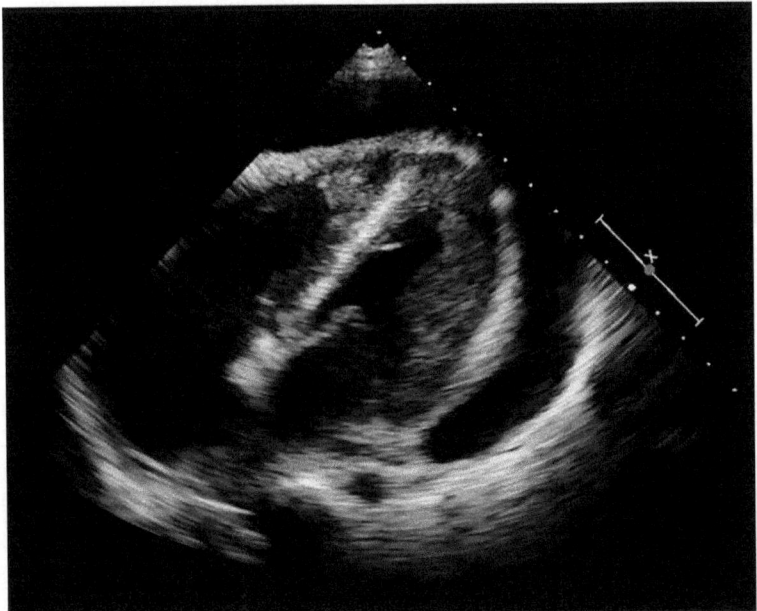

Courtesy by: Sarvesh Pal Singh, Sonali Bansal

Fig. 3: Echocardiogram showing pericardial effusion only with "NO" collapse of RV; therefore no tamponade.

sounds (to differentiate with tension pneumothorax). Ultrasound has found an important and definitive role in identifying pericardial effusion in trauma victims. The parasternal views or the subxiphoid view could easily diagnose pericardial effusion as a hypoechoic image between the hyperechoic pericardial blades and collapse of right chamber of heart (Figs. 2 and 3).

The ultrasound reveals noncollapsible inferior vena cava, heart sways like pendulum in effusion in pericardium "swinging heart" in patients with pericardial effusion. Cardiac tamponade requires an invasive intervention like pericardiocentesis or open thoracotomy. Use of ultrasound provides definitive therapy and thus avoidance of unnecessary or delayed interventions. It is to emphasize to correlate the ultrasound findings clinically with patient status during management.

Hypovolemia

Hypovolemia remains an important cause of cardiac arrest especially in trauma victims. The clinical signs like pulse quality are usually identified for hypovolemia and patients receive fluid replacements for the same. In severe hypovolemia patient may manifest as cardiac arrest with PEA rhythm. The use of ultrasound to identify hypovolemia would lead to increased diagnostic clarity in a timely manner. The early definitive diagnosis will narrow down the differentials and thus appropriate timely management. Also, etiologies with contradictory management like thrombolytics in pulmonary embolism in patients with aneurysm rupture could be managed conclusively.

The ultrasound is good modality to identify flattened right and left ventricles (Figs. 4A and B). The short axis parasternal of ultrasound view will aid in measuring left ventricular end-diastolic (LVED) area. The presence of hyperkinetic left ventricular wall along with close approximation of ventricular walls (kissing walls) during contraction identifies hypovolemia. Not only identifying hypovolemia but also its probable cause may be picked up using ultrasound scan. The ruptured abdominal aortic aneurysm is one of the causes of acute hypotension and fluid replacement may not serve the optimal outcome as it requires immediate surgical intervention. Ultrasound can pick up the aneurysm (Fig. 5).

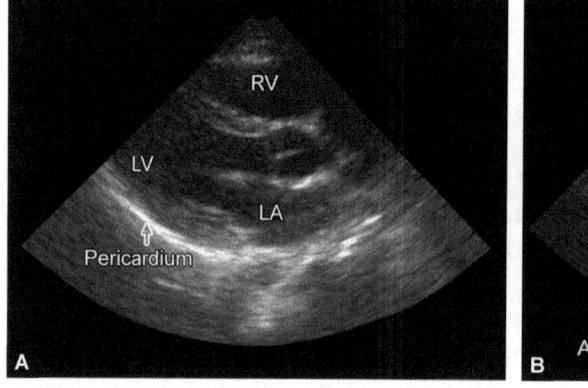
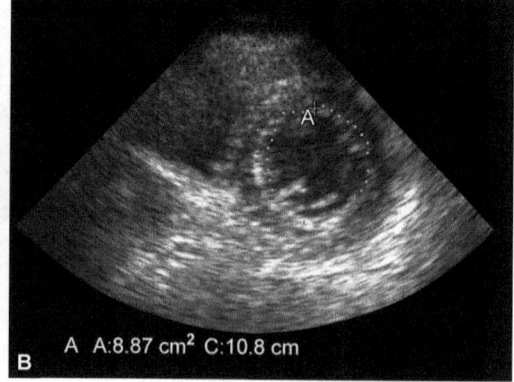

Courtesy by: Tanvir Samra

Figs. 4A and B: (A) Parasternal long-axis view [Rapid Ultrasound for Shock and Hypotension (RUSH) exam]; (B) Parasternal short-axis view [Rapid Ultrasound for Shock and Hypotension (RUSH) exam] at level of papillary muscle.

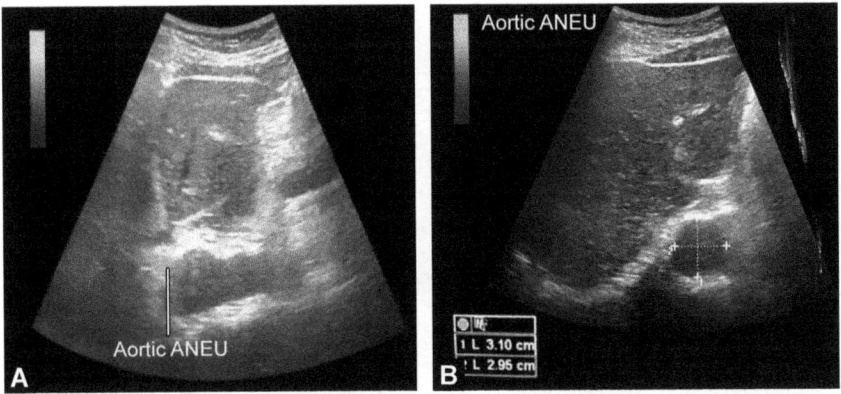

Figs. 5A and B: Ultrasound image abdominal aortic aneurysm in abdominal transverse view.

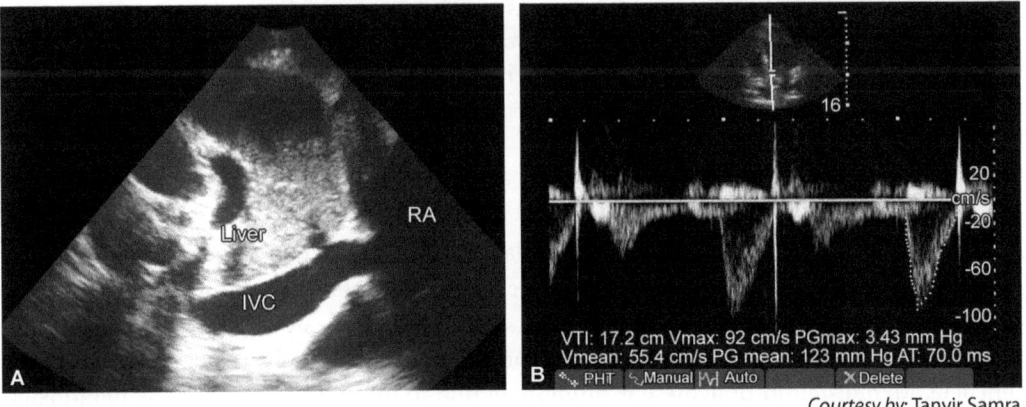

Courtesy by: Tanvir Samra

Figs. 6A and B: (A) Subcostal transthoracic echocardiography (TTE) with inferior vena cava (IVC) draining into right atrium (RA); (B) Calculation of velocity time integral (VTI) on transthoracic echocardiography (TTE).

Inferior vena cava (IVC) can easily be scanned as subxiphoid space using liver window for evaluation of fluid status which is one of the causes of cardiac arrest (Figs. 6A and B). The IVC diameter and its collapsibility correlate well with the volume status. Presence of collapsed or IVC diameter less than 5 mm indicates hypovolemia and should prompt aggressive fluid resuscitation. Though these correlations are better in spontaneously breathing patients and may not be appropriate in patients with cardiac arrest receiving positive pressure ventilation. The perfusion deficit in a patient may not only be due to hypovolemia but also be related to cardiac failure conditions which may occur due to cardiac and pulmonary conditions. The cardiac conditions include congestive heart failure pericardial tamponade, myocardial infarction and cardiac arrest. The pulmonary condition include pulmonary embolism. In such cases, IVC would be dilated and full with diameter more than 20 mm.

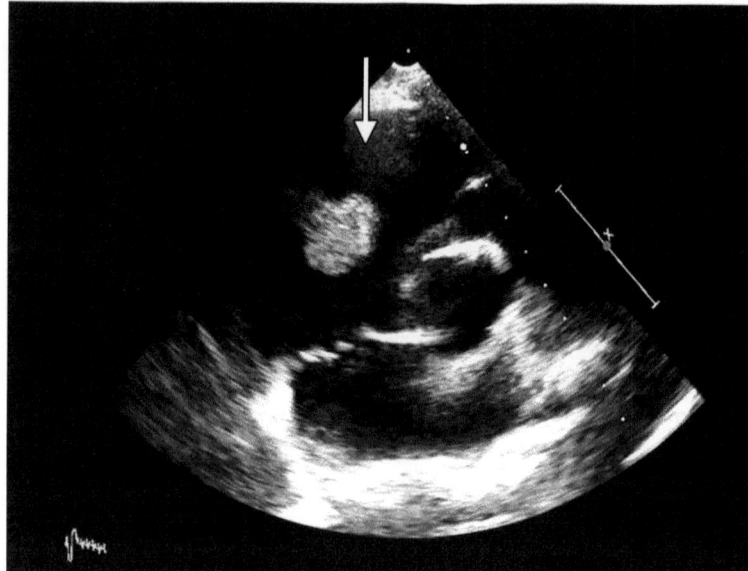

Courtesy by: Sarvesh Pal Singh, Sonali Bansal

Fig. 7: Parasternal short-axis view showing thrombus in right ventricle.

Pulmonary Embolus

Acute pulmonary embolus may lead to cardiac arrest. The cardiac rhythm could either be PEA or asystole. Early identification and specific management using thrombolytic therapy or embolectomy improves the outcome. Ultrasonography reveals engorged right ventricle, right ventricular hypokinesis or dysfunction, or pulmonary hypertension with a flattened left ventricle. Manifestation of acute cor pulmonale due to embolism manifests as paradoxical septal wall motion (D-sign). The right heart thrombus could also be observed in 10% of patients. Ultrasound has moderate sensitivity and good to excellent specificity as a screening test for diagnosing pulmonary emboli. Higher specificity may be considered to be more important in cardiac arrest victims. In view of these benefits, ultrasound should be a part of assessing the cardiac arrest victims suspicious of large acute embolus as a real time bedside tool for early diagnosis and appropriate management.

Pneumothorax

Tension pneumothorax is an emergency and diagnosed based on clinical signs and mandating immediate intervention. This clinical situation may present later once positive pressure ventilation is initiated during cardiopulmonary resuscitation. It may be confused with cardiac tamponade. The clinical sign of absent breath sounds or percussion note may not be practically feasible in busy or noisy place like that of emergency department. Ultrasound proves to be bedside tool for immediate recognition of pneumothorax. It remains very accurate, repeatable and can also differentiate tension pneumothorax and cardiac tamponade which may

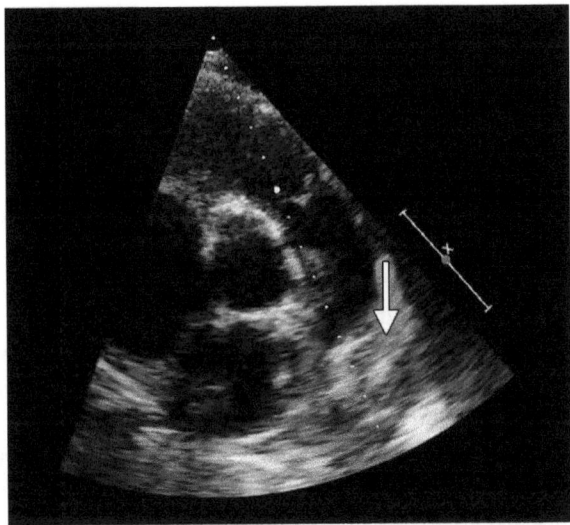

Fig. 8: Parasternal short-axis view showing thrombus in main pulmonary artery.

Courtesy by: Sarvesh Pal Singh, Sonali Bansal

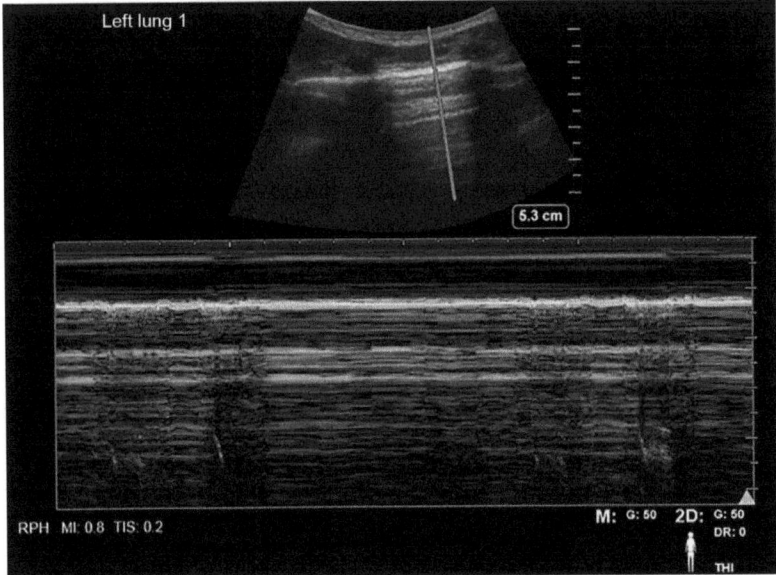

Courtesy by: Sarvesh Pal Singh, Sonali Bansal

Fig. 9: Barcode sign. Apical pneumothorax on left side displaying the barcode sign on M mode.

have similar clinical signs and thus confusion in early diagnosis. Pneumothorax is assessed with ultrasound as "absence of sliding" when scanned at second intercostal interspace in mid clavicular line (Fig. 9). This sign is indicated as absence of pleural movements of both the pleura that usually happens in an expanded lung and can be confirmed with M mode of ultrasound.

Cardiac Standstill (True Asystole)

In patients with no cardiac activity on ultrasound view during resuscitation is good predictor of death. Ultrasound has been found to be a good tool to differentiate true asystole from other rhythms of cardiac arrest. Echocardiogram would reveal "no wall motion in asystole with cardiac stand still or pulseless electrical activity". True asystole is detected on ultrasound as absence of heart motion including chambers, walls and cardiac valves. This is important for management plan and for prognostication as well. Victim with true asystole, outcome is meager and prognostication needs to be done accordingly. In situations of asystole, but with some kinetic activity as assessed with ultrasound may be considered for further continued resuscitation. Also repeatability benefit for assessment of cardiac activity is useful tool to assess the outcome of resuscitation. It is prudent to terminate the resuscitative efforts, if ultrasound reveals persistent absent cardiac activity even after optimal resuscitative efforts and interventions. It also remains an easy tool to differentiate fine ventricular fibrillation with asystole as it can be missed in electrocardiographic tracing. In patients with fine ventricular fibrillation, the ultrasound images would reveal atrial and ventricular movements which would be absent in patients with true asystole.

Airway Management during Resuscitation

Ultrasound has been shown to be beneficial in airway management with regards to airway assessment, real time endotracheal intubation and assessing its correct placement including endobronchial or esophageal placement. It would also be beneficial to identify airway related pathology like laryngeal injuries. In case of pneumothorax with positive pressure ventilation, it would pick pneumothorax as well. It remains a useful tool for surgical airway access and thoracentesis, which may be required as management of etiology for cardiac arrest.

Cannulation during Resuscitation

Presence of intravenous access is essential during cardiopulmonary resuscitation to administer emergency drugs as recommended by various resuscitation councils. However, the vein collapses in patients with absent pulses. They may also be vasoconstricted due to hypothermia. Alternate access like intraosseous access has been suggested but either the specific needle or the expertise may not be available. Ultrasound has emerges as an easy bedside tool for securing an intravenous access. It has higher success rate with lesser complications. The identification of cubital vein, femoral or internal jugular vein can easily be identified and real time cannulation be done using ultrasound.

Ultrasound as an Aid to Assess Quality of Resuscitation

The outcome of resuscitation in a cardiac arrest victim requires includes many components and early and effective chest compressions is one of them. It requires correct placement of hand and technique of compression. The usual method is based on anatomical landmark. Ultrasound could be useful adjunct by adjusting the hand position based on effective ventricular compression during ongoing cardiac compression. Thus, ultrasound could demonstrate the effectiveness of chest compressions. The placement of ultrasound probe subxiphoid site is useful during resuscitation as it would not interrupt chest compressions. Placement of ultrasound probe just prior to completion of chest compression cycles during rhythm analysis can provide an optimal view to make diagnosis related to the cardiac status.

Reported Approaches of Ultrasound in Cardiac Arrest Victim

Various approaches have been reported in literature for incorporation of ultrasound as an adjunct in management of cardiac arrest victim. The Cardiac Arrest Ultrasound Exam (CAUSE) approach has been reported in the literature for use of ultrasound in patient with cardiac arrest. This approach uses chest imaging for heart/pericardium assessment using four chamber view and lung or pleura assessment using anteromedial view (Flowchart 1). Four chamber view for heart and pericardium assessment can be done using ultrasound using thoracic windows including subcostal, parasternal or apical. Four chamber views provide clue related to cause of cardiac arrest like hypovolemia, massive pulmonary embolism and cardiac tamponade. The lung or pleura assessment using ultrasound requires imaging on anterior chest in second intercostal space in the midclavicular line. The pneumothorax is classically identified on ultrasound imaging as absence of normal lung sliding sign and presence of comet-tail artifact. The CAUSE protocol is applicable to nonarrhythmogenic groups to rule out the potential cause and manages them (Flowchart 1).

Another reported protocol for use of ultrasound during resuscitation includes focused echocardiographic evaluation in resuscitation (FEER). The timely integration of ultrasound point of care during cardiopulmonary resuscitation helps in identifying the probable etiology and response to resuscitation attempts. The myocardial infarction (MI) remains one of the major causes of nontraumatic cardiac arrest and its diagnosis remains a big pandora box in situation of associated cardiac arrest secondary to MI. The assessment of other reversible cause also needs to be diagnosed and managed. The authors proposed ten-step procedure done along with other steps of conventional resuscitation without undue interruptions of cardiac compressions (Tables 1 and 2). The integration of echocardiographic scan is time bound. Initially the standard protocol of initiation of resuscitation is followed. At this juncture the infrastructure and expert for scanning is initiated by the expert team. During ongoing cardiac compression, identification of xiphoid is done and probe is kept ready at the subxiphoid site. During interruptions of the chest compression after 5 cycles, the probe is positioned and aligned at the subcostal window at assess the cardiac status by visualizing ultrasound image of four-chamber view. The status of the hear contraction and valves is commented and cardiac compression is started within 10 seconds for further 5 cycles. This is repeated every 5 cycles till return of cardiac activity and then a postresuscitation echocardiography would comment on further cardiac functioning and the insult occurred during the incident.

The use of ultrasound for cardiac assessment is reported by other groups where assessments of etiological factors for shock which may lead to cardiac arrest were evaluated. These include SIMPLE and RUSH protocols. The rapid ultrasound for shock and hypotension (RUSH)

Flowchart 1: The use of ultrasound in patient with cardiopulmonary arrest following CAUSE protocol.

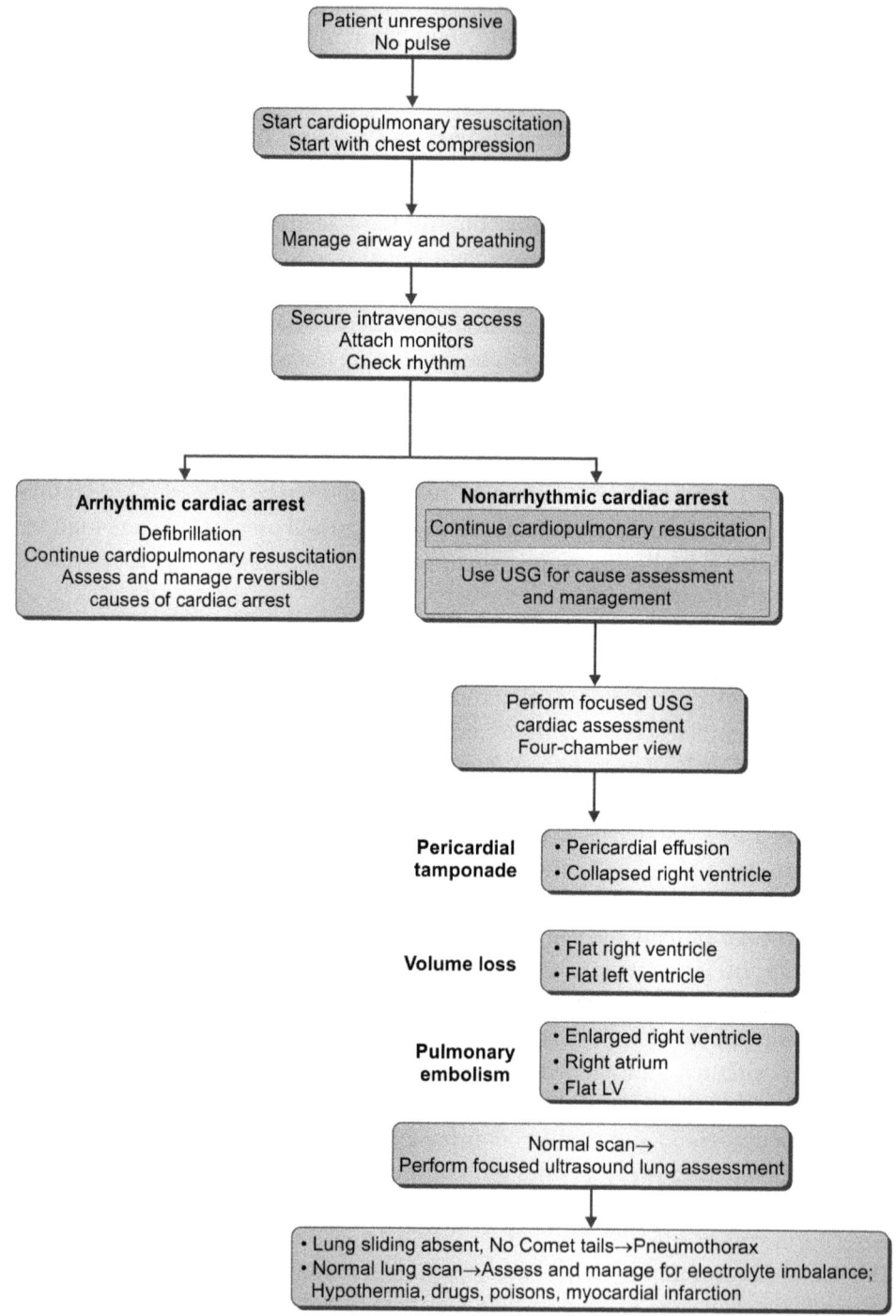

Source: (Adapted from: Hernandez C, Klaus Shuler, Hannan H, Sonyika C, Likourezos A, Marshall J. CAUSE: Cardiac arrest ultrasound exam-A better approach to managing patients in primary non-arrhythmogenic cardiac arrest. Resuscitation. 2008;76:198-206).

Table 1: Focused echocardiographic evaluation in resuscitation (FEER).	
Cardiopulmonary resuscitation	• Start CPR as per standard National guidelines • Initiate the process of arranging and prepare the ultrasound machine for use • Test the machine • Prepare the patient for assessment using ultrasound scan
Integration of ultrasound echocardiography during resuscitation	• After completion of 5 cycles, inform team for pulse check or rhythm check • During this time of 10 seconds, initiate scan • Place the probe for a subxiphoid view and assess • Assess via subcostal long-axis view. If diagnosis made within the time of 10 second interruption, manage accordingly or otherwise resume chest compression again and assess after 5 cycles.
Outcomes	• Close loop communication regarding the assessment findings and integration of management plans accordingly.

Source: (Adapted from: Breitkreutz R, Walcher F, Seeger FH. Focused echocardiographic evaluation in resuscitation management: concept of an advanced life support—conformed algorithm. Crit Care Med. 2007;35(Suppl.):S150-61).

protocol includes assessment of IVC, abdominal aorta, lungs, leg veins. Though these are typically described for management of shock in critically ill-patients. These could be extrapolated to understand the utility of ultrasound for assessment of various etiologies of cardiac arrest. They described various findings that could be assessed using ultrasound for making a diagnosis (Table 3).

The protocol has also been reported related to termination of resuscitation. The patient assessment and prediction of the outcome including steps to stop further resuscitation can be made using the protocol of serial focused echocardiographic evaluation in life support (FEEL), especially when a patient is shifted from out of hospital with cardiac arrest. The authors propose that cardiac standstill as assessed with ultrasound for a given period (>10 minutes) may predict failure of return of cardiac activity.

Training for Ultrasound for Resuscitation

The main concern of identifying the etiology with ongoing cardiopulmonary resuscitation remains a challenge. The other challenge the need of trained ultrasonologists for performing ultrasound and its immediate availability at the site of cardiac arrest which could be any part of the hospital. It has been found that emergency sonographers with training can effectively use ultrasound during resuscitation without interruptions [a10]. The identification of the underlying cause cardiac tamponade has been found to be comparable among emergency physicians with focused training and cardiologists with high accuracy. The training for learning the ultrasound is to understand the basic concepts and diagnosis of relevant pathology rather than a detailed structural evaluation as done by conventionally trained ultrasonologists or radiologists.

Phase	Step with command, element
Table 2: Resuscitation steps with integration of focused echocardiography.	
High-quality CPR, preparation, team information	• Perform immediate and accurate BLS and ACLS according to AHA/ERC/ILCOR guidelines, at least five cycles of chest compression/ventilation • Tell the CPR team: I am preparing an echocardiogram • Prepare portable ultrasound (let prepare) and test it • Accommodate situation (e.g. best position of patient and doctor, removal of clothes), be ready to start
Execution, obtaining the echocardiogram	• Tell CPR team to count down 10 secs and to undertake a pulse check simultaneously • Command: Interrupt at the end of this cycle for echocardiography • Put the probe gently onto the patients subxiphoidal region during chest compressions • Perform a subcostal (long axis) echocardiogram as quickly as possible. If you cannot identify the heart after 3 secs, stop the interruption and repeat again five cycles later and/or with the parasternal approach
Resuming CPR	• Command after 9 secs at the latest: "Continue CPR" and control it
Interpretation and consequences	• Communicate (after continuation of chest compressions only) the findings to the CPR team (e.g. wall motion, heart is squeezing, cardiac stand still, (massive) pericardial effusion, no conclusive finding, suspected pulmonary artery embolism, hypovolemia) and explain consequences and follow-up procedure

Source: (Adapted from: Breitkreutz R, Walcher F, Seeger FH. Focused echocardiographic evaluation in resuscitation management: concept of an advanced life support—conformed algorithm. Crit Care Med. 2007;35(Suppl.):S150-61).

LIMITATIONS

The ultrasound has proven its important role as point of care in critical situations. It is just to remind that finally patients need to be treated and not the imaging. So the ultrasound imaging needs to be correlated well with clinical situations and managed accordingly. Ultrasound should be considered as an aid to clinical management and should not be considered as replacement for appropriate history and clinical examination of the patient. Ultrasound use has inherent caveats and needs to be understood while using for managing cardiac arrest victims. For example, presence of pericardial effusion may be a finding on ultrasound scan in patients of chronic renal disease but the cause of cardiac arrest in this patient may

Table 3: Ultrasound findings in various causes of shock.

	Type of shock	Hypovolemic	Cardiogenic	Septic	Distributive	Pulmonary embolism	Cardiac tamponade	Aortic dissection
S	Chamber size	Small LV	Dilated LV	Early: small LVESA Late: normal: dilated	Near normal LVEDA but small LVESA	Dilated RV, small/normal LV	Diastolic collapse of RA and RV; normal LV	Usually normal
I	IVC thickness	Collapsed	Distended <50% respiratory collapse	Early: collapsed Late: distended	Collapsed	Distended and loss of respiratory collapse	Distended and loss of respiratory collapse	Normal when no cardiac tamponade
	IVS movement	Normal	Reduced	Early: normal Late: reduced	Normal	Paradoxical IVS and D-shaped LV	Normal	Normal
	Intimal flap	Absent	Absent	Absent	Absent	Absent	Absent	Present
M	Myocardial thickening/motion	Hyperdynamic	Hypokinetic	Early: hyperdynamic Late: hypokinetic	Hyperdynamic or normal	McConnell's sign, LV hyperdynamic	Diastolic collapse of RA and RV	Normal, if coronary ostia not involved
	Masses in heart	Absent	Intramural thrombi, if AF/AMI	Absent	Absent	Thrombi in RA, RV and IVC	Absent	Absent
P	Pericardial effusion	Absent	Small amount, if inflammatory cause	Absent	Absent	Absent	Moderate to large but can be small, if acutely collected	Present, if retrograde dissection and echogenic
	Pleural effusion	Absent	Present	Present, if pneumonia	Absent	Usually absent	Absent	Present, if hemothorax
L	LV systolic function	Hyperdynamic	Poor	Early: normal or hyperdynamic Late: impaired	Normal or hyperdynamic	Normal or hyperdynamic	Normal	Normal
E	Abdominal aorta in epigastrium	Aneurysmal, if due to AAA rupture	Normal	Normal	Normal	Normal	Normal	Intimal flap seen

Abbreviations: AF: Atrial fibrillation; AMI: Acute myocardial infarct; LV: Left ventricle; LVEDA: Left ventricular end-diastolic area; LVESA: Left ventricular end-systolic area; RA: Right atrium; RV: Right ventricle. Make it SIMPLE: enhanced shock management by focused cardiac ultrasound.

Source: (Adapted from: Mok KL. Make it SIMPLE: enhanced shock management by focused cardiac ultrasound. J Intensive Care. 2016;4:51).

be hyperkalemia. So, the overall assessment including clinical differential diagnosis needs to be sought while managing such patients.

SUMMARY

The cardiopulmonary resuscitation requires timely assessment and management of various reversible causes for a good outcome. The critical situation prevents shifting patient to specialty medical and imaging suites. The availability of ultrasound as bedside point of care is emerging as an important tool for managing patient of cardiac arrest. It aids in timely decision making in victims of cardiac arrest where time is essence. It is easy to use, can be performed simultaneously along with resuscitation and repeatability proves useful during resuscitation. So use of ultrasound may be considered as in important adjunct tool and as an extension to conventional advanced cardiac resuscitation techniques.

FURTHER READING

1. Amaya SC, Langsam A. Ultrasound detection of ventricular fibrillation disguised as asystole. Ann Emerg Med. 1999;33:344-6.
2. Anderson KL, Castaneda MG, Boudreau SM, et al. Ultrasound guided chest compressions over the left ventricle during cardiopulmonary resuscitation increases coronary perfusion pressure and return of spontaneous circulation in a swine model of traumatic cardiac arrest. Circulation. 2014;130:A15853.
3. Arntfield RT, Millington SJ. Point of care cardiac ultrasound applications in the emergency department and intensive care unit-a review. Curr Cardiol Rev. 2012;8:98-108.
4. Ball CG, Kirkpatrick AW, Feliciano DV. The occult pneumothorax: what have we learned? Can J Surg. 2009;52:173-9.
5. Blaivas M, Fox JC. Outcome in cardiac arrest patients found to have cardiac standstill on the bedside emergency department echocardiogram. Acad Emerg Med. 2001;8:616-1.
6. Blaivas M, Fox JC. Outcome in cardiac arrest patients found to have cardiac standstill on the bedside emergency department echocardiogram. Acad Emerg Med. 2001;8:616-21.
7. Blyth L, Atkinson P, Gadd K, et al. Bedside focused echocardiography as predictor of survival in cardiac arrest patients: a systematic review. Acad Emerg Med. 2012;19:1119-26.
8. Breitkreutz R, Price S, Steiger HV, et al. Emergency Ultrasound Working Group of the Johann Wolfgang Goethe-University Hospital, Frankfurt am Main. Focused echocardiographic evaluation in life support and peri-resuscitation of emergency patients: a prospective trial. Resuscitation. 2010;81:1527-33.
9. Breitkreutz R, Walcher F, Seeger FH. Focused echocardiographic evaluation in resuscitation management: Concept of an advanced life support—conformed algorithm. Crit Care Med. 2007; 35:S150-61.
10. Byhahn C, Bingold TM, Zwissler B, et al. Prehospital ultrasound detects pericardial tamponade in a pregnant victim of stabbing assault. Resuscitation. 2008;76:146-8.
11. Cebicci H, Salt O, Gurbuz S, et al. Benefit of cardiac sonography for estimating the early term survival of the cardiopulmonary arrest patients. Hippokratia. 2014;18:125-9.
12. Corbett SW, O'Callaghan T. Detection of traumatic complications of cardiopulmonary resuscitation by ultrasound. Ann Emerg Med. 1997;29:317-22.
13. Cureton EL, Yeung LY, Kwan RO, et al. The heart of the matter: utility of ultrasound of cardiac activity during traumatic arrest. J Trauma Acute Care Surg. 2012;73.

14. Dulchavsky SA, Schawarz KL, Kirkpatrick AW, et al. Prospective evaluation of thoracic ultrasound in the detection of pneumothorax. J Trauma. 2001;50:201-5.
15. Dulchavsky SA, Schwarz KL, Kirkpatrick AW, et al. Prospective evaluation of thoracic ultrasound in the detection of pneumothorax. J Trauma. 2001;50:201-5.
16. Hayhurst C, Lebus C, Atkinson PR, et al. An evaluation of echo in life support (ELS): is it feasible? What does it add? Emerg Med J. 2011;28:119-21.
17. Hendrickson RG, Dean AJ, Costantino TG. A novel use of ultrasound in pulseless electrical activity: the diagnosis of an acute abdominal aortic aneurysm rupture. J Emerg Med. 2001;21:141-4.
18. Hernandez C, Klaus Shuler, Hannan H, et al. CAUSE: Cardiac arrest ultrasound exam—A better approach to managing patients in primary non-arrhythmogenic cardiac arrest. Resuscitation 2008;76:198-206.
19. Hilty WM, Hudson PA, Levitt MA, et al. Real-time ultrasound guided femoral vein catheterization during cardiopulmonary resuscitation. Ann Emerg Med. 1997;29:331-7.
20. Hoppmann RA, Bell FE, Hoppmann NA, et al. Hand-held ultrasonography to assess external chest compressions on a fresh cadaver. Resuscitation. 2013;84(8):e93.
21. Hwang SO, Zhao PG, Choi HJ, et al. Compression of the left ventricular outflow tract during cardiopulmonary resuscitation. Acad Emerg Med. 2009;16:928-33.
22. Kirkpatrick AW, Sirois M, Laupland KB, et al. Hand-held thoracic sonography for detecting post-traumatic pneumothoraces: the extended focused assessment with sonography for trauma (EFAST). J Trauma. 2004;57:288-95.
23. Kirkpatrick W, Sirois M, Laupland KB, et al. Hand-held thoracic sonography for detecting post-traumatic pneumothoraces: the extended focused assessment with sonography for trauma (EFAST). J Trauma. 2004;57:288-95.
24. Knowles P. Transthoracic echocardiography during cardiac arrest due to massive pulmonary embolism. Emerg Med J. 2003;20:395-6.
25. Labovitz AJ, Noble VE, Bierig M, et al. Focused cardiac ultrasound in the emergent setting: a consensus statement of the American Society of Echocardiography and American College of Emergency Physicians. J Am Soc Echocardiogr. 2010;23:1225-30.
26. Legome E, Pancu D. Future applications for emergency ultrasound. Emerg Med Clin North Am. 2004;22:817-27.
27. MacCarthy P, Worrall A, McCCarthy G, et al. The use of transthoracic echocardiogram to guide thrombolytic therapy during cardiac arrest due to massive pulmonary embolism. Emerg Med J. 2002;19:178-9.
28. Mansencal N, Vileillard-Baron A, Beauchet A, et al. Triage patients with suspected pulmonary embolism in the emergency department using a portable ultrasound device. Echocardiography. 2008;25:451-6.
29. Matsushima K, Frankel HL. Beyond focused assessment with sonography for trauma: ultrasound creep in the trauma resuscitation area and, beyond. Curr Opin Crit Care. 2011;17:606-12.
30. Mok KL. Make it SIMPLE: enhanced shock management by focused cardiac ultrasound. J Intensive Care. 2016;4:51.
31. Niendorff D, Rassias AJ, Palac R, et al. Rapid cardiac ultrasound of inpatients suffering PEA performed by nonexpert sonographers. Resuscitation. 2005;67:81-7.
32. Niendorff DF, Rassias AJ, Palac R, et al. Rapid cardiac ultrasound of inpatients suffering PEA arrest preformed by nonexpert sonographers. Resuscitation. 2005;67:81-7.
33. Osman A, Sum KM. Role of upper airway ultrasound in airway management. J Intensive Care. 2016;4:52.
34. Perera P, Mailhot T, Riley D, et al. The RUSH Exam: rapid ultrasound in shock in the evaluation of the critically III. Emerg Med Clin North Am. 2010;28:29-56.
35. Price S, Uddin S, Laupland KB, et al. Echocardiography in cardiac arrest. Curr Opin Crit Care. 2010;16:211-5.

36. Pérez-Coronado JD, Franco-Gruntorad GA. Utility of ultrasound in resuscitation. Rev Colomb Anestesiol. 2015;43:321-30.
37. Salen P, Melniker L, Chooljian C, et al. Does the presence or absence of sonographically identified cardiac activity predict resuscitation outcomes of cardiac arrest patients? Am J Emerg Med. 2005;23:459-62.
38. Salen P, O'Connor R, Sierzenski P, et al. Can cardiac sonography and capnography be used independently and in combination to predict resuscitation outcomes? Acad Emerg Med. 2001;8:610-5.
39. Scalea TM, Rodriguez A, Chiu WC, et al. Focused assessment with sonography for trauma (FAST): results from an international consensus conference. J Trauma. 1999;46:466-72.
40. Schuster KM, Lofthouse R, Moore C, et al. Pulseless electrical activity, focused abdominal sonography for trauma, and cardiac contractile activity as predictors of survival after trauma. J Trauma. 2009;67:1154-7.
41. Shin J, Rhee JE, Kim K. Is the inter-nipple line the correct hand position for effective chest compression in adult cardiopulmonary resuscitation? Resuscitation. 2007;75:305-10.
42. Tayal VS, Kline JA. Emergency echocardiography to detect pericardial effusion in patients in PEA and near-PEA states. Resuscitation. 2003;59:315-8.
43. Tayal VS, Kline JA. Emergency echocardiography to detect pericardial effusion in patients in PEA and near-PEA states. Resuscitation. 2003;59:315-8.
44. Volpicelli G, Elbarbary M, Blaivas M, et al. International Liaison Committee on Lung Ultrasound (ILC-LUS) for International Consensus Conference on Lung Ultrasound (ICC-LUS). International evidence-based recommendations for point-of-care lung ultrasound. Intensive Care Med. 2012;38:577-91.
45. Walcher F, Kortüm S, Kirschning T, et al. Optimized management of polytraumatized patients by prehospital ultrasound. Unfallchirurg. 2002;105:986-94.
46. Walcher F, Weinlich M, Conrad G, et al. Prehospital ultrasound imaging improves management of abdominal trauma. Br J Surg. 2006;93:238-42.

Chapter 10

Ultrasound-guided Therapeutic Procedures in Intensive Care Unit

Nitish Parmar, Puneet Khanna, Devasenathipathy Kandasamy

INTRODUCTION

Since its advent ultrasonography (USG) has increasingly found its application both as diagnostic modality and as an aid to therapeutic interventions in clinical medicine. It is noninvasive, relatively inexpensive, radiation free and highly versatile modality, and is being increasingly used as a point-of-care device. Advance in the ultrasound technology has led to the development of smaller and portable devices which can be used at bedside. These characteristics make it particularly suitable for use in critically ill since often these patients are difficult to shift to radiology suite. Also this patient population has a higher morbidity and mortality, therefore, early identification and treatment of pathology becomes vital. With increasing appreciation of the varied application of this technology there is increased interest in ultrasound usage among various specialties including critical care. With the publication of guidelines drawn up by a joint working party of the Association of Anesthetists of Great Britain and Ireland, the Royal College of Anesthetists, and the Intensive Care Society in November 2010, it is inevitable that formal USG training will become part of the anesthesia and intensive care curriculum.[1]

Though USG has found numerous applications in intensive care unit (ICU) this chapter shall deal with ultrasound-guided therapeutic procedures in ICU.

GENERAL PRINCIPLES

Preprocedural Assessment

As with any other intervention the safety and success of ultrasound-guided procedures depend upon appropriate patient selection and preprocedural workup. Prior to performing the ultrasound-guided intervention the indication for procedure should be clearly documented. Appropriate medical history and focused physical examination should be performed and relevant laboratory investigations should be sought prior to procedure to rule out any contraindications to the procedure. Common contraindications will include lack of consent, bleeding disorders, and local infection on the overlying skin.

Equipment

Selection of appropriate equipment is vital to the success of the planned procedure. The following equipments are common to all the therapeutic procedures discussed in the chapter:
- *Portable ultrasound with transducer (Fig. 1):* Primary determinant of the choice of ultrasound transducer is the depth of the target organ. Linear transducer has a higher

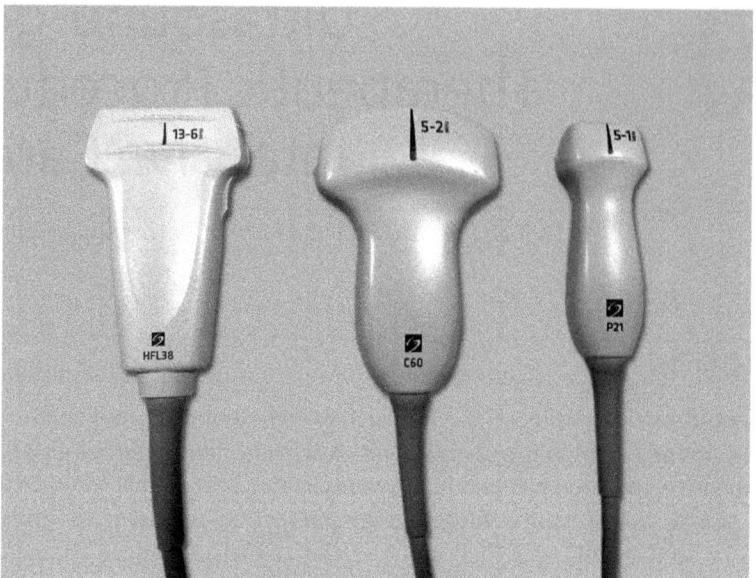

Fig. 1: (From left) Linear, curvilinear, and phased array transducers.

frequency (>10 MHz) and is usually used to visualize superficial structures like vessels, pleura, etc. For abdominal ultrasound a low frequency (3–5 MHz) curvilinear probe is preferred since it offers visualization of greater depths. Application of color Doppler allows visualization of blood vessels which helps in minimizing vascular injury.
- Material for maintaining asepsis like sterile gloves, cap, mask, gowns, skin disinfectant solution, sterile drapes, etc. should be available.
- All equipment required for the planned procedure like needles, syringes, local anesthetic, guidewires, dilators, catheters, sutures, drainage bags, and vials for sample collection should be available as warranted.
- *Needle selection:* Typical fluid collections are readily drained through an 18-G needle. There is no significant difference in needle trauma between an 18-G and 22-G needle, however, using a smaller-sized needle may cause negative aspiration in cases of abscesses and pleural collections with presence of debris and pus. Also usage of 18-G needle allows introduction of 0.038 inch guidewire.
- *Catheter selection:* There are various types of catheters of various sizes available. The most commonly used catheters are Malecot catheter and Pigtail catheter.
  - *Malecot catheter (Fig. 2A):* It is radiopaque catheter with a flower tip which prevents catheter dislodgment from the cavity and facilitates enhanced drainage.
  - *Pigtail catheter (Fig. 2B):* Pigtail catheters are also radiopaque catheters which get their name from characteristic pigtail-shaped loop at the distal end. These catheters have multiple holes typically on inside of loop for more effective drainage. The locking variety of this catheter has a small string which is fixed to distal curved end. The other end of string exits below the hub of the catheter where it is tied after insertion which causes catheter to curl and lock. This prevents catheter dislodgment and facilitates drainage. The basic principle in selecting the catheter size remains that the catheter should be

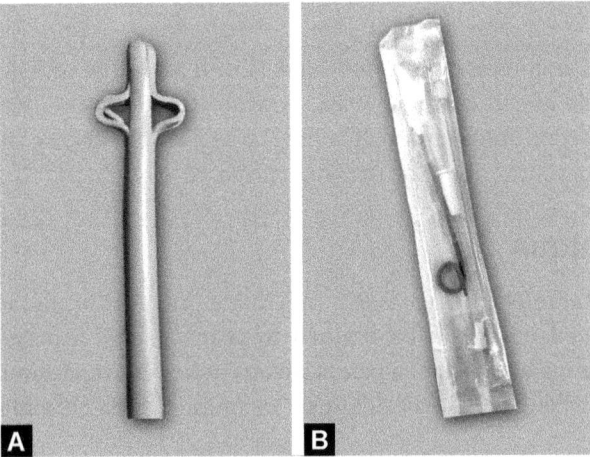

Figs. 2A and B: (A) Malecot catheter. The flower tip prevents catheter dislodgment from the drainage cavity; and (B) Pigtail catheter. Notice the characteristic pigtail-shaped distal loop.

large enough to drain the nature of fluid for which it is inserted. If the fluid is easily aspirated through an 18-G needle then an abscess can be easily drained through a 6–8 Fr catheter. Similar size is sufficient to drain pleural and ascitic fluid whereas larger sizes of 12–16 Fr will be required to drain infected fluid or abscess.

- Proprietary kits are also available for drainage of collections, e.g. Arrow-Clarke™ Pleura-Seal® Thoracentesis Kit.

Consent

An informed consent should be obtained from all the patients or next to kin prior to performing any procedure. Consent should involve the description of the procedure in a manner and language that patient and the immediate family understands. All the potential benefits and risks associated with the procedure should be carefully outlined.

Sedation and Analgesia and Antibiotics

A plan for providing adequate sedation and analgesia should be made prior to embarking upon the procedure. Superficial drainage procedures like thoracocentesis or paracentesis will usually only require administration of local anesthetics, however, drainage of deep-seated collections or insertion of drains and catheters will also require systemic analgesia and sedation. Throughout the procedure it is important to closely monitor the vitals of the patient.

Patient Positioning and Preliminary Scan

Proper positioning of the patient is of vital importance and may be the difference between success and failure of the procedure. Also prior to starting the procedure the correct site, size, and nature of the collection must be verified by a preliminary ultrasound scan. The ultrasonographic appearance of various pathologies is discussed in the relevant sections of the chapter.

Asepsis

Proper universal precautions should be followed as with any other invasive procedure. Operator should don sterile gown, cap, and mask prior to the procedure. Patient skin should be prepared with disinfectant solution and area draped with sterile drapes. Ultrasound transducer provides an important source of contamination and sterile jelly and probe cover should be used.

Drainage Technique

Evacuation and continuous drainage is the underlying principle for management of any fluid collection. Smaller collection can be drained by using needle aspiration, however, larger collections will require placement of a percutaneous drain. Two techniques of catheter placement have been described: (1) Classic Seldinger technique, and (2) Tandem trocar technique.

Seldinger Technique (Figs. 3A to E)

Seldinger first described this technique for vascular access in 1953. The technique has remained relatively unchanged since then and now has been adapted for percutaneous access to almost every organ. In this technique the skin is punctured using an 18-G needle. The needle is then advanced under ultrasound guidance until the rim of collection is punctured which is identified by aspirating a small quantity of fluid or by direct visualization of needle in the collection via USG. A 0.035/0.038 inch guidewire is then advanced through the introducer needle under ultrasound guidance. The guidewire is then left in situ and needle is withdrawn. The tract is then dilated using serial dilators along the intended catheter tract. Care should be taken to avoid kinking of guidewire by the dilator. Once the tract is sufficiently dilated to accommodate an appropriate sized drainage catheter, it is advanced over the guidewire. Once in the cavity the guidewire is removed and formation of a pigtail is visualized. The catheter is then sutured to the skin. Throughout the procedure it is important to never loose guidewire access since reaccess may be difficult once the collection gets punctured. Since the introducer needle is small caliber there is lower risk of injuring the overlying vessels and nerves. However, usage of serial dilators can often be more painful and can cause leakage of contents around the catheter. Serial dilation can cause leakage around the catheter. Kinking and buckling of the guidewire can be problematic in this technique. Seldinger technique allows manipulation of guidewire in the fluid cavity which may help in breaking down loculations and allow more precise placement of catheter.

Trocar Technique (Figs. 4A and B)

Fluid collection is punctured using a small caliber needle which allows localization of collection, assess the depth, and also functions as a tandem localizer. An appropriate sized trocar mounted catheter is then introduced parallel to the introducer needle and advanced into the collection. The catheter position is optimized using ultrasound guidance and the trocar is then withdrawn, position of the catheter ascertained, and secured to skin with sutures. Trocar technique allows rapid insertion of the catheter, however, it is difficult to reposition a suboptimally positioned catheter. Since there is usage of large bore sharp trocar for insertion there is greater risk of injuring overlying vessels and nerves as compared to Seldinger technique.

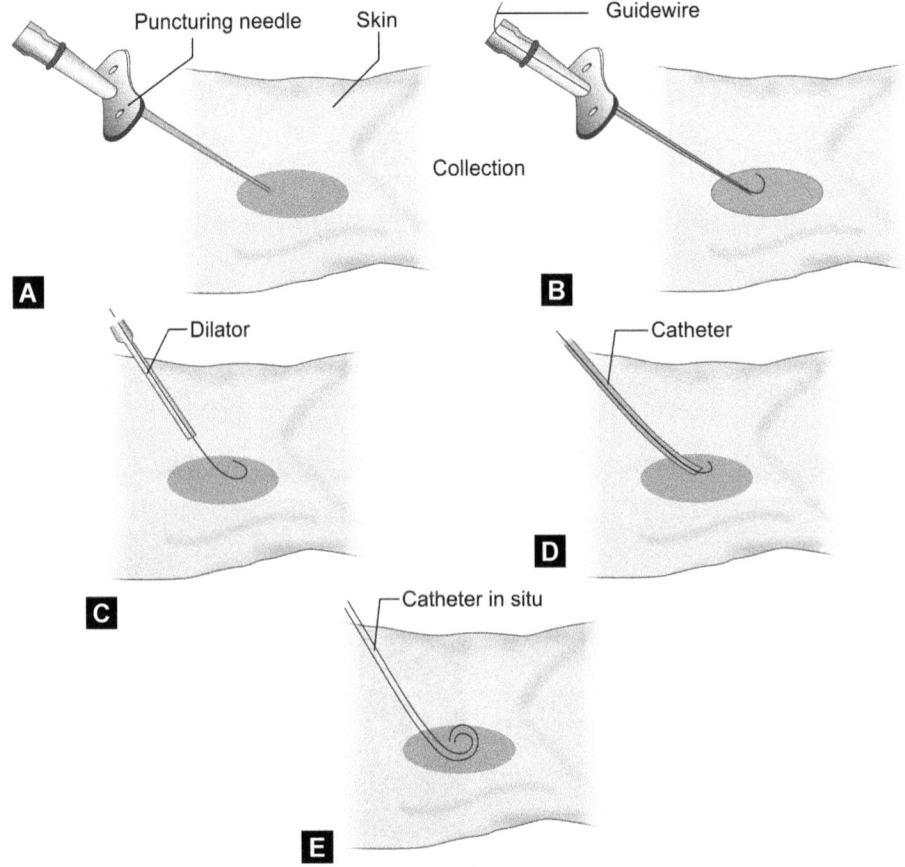

Figs. 3A to E: Seldinger technique of catheter insertion. (A) Collection is punctured with an introducer needle; (B) A guidewire is then advanced through the needle and needle withdrawn leaving guidewire in situ; (C) The tract is then dilated using a dilator; and (D) Once tract is sufficiently dilated a drainage catheter is railroaded over guidewire; and (E) Guidewire is then removed.

Needle Tracking

Ultrasound scanners generate a two-dimensional (2D) image of a three-dimensional (3D) structure. With practice and development of trained eye identification of anatomical structures is relatively easy, however, tracking the movement of advancing needle under ultrasound guidance is a difficult skill to acquire.[2] Failure to do so often results in failed procedure or unintentional vascular, neural, or visceral injury. Once the localization of fluid collection is done the point of entry should be marked with an indelible skin marker. Needle should be advanced during the procedure under direct ultrasound visualization using the following approaches.

In-plane Approach (Figs. 5A and B)

This approach refers to advancing the needle parallel to the long axis of the ultrasound transducer in such a way that the ultrasound beam and the needle shaft are collinear. This requires movement of needle in 3D space while visualizing on a 2D screen which coupled

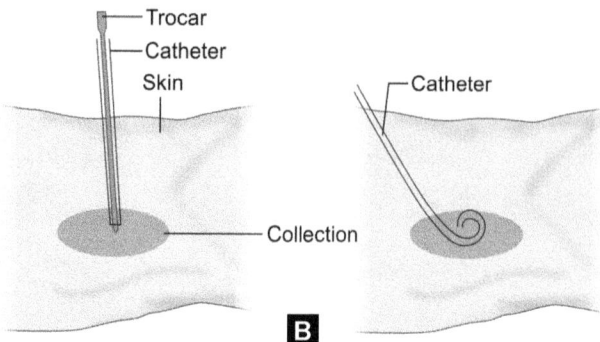

Figs. 4A and B: Trocar technique. (A) Trocar mounted catheter introduced into the collection; and (B) The trocar is then withdrawn leaving catheter in situ.

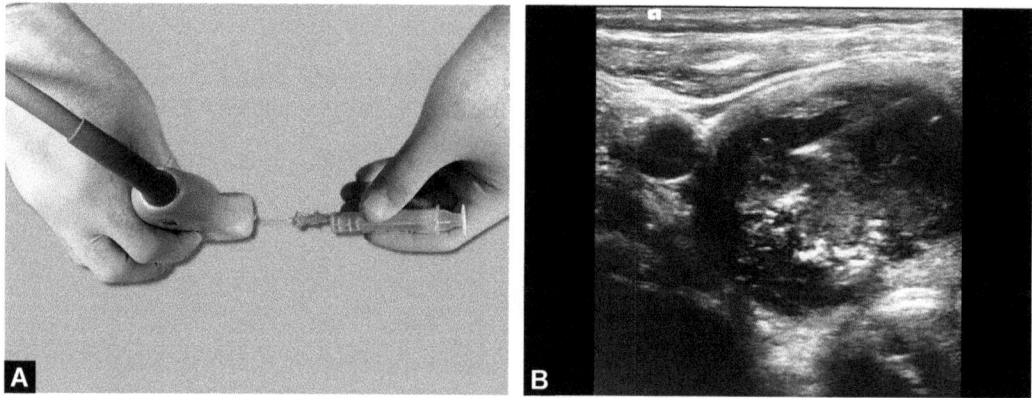

Figs. 5A and B: In-plane approach. (A) The advancing needle is kept parallel to long axis of ultrasound transducer; and (B) Ultrasonographic appearance of needle in the in-plane approach.

with narrow ultrasound beam (1 mm) that makes it difficult to maintain the beam-needle alignment. This can be reduced with usage of mechanical needle guides which are attached with the transducer. However, they restrict needle redirection. Appropriate stability of the transducer should be maintained during the procedure and attempt to advance the needle should not be done, if needle visualization is not there. Visual inspection of needle and transducer should be done to rule out gross misalignment. The transducer then should be slowly manipulated by sliding, tilting, or gently rotating to bring the needle under view. The advantage of using this technique is that during the procedure the entire needle is under direct visualization.

Out-of-Plane Approach (Figs. 6A and B)

This approach refers to advancing the needle perpendicular to the long axis of the ultrasound transducer (Fig. 6A). The ultrasound beam cuts a cross-section of the needle and the needle appears as an echogenic dot on the screen (Fig. 6B). The disadvantage of this technique is that it is not possible to tell whether the echogenic dot represents the tip or the shaft since

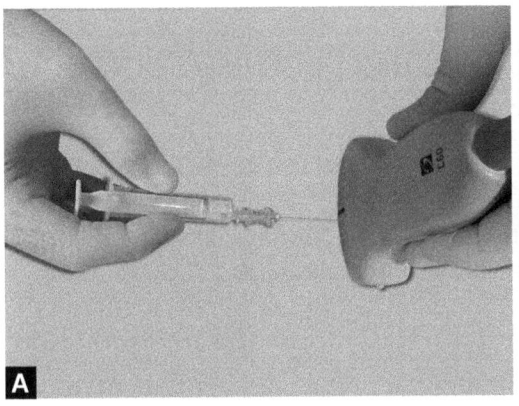 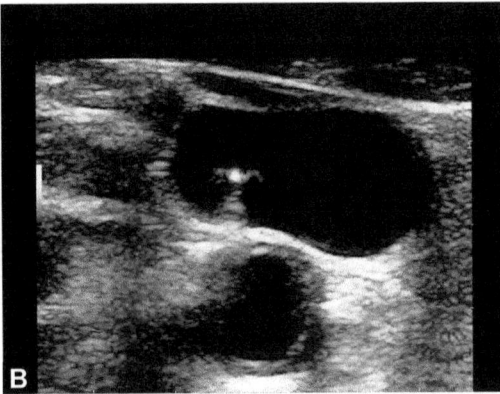

Figs. 6A and B: Out-of-plane approach. (A) Advancing needle is kept perpendicular to the long axis of the ultrasound transducer; and (B) Ultrasonographic appearance of needle in the out-of-plane approach.

both have a similar cross-section. To avoid the needle passing beyond the ultrasound transducer it is important that the needle should be advanced at an angle steep enough to keep the needle under the ultrasound transducer throughout the procedure. Out-of-plane approach is typically suited for accessing superficial structures, e.g. placement of a central venous catheter.[3]

Though any technique can be used for needle advancement, it is a good idea to be proficient in both the approaches and can be both used during the procedure to ascertain the correct position of the needle.

In the following section of the chapter various commonly performed ultrasound-guided interventional procedures in intensive care units is described.

THORACOCENTESIS

Over 60% patients admitted to medical ICU will develop radiographic evidence of pleural effusion at some point in their hospital stay.[4] The common causes include parapneumonic effusions, atelectasis, cardiac failure, and neoplasms. Infective pathologies can further lead to development of empyema and traumatic injury can lead to hemothorax. Thoracocentesis refers to removal of fluid from the pleural space. Small effusions rarely need drainage, however, infected effusions and effusions large enough to cause respiratory compromise that need drainage. Ultrasound allows detection of effusions as small as 20 mL and higher accuracy as compared to bedside chest X-ray.[5,6] Though thoracocentesis is a relatively safe procedure the use of ultrasound allows direct visualization and safe access of the pleural cavity reducing the procedural complication rates and is strongly recommended.[7] It can be used both for a single time aspiration and placement of pleural catheter. In this section, we shall review the technique of ultrasound-guided thoracocentesis.

Indications

Diagnostic

To evaluate the nature and etiology of pleural effusion.

Therapeutic

- Aspiration in symptomatic patients with large pleural effusion.
- Cather insertion for continuous therapeutic drainage.

Contraindications

There are no absolute contraindications for thoracocentesis, however, bleeding diathesis, and infection at the site of needle puncture are relative contraindications.

Technique

Localization of the Collection

The details of lung ultrasound for diagnosis of pleural effusion have been discussed in relevant chapter. However, we will briefly discuss about the localization of pleural collection. In any patient when there is clinical or radiological suspicion of pleural effusion a preliminary scan should be performed and the side, site, and nature of the collection should be documented. Ultrasound allows rapid identification and localization of even small effusions. Pleural effusions are identified in the dependent areas of lung between chest wall and diaphragm. Patient should be kept in semirecumbent position which allows effusion to gravitate in the dependent region and detect even a small volume of effusion. Using the low frequency (3–5 MHz) curvilinear probe the scanning should begin from flank, identifying the splenorenal (left side), and hepatorenal (right side) recess. The probe should be then moved in a cephalad direction and once fluid collection is identified above the diaphragm the area should be marked (Figs. 7A to C).

The fluid appears as a dark anechoic shadow displacing the lung from its apposition to the chest wall and diaphragm (Figs. 8A and B). If the effusion is large the lung is often observed to swim in the effusion. Complex effusions will often be observed as presence of septations, loculations, and fibrin strands. Ultrasound can also be used to identify the depth of needle penetration by freezing the ultrasound image. Ultrasound allows estimation of volume of effusion. One study described an equation between the volume of the effusion and the separation distance between the lung and the outer parietal pleura as effusion size (cc) = 20 × separation in mm.[8] It is recommended that an interpleural distance of at least 10 mm with effusion visible at the adjacent superior and inferior intercostal spaces is necessary in order to perform a safe pleural tap.[7] The optimal puncture site is determined by the identification of the largest fluid pocket. Conventionally it lies between the 7th–9th intercostals space in the posterior axillary line. Once the site of maximum interpleural distance is identified the point of entry should be marked with indelible skin marker (Fig. 9).

A well-informed consent, plan for adequate analgesia and sedation, and proper asepsis should be maintained prior to procedure as described earlier. The procedure is ideally described with the patient sitting with arms elevated and clasped behind the head. However, in the ICU setting it is frequently not possible to achieve this position. In ventilated patients the procedure can be performed in supine position with the head end elevated or also in lateral position with the affected side up. After localization of the effusion the site should

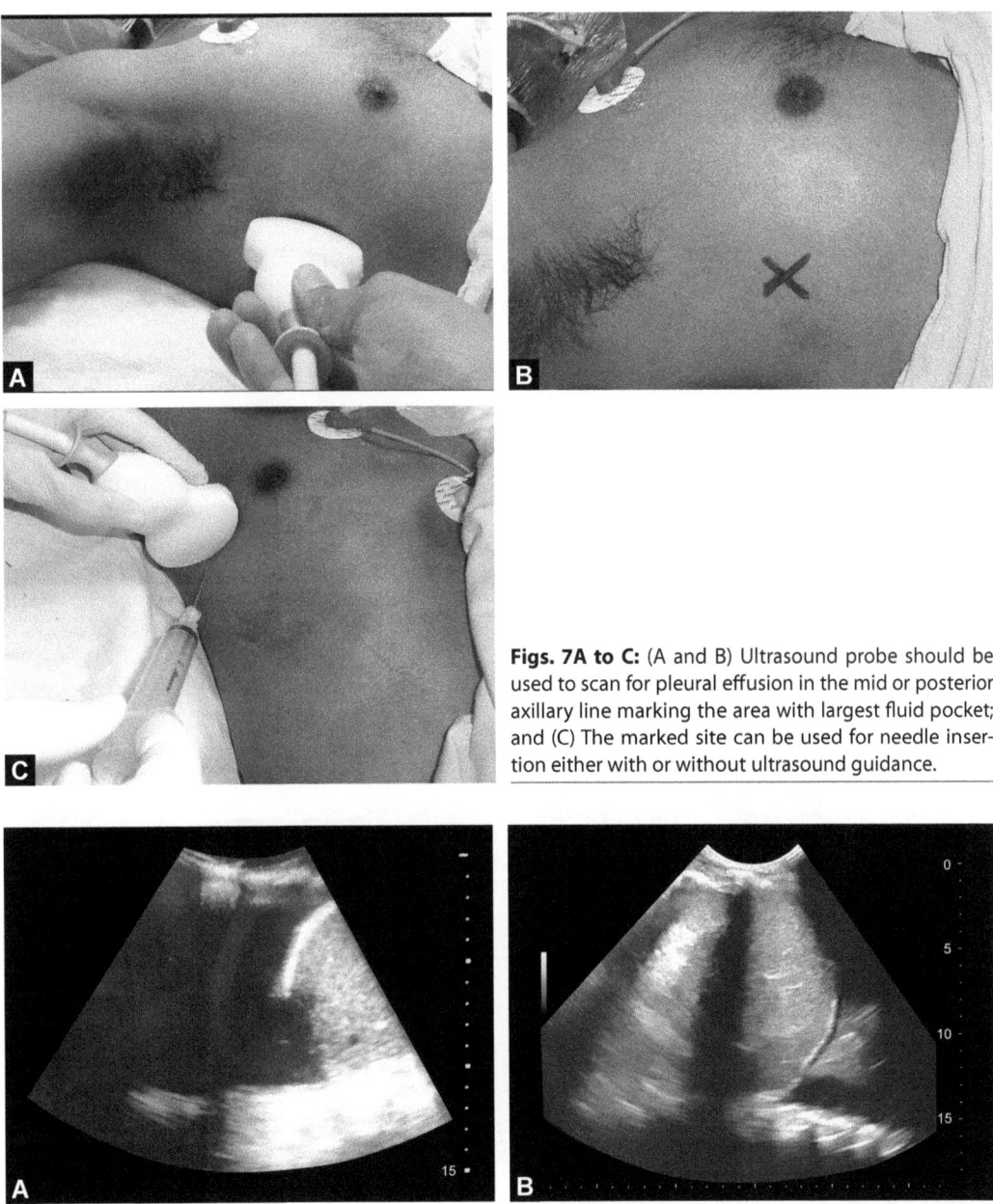

Figs. 7A to C: (A and B) Ultrasound probe should be used to scan for pleural effusion in the mid or posterior axillary line marking the area with largest fluid pocket; and (C) The marked site can be used for needle insertion either with or without ultrasound guidance.

Figs. 8A and B: (A) Large right-sided pleural effusion; and (B) Right-sided pleural effusion showing floating lung.

be then prepped with a disinfectant solution and a sterile drape should be placed. Two ultrasound-guided techniques are described: (1) Static—In this technique the ultrasound is merely used to mark the optimal needle insertion site and needle is not guided continuously using the ultrasound, and (2) Dynamic—The procedure uses continuous real-time ultrasound guidance for needle placement. Once the point of needle entry is identified, local

176 *Ultrasound in Critical Care*

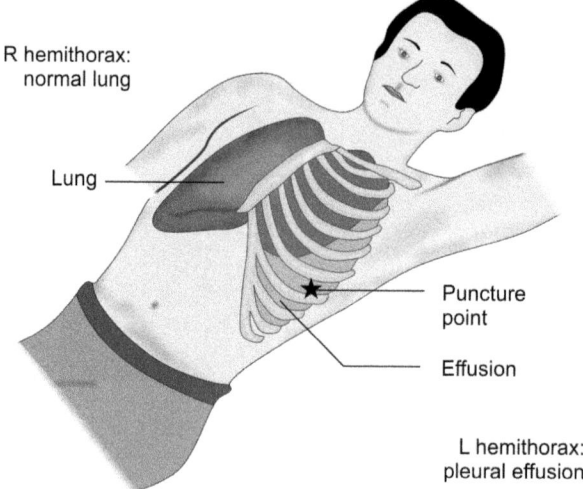

Fig. 9: Patient positioning for thoracocentesis. Patient should be kept in semirecumbent position with arm extended behind the neck.

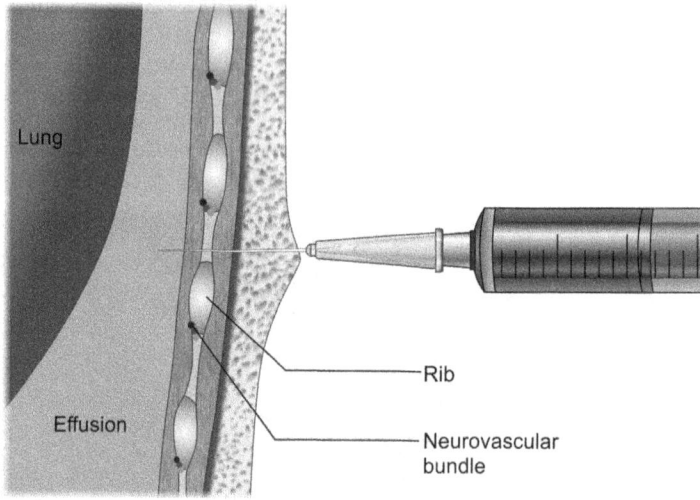

Fig. 10: Needle should be advanced over the top of the rib to avoid injury to neurovascular bundle.

anesthetic should be infiltrated up to parietal pleura to minimize patient pain and discomfort. The needle is then advanced in plane to the ultrasound transducer while maintaining steady aspiration. It is important to remember that the neurovascular bundle lies in a groove in the inferior border of the rib therefore the needle should be advanced over the top of the rib to avoid injury to neurovascular bundle (Fig. 10). Direct visualization of needle in the pleural space and aspiration of fluid in the syringe confirms the presence of needle in the pleural space. For diagnostic purposes adequate amount of fluid is aspirated and the needle is

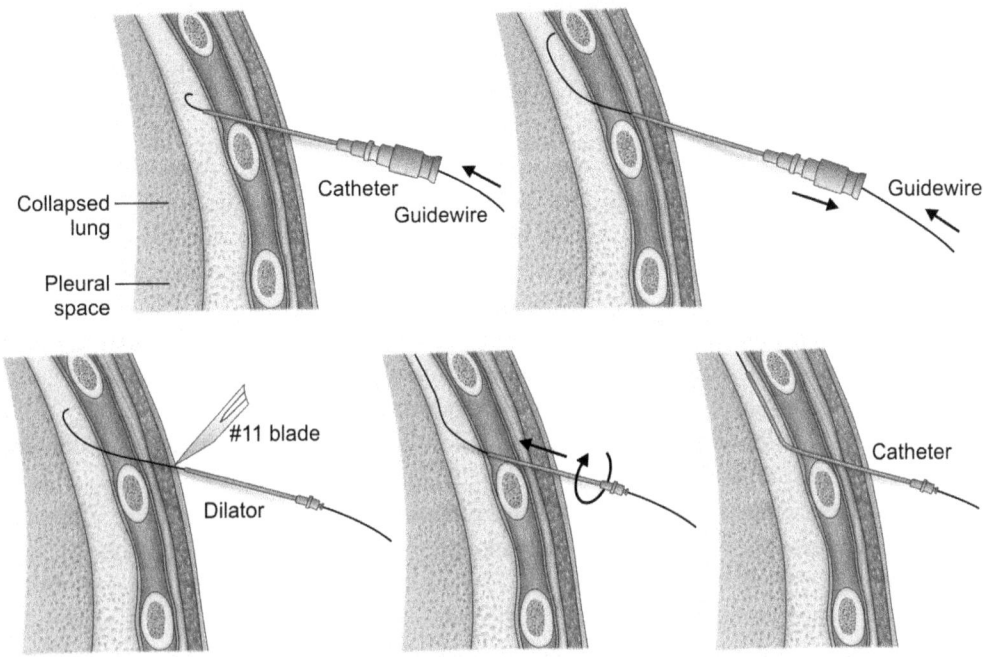

Fig. 11: Diagrammatic representation of pigtail thoracocentesis.
Source: (Adapted from: Reichman EF, Simon RR. Emergency Medicine Procedures, 2nd edition. New York: McGraw-Hill Professional; 2003).

then withdrawn. For therapeutic drainage the needle is stabilized and repeated aspiration is done using a commercially available kit or repeated aspiration using a syringe.

A drainage catheter can also be placed using this technique. Both trocar technique and Seldinger technique have been described, however, Seldinger technique is recommended due to lesser chance of injury to underlying structures. The syringe is removed and a flexible guidewire is introduced through the needle into the pleural cavity. Once the guidewire is in the pleural cavity its position is conformed using the ultrasound and the needle is removed. A small skin incision is made to accommodate the dilator and the catheter. The tract is then dilated using a dilator. After dilation a pigtail catheter with its introducer is advanced into the pleural cavity over the guidewire. The catheter is then connected to drainage system and catheter is secured to the skin using sutures (Fig. 11).

Postprocedure Care

Observe patient for worsening of respiratory function. Immediate postprocedure ultrasound can help to diagnose pneumothorax. Though chest X-ray (CXR) is not routinely recommended after simple pleural aspiration it is recommended, if the procedure is difficult or has required multiple attempts or if the patient becomes asymptomatic. Also CXR should be performed to evaluate the resolution of effusion and re-expansion of lung.

Complications

Ultrasound-guided thoracocentesis is a very safe procedure with low rate of complications, however, following complication can occur. Complications are most commonly seen after insertion of large bore catheters.
- *Pneumothorax:* It is the most commonly reported complication of thoracocentesis, with incidence ranging from 4.3% to 30%. However, with the use of ultrasound it was 0–9.1%.[9] It occurs due to accidental puncture of visceral pleura or entry of atmospheric air into the pleural cavity. Pneumothorax can be avoided by continuous visualization of the needle and maintenance of closed drainage at all times.
- *Vascular injury and hemothorax:* It a potentially life-threatening complication and should be suspected, if the patient develops worsening of respiratory function or CXR suggests significant increase in pleural effusion or bloody drainage.
- *Re-expansion pulmonary edema:* Re-expansion pulmonary edema can develop due to rapid or large volume drainage of pleural collection. Limiting drainage volume to less than 1.5 L in day decreases the incidence of this complication.
- *Infection:* Empyema can develop after drainage.
- *Accidental removal and drain blockage:* It can be avoided by securely suturing the drain and careful postprocedure care.
- *Others:* Pain, shortness of breath, cough and vagal reactions, and injury to the underlying organs (lung, liver, and spleen).

Clinical Pearls

- Diaphragm can be confused with hepatorenal and splenorenal recess as both are curvilinear structures. Wrong identification can lead to injury of these organs.
- Insert the needle in horizontal direction, severe cephalad angling can lead to injury of neurovascular structures.
- Avoid rapid removal of pleural fluid and limit drainage to less than 1.5 L in a day to reduce the incidence of re-expansion pulmonary edema.
- Color Doppler should be used to identify blood vessels in the needle path to avoid injuring the vascular structures.

PERICARDIOCENTESIS

Pericardium is a sac-like structure surrounding the heart (Fig. 12). It is formed by two layers, viz. (1) Outer fibrous pericardium, and (2) Inner serous pericardium. The two layers of serous pericardium (parietal and visceral) form a potential pericardial space which contains 15–30 mL of fluid and functions as a lubricant. Fluid can accumulate in the pericardial space leading to pericardial effusion which may lead to cardiac tamponade. Cardiac tamponade is a dreaded life-threatening condition that requires prompt diagnosis and management. It is diagnosed by the classic Beck's triad of hypotension, muffled heart sound, and raised jugular venous pulse, though studies have shown that only a minority of patients develop all the features and thus is a difficult diagnosis to make based on clinical findings alone.[10]

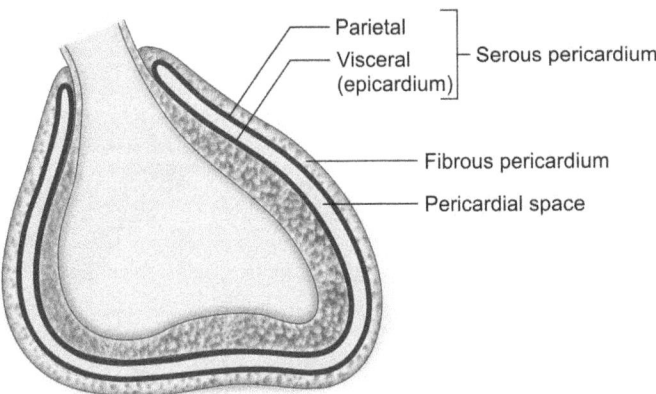

Fig. 12: Pericardial layers.

Pericardiocentesis refers to aspiration of fluid from the pericardial space surrounding the heart via needle aspiration or placement of a percutaneous drain. In this chapter, we shall review the ultrasound-guided drainage of pericardial effusion.

Indications

- *Diagnostic:* Aspiration pericardial effusion to identify the etiology (infective, neoplastic).
- *Therapeutic:* To relieve cardiac tamponade.[11]

Contraindications

There is no absolute contraindication to pericardiocentesis in hemodynamically unstable patient. Draining even a small amount of fluid can significantly improve the hemodynamic status of the patient. Relative contraindications include coagulation disorders, aortic dissection and myocardial rupture, patient on anticoagulation therapy, or following thrombolysis.[11]

Technique

The basics of echocardiography is beyond the scope of this chapter. However, we will briefly review the echocardiographic findings of pericardial effusion here. In a patient with clinical and radiological suspicion of pericardial effusion an echocardiography should be performed in all cases to confirm the presence of pericardial effusion and ascertain its size. The commonly used views are the subxiphoid, parasternal long axis, and apical four-chamber view. Pericardial effusion is an anechoic space between the ventricular free wall and the pericardium posteriorly and chest wall anteriorly (Figs. 13A and B). It is possible to roughly estimate the volume of pericardial effusion from the thickness of effusion. A posterior thickness of less than 1 cm, 1–2 cm, and more than 2 cm usually corresponds to volume of less than 200 mL, 200–500 mL, and more than 500 mL. However, there is no specific volume associated with development of cardiac tamponade which can either be due to large effusion or a rapid

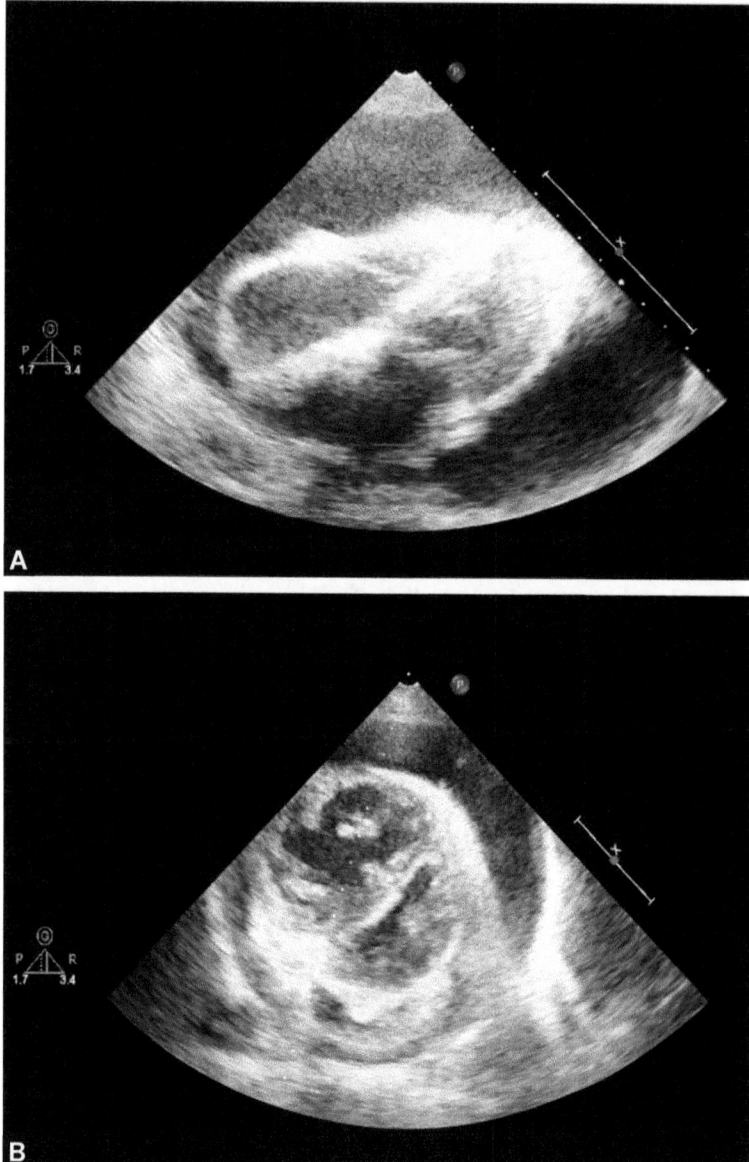

Figs. 13A and B: (A) Large pericardial effusion subxiphoid view; and (B) Apical view.

accumulation of a smaller one. A large accumulation can develop insidiously and not lead to cardiac tamponade due to stretching of pericardial sac. On echocardiography cardiac tamponade can be identified by diastolic collapse of right atrium and right ventricle. This is best observed in the apical four-chamber view.

The general principle of invasive procedures remains the same including consent, analgesia, and sterile precautions. Ultrasound or echocardiography-guided procedure is recommended since it has a lower complication rate than blind approach.[12]

Often pericardiocentesis is performed in emergency setting, therefore, equipment and drugs for resuscitation should be immediately available. Identify the pericardial effusion in different views and identify the best possible route of access. The patient is positioned in supine position with head end elevated by 30–45°. This allows the fluid to gravitate inferiorly and brings the heart closer to the anterior chest wall. Subxiphoid, parasternal, and apical approach have been described for pericardiocentesis.

Subxiphoid Approach

Xiphoid process is identified and site for insertion is marked just caudal to the xiphoid process using a skin marker (Figs. 14A and B). If available a pericardiocentesis kit can be used otherwise an 18–22-G spinal needle is attached to 20 mL syringe filled with 5 mL of normal saline. The needle is inserted under ultrasound guidance using in-plane technique at 45° angle to the abdominal wall and 45° of the midline sagittal plane, directed towards left shoulder. The needle is advanced under continuous ultrasound guidance while maintaining a negative pressure on the syringe. Note that entering the thoracic cavity via parasternal approach can put the patient at risk for liver and bowel injury.

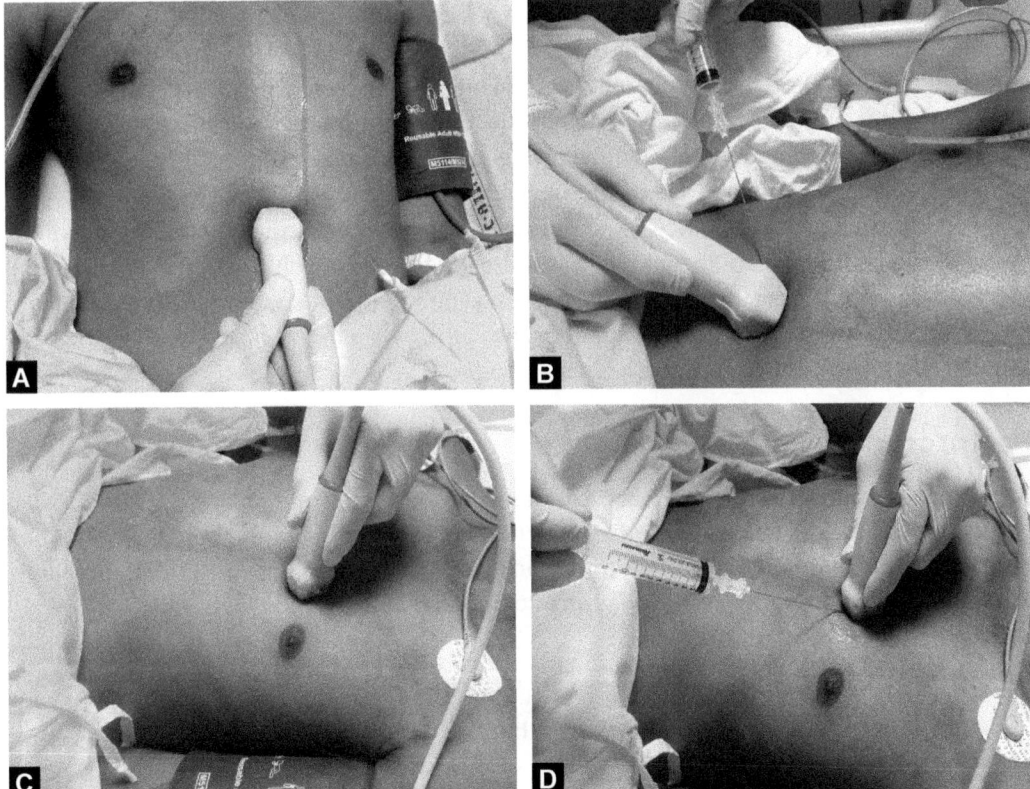

Figs. 14A to D: (A and B) Subxiphoid approach to pericardiocentesis; and (C and D) Parasternal approach to pericardiocentesis.

Alternatively parasternal approach may be used in which the needle is inserted just lateral to the sternum in the 3rd or 4th intercostals space perpendicular to the chest wall (Figs. 14C and D). Site of largest collection is identified and needle is introduced just lateral to the end of transducer closest to the heart and directed towards the patient's spine using ultrasound guidance the needle is advanced into the pericardial space to drain the collection. Parasternal approach offers advantage of being closer to the pericardial effusion and avoids liver and lung injury.

Apical Approach

Under ultrasound guidance apical approach is fast becoming the most commonly employed method for pericardiocentesis. After ascertaining the presence of effusion, insert the needle just lateral to the transducer. Visualize the needle as it enters the pericardial space at the apex of the heart. If the lung is visualized overlying the heart, movement of probe in the medial and caudal direction may be a more optimal site for needle entry.

After entering the pericardium the position of the needle is confirmed by aspiration of pericardial fluid or observing bubbles in pericardial sac after injecting agitated saline. During the entire procedure electrocardiogram (ECG) should be carefully monitored. If there is presence of ECG changes it reflects direct contact with the myocardium. Gradually withdraw the needle till the ECG rhythm returns to normal.

Though single time aspiration may significantly improve the hemodynamic status of the patient it may not completely evacuate the effusion and reaccumulation can occur. Placement of a percutaneous drain is often necessary to continuously drain the effusion. Placement of catheter is recommended using the Seldinger technique. After placement of needle in the pericardial space a guidewire is introduced in the pericardial space, the needle is then removed, and a percutaneous drain is railroaded over the guidewire in the pericardial space. The position of catheter is confirmed by aspirating pericardial fluid through it and is secured with sutures and dressing applied. The catheter can then be used to drain the pericardial collection by attaching it to a collection bag.

Postprocedure Care

Following pericardiocentesis visualize the heart to evaluate for reduction in pericardial collection and improvement in cardiac function. A chest radiograph should be obtained to rule out pneumothorax. Continue monitoring the patient for new onset hemodynamic instability. Consider specialist consultation after initial management.

Complications

Complications include hemothorax, pneumothorax, coronary artery puncture or aneurysms, and dysrhythmias. Subxiphoid approach may also result in hepatic or stomach injury, peritoneal puncture, and diaphragmatic injury. Parasternal approach may be associated with puncture of internal thoracic artery.[13]

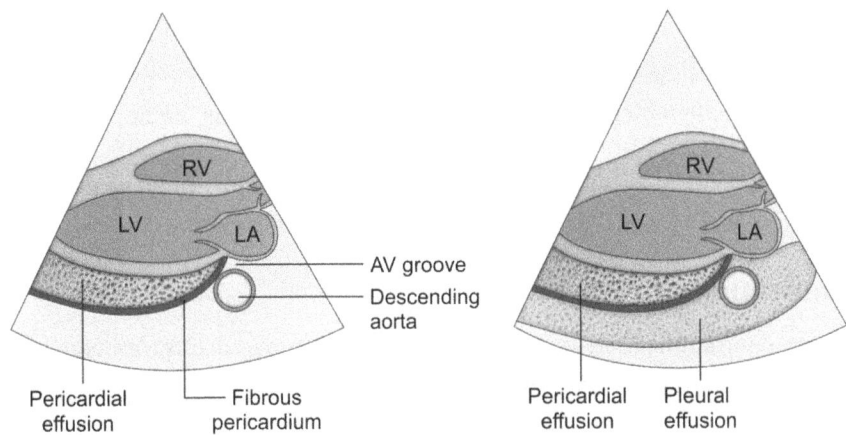

Fig. 15: Differentiating between pleural and pericardial effusion. Fluid anterior to aorta is pericardial effusion whereas posterior to aorta is pleural effusion.

Clinical Pearls

- Epicardial fat pad is often confused with pericardial effusion. Remember the fat pad is an anterior structure and effusions are usually circumscribed.
- If there is suspicion that the needle is placed in the ventricles it can be confirmed using the fact that intracardiac blood forms a clot whereas pericardial aspirate does not.
- If there is confusion between pericardial and pleural effusion look at descending aorta in parasternal long axis. Fluid anterior to aorta is pericardial effusion whereas posterior to aorta is pleural effusion (Fig. 15).

PARACENTESIS

Ascites is a common endpoint of many pathological conditions like hepatic cirrhosis, renal failure, cardiac failure, and severe hypoalbuminemia. It may develop in conditions like abdominal and gynecological malignancies. Small collections of fluid are usually asymptomatic, however, with increasing volumes there is progressive increase in intra-abdominal pressure which can lead to severe patient discomfort. Elevation of diaphragm can lead to respiratory distress especially in supine position. Presence of ascites can also lead to development of spontaneous bacterial peritonitis which is a potentially life-threatening condition. Paracentesis refers to drainage of fluid from the peritoneal cavity. With the use of ultrasound it is possible to detect as little as 100 mL of free fluid in the peritoneal cavity.[14] Paracentesis forms the backbone of diagnosis and management of conditions like spontaneous bacterial peritonitis. It also allows assessment of presence of intraperitoneal blood in cases of trauma. Use of ultrasound guidance for paracentesis has lead to significant increase in the success rate as well as reduction in the complications associated with blind technique.[15,16] In this section, we shall review the technique of ultrasound-guided paracentesis.

Indications

- *Diagnostic:* For evaluation of ascitic fluid and peritoneal free fluid to ascertain the cause (liver disease, malignancies, spontaneous bacterial peritonitis, hemoperitoneum due to trauma, etc.).
- *Therapeutic:* Drainage of peritoneal collection in symptomatic patients due to increased intra-abdominal pressure (dyspnea, decreased urine output).

Contraindications

Only absolute contraindication to paracentesis is disseminated intravascular coagulation, however, studies do not support routine evaluation and correction of coagulation studies prior to paracentesis. Relative contraindications include infection at point of puncture, pregnancy, severe bowel distension, or intra-abdominal adhesions.

Technique

Basics of abdominal ultrasound is discussed in the relevant chapter, however, in this section we shall review the ultrasonographic findings of peritoneal fluid collection. In all the patients with clinical suspicion of peritoneal collection bedside abdominal USG should be performed with a low frequency curvilinear transducer (3–5 MHz) which allows a greater visualization of deeper structures. The dependent regions of the abdomen (Figs. 16A and B) are usually scanned. Ascites appears as an anechoic shadow within the peritoneal cavity and frequently bowel loops and mesentery may be seen floating in the ascitic fluid (Figs. 17A and B). Ascitic fluid may often be complicated with the presence of fibrin strands, loculations, and hemoperitoneum. Following identification of the site with maximum collection abdomen wall should be scanned in that area with a high frequency probe with Doppler to rule out overlying vessels. It is important to remember that the inferior epigastric artery runs along the rectus abdominis muscle therefore needle insertion site should be lateral to rectus abdominis. Conventionally

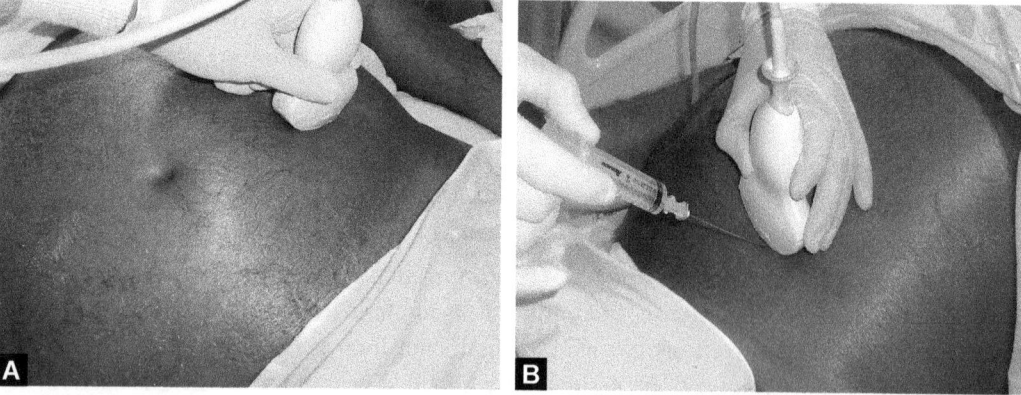

Figs. 16A and B: (A) Dependent regions of the abdomen should be scanned for ascites; and (B) The site with largest fluid collection should be used for puncture.

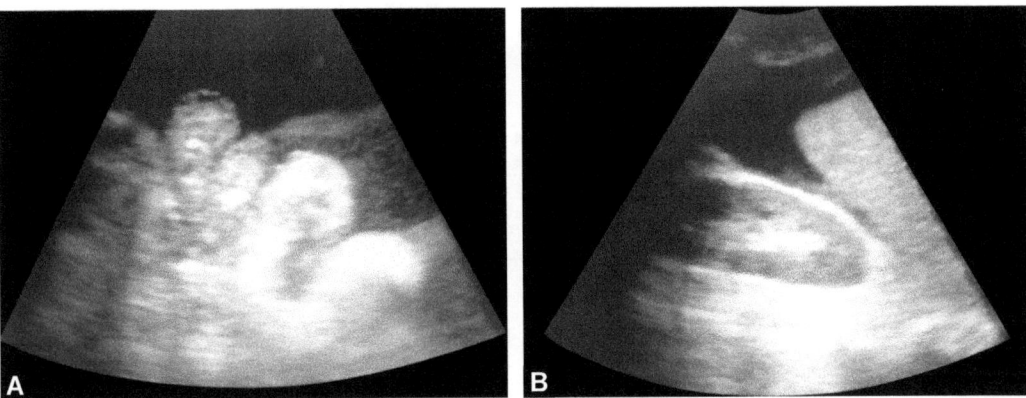

Figs. 17A and B: Ultrasonographic appearance of ascites.

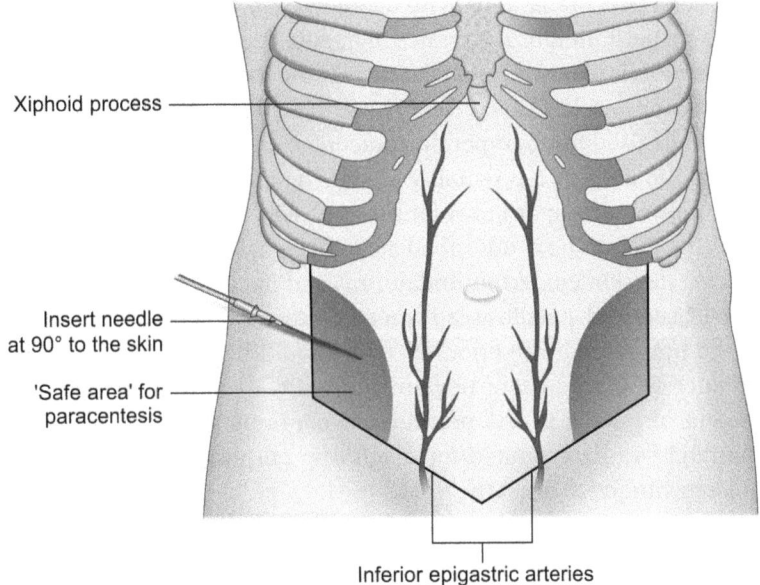

Fig. 18: Safe area of needle insertion for paracentesis.

left lower quadrant is preferred over right lower quadrant to avoid injuring a distended cecum. Alternatively a midline entry can be performed 2 cm below the umbilicus since this corresponds to linea alba which is an avascular structure (Fig. 18). Patients with portal hypertension due to cirrhosis have significant collaterals and puncturing these vessels should be avoided.

A well-informed consent, plan for adequate analgesia and sedation, and proper asepsis should be maintained prior to procedure as described earlier. Patient should be positioned supine with head end elevated by 30–45°. This allows collection of free fluid in the pelvis. The largest point of collection is noted and marked with a skin marker. Select an appropriate sized needle depending upon the depth and whether a diagnostic or therapeutic drainage is

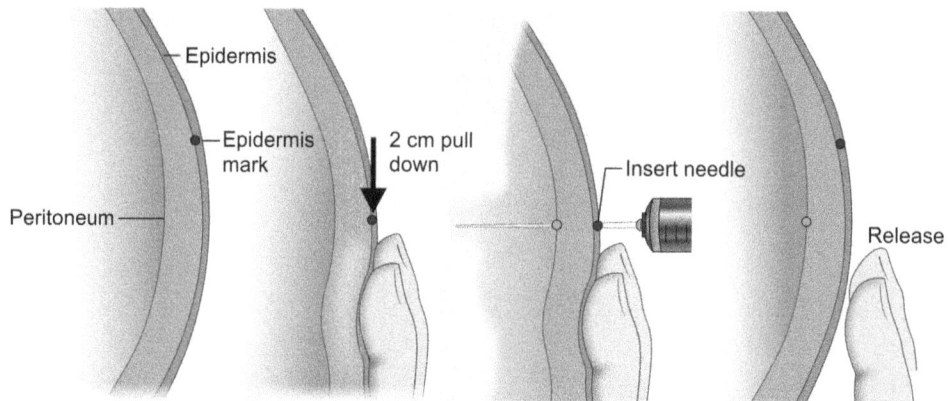

Fig. 19: Z-track technique of needle insertion for paracentesis.

being performed. After part preparation with disinfectant solution local anesthesia should be applied up to parietal peritoneum. The choice of needle depends upon whether diagnostic or therapeutic paracentesis is planned. As a general rule the narrowest needle should be selected to minimize complication. A 22-G needle may be sufficient for diagnostic tap whereas a larger 18-G needle should be used for therapeutic paracentesis. Avoid using sheathed needle since with Z-track technique there is a possibility of shaving off part of plastic sheath into abdominal wall which may require laparotomy for removal. Needle should be introduced using the "Z-tract method". The technique is described as applying traction to the skin in caudal direction and puncturing the skin epidermis and dermis and once these structures are penetrated the traction is released and needle is advanced further (Fig. 19). This prevents the leakage of fluid in the postprocedure period. Needle should be advanced continuously under ultrasound guidance with a constant negative pressure. Aspiration of fluid or direct ultrasound visualization of needle in the peritoneum confirms the needle position in peritoneal cavity. The fluid can be aspirated for diagnostic purpose or a catheter can be placed to using Seldinger technique to drain the fluid.

Postprocedure Care

Remove the catheter after the desired volume of ascitic fluid has been drained and apply a sterile gauze dressing over the puncture site. Monitor hemodynamics of the patient and also observe for any leakage from the puncture site.

Complications[17]

- *Hemorrhage (1–2%):* Bleeding in the peritoneal cavity is the most dreaded complication of paracentesis especially in patients with liver disease who may have concomitant coagulation defects and thrombocytopenia. Most commonly occurs due to injury to epigastric vessels. Risk can be minimized with ultrasound guidance. Coagulopathy should be corrected, if present.

- *Bowel injury and infections:* Infection is rare unless bowel is perforated. Patient may develop abdominal pain and signs of peritonitis.
- *Persistent leakage of ascitic fluid:* Leaks typically arise when Z track has not been properly employed, or large bore needle has been used. Pressure dressing over the puncture site will decrease the leak over few days, however, suture can be used, if the leak still persists.
- *Hemodynamic instability:* Large volume paracentesis (>5 L) can lead to hemodynamic instability. It can generally be managed with intravenous fluids, however, occasionally vasopressor infusion may be required. If large volume of paracentesis is planned intravenous albumin (6–8 g/L of ascitic fluid) should be administered to the patient.

Clinical Pearls

- Identify the largest fluid pocket. Do not reposition the patient after this step, as fluid pockets may move with repositioning.
- Avoid subcutaneous varices by using color flow Doppler on abdominal wall. Varices appear anechoic and are compressible.
- Bowel may get sucked into the pigtails fenestrations leading to impaired drainage. Flush the catheter or reposition the patient.
- Always confirm that bladder is empty since filled bladder can be misdiagnosed as ascites.

PERCUTANEOUS DRAINAGE OF INTRA-ABDOMINAL COLLECTIONS

Abdominal surgery and other abdominal pathologies are frequently complicated with development of intra-abdominal abscesses and collection. A delay in diagnosing and management of intra-abdominal infections may lead to sepsis and multiorgan failure. Mortality from intra-abdominal sepsis can be as high as 30–35%, with mortality in patients requiring a second operation reaching 50% and in those with an undrained abscess exceeding 90% often as result of multiple organ failure.[18] Historically intra-abdominal abscess were drained surgically, however, evidence has emerged that operative drainage is associated with high morbidity and mortality and therefore the management protocol has shifted towards less invasive methods.[18] Though computed tomography (CT)-guided percutaneous drainage of intra-abdominal collections is the standard technique, ultrasound-guided drainage offers a good alternative since it is portable and patient transport can be avoided, also it provides real-time visualization of needle, guidewire, and catheter. However, it is often technically difficult to locate deep-seated intra-abdominal collections using ultrasound due to overlying bowel and is well suited for organ specific and superficially located intraperitoneal abscesses. In this chapter, we shall review the technique of ultrasound-guided percutaneous drainage of intra-abdominal collections.

Indications

Presence of abnormal fluid collection in the abdomen which is suspected to be infected (postsurgical fluid collections) or presence of organ specific abscess (renal, hepatic, and pancreatic). Based upon the size, nature, and the clinical condition of the patient either single time aspiration or placement of drainage catheter can be performed.

Contraindications

Though there are no absolute contraindications, commonly encountered relative contraindications are: Bleeding diathesis, overlying bowel, and vascular structure.

Antibiotic Prophylaxis

Society of interventional radiology considers percutaneous abscess drainage as a dirty procedure and recommends preprocedural antibiotic prophylaxis 1 hour prior to the anticipated start of procedure.[19] Though there is no consensus on the ideal antibiotic, broad-spectrum antibiotics such as third-generation cephalosporin are recommended since abscesses tend to be polymicrobial.

Technique
Localization of the Pathology

In a patient with suspected intra-abdominal collection prior to aspiration or placement of drain the patient should undergo scanning with an ultrasound as it provides information regarding feasibility, best possible approach, and associated risks. It is often prudent to obtain a CT scan to better identify the location, extent, and nature of collection. During preliminary scan the size of the collection should be measured using the ultrasound, the depth from the insertion site to the collection should be measured. Target should be to place the catheter in the most dependent position since it helps to effectively drain the collection, also shortest possible route should be used to minimize catheter length. Also presence of overlying bowel loops and vascular structures should be evaluated since this may lead to viscus perforation or hematoma/aneurysm formation. Based on the measurement the length of the guidewire insertion (should be slightly longer to ensure adequate access to the cavity) as well as drain length should be determined (measure from skin to beginning of the cavity and second measurement to end of cavity). Holes should be embedded into the cavity; no proximal side holes should be external to the cavity.

A well-informed consent, plan for adequate analgesia and sedation, and proper asepsis should be maintained prior to procedure as described earlier. The procedure should be performed in supine position. After aseptic preparation the area should be infiltrated with local anesthetic. The needle should be introduced and moved towards the abdominal collection under ultrasound guidance. Continuous negative pressure should be applied on the needle and once there is presence of aspiration the needle position is confirmed. The attached syringe should be removed and a guidewire is introduced through the needle. The needle is then removed while keeping guidewire in situ. A small incision is made at the puncture site and the tract of the guidewire is dilated using serial dilators to appropriate size. Following dilatation a pigtail catheter of selected size is railroaded over the guidewire. If the trocar technique is used the catheter mounted trocar is introduced parallel to the plane of the needle up to a predetermined depth. Either trocar or Seldinger technique can be used for catheter placement, however, Seldinger technique is frequently utilized for small deeper collections whereas superficial large collections are more amenable to drainage with trocar technique.

Correct placement of the drain is confirmed by free flow of fluid through the catheter or by ultrasound visualization. The catheter should be attached to the closed drainage system and catheter is secured to skin with sutures and dressing is applied.

Specific Pathologies (Hepatic Abscess, Percutaneous Cholecystostomy, Percutaneous Nephrostomy, and Pelvic Abscess)

Hepatic abscess: Though rare hepatic abscess is a life-threatening condition associated with mortality approaching 100% in undrained cases.[20] Advent of ultrasound or CT-guided percutaneous drainage has significantly reduced the mortality and morbidity associated with this condition.[20] Pyogenic abscess forms greater than 80% of the cases of liver abscess. Amebic and fungal abscess form the other major causes. With the help of ultrasound abscess as small as 1 cm can be located and drained. Preprocedure preparation and equipment is similar to drainage of any other abdominal collection. Ultrasound scanning should be performed prior to the procedure with 3–5 MHz curvilinear transducer and size and location of the collection should be recorded. Liver abscesses are typically poorly demarcated with a variable appearance, ranging from predominantly hypoechoic (still with some internal echoes however) to hyperechoic (Figs. 20A and B). A record should also be made of number of collections and the best approach for drain placement (subcostal or intercostal). With patient in supine position the abscess can be accessed via subcostal approach or depending upon the location of the collection intercostal approach can also be used with patient in the lateral decubitus position. Care should be taken to avoid pleural, bowel, and intrahepatic vascular injury by continuous visualization of the entire procedure under ultrasound. Both trocar as well as Seldinger techniques are acceptable for drain placement. Usually a collection less than 5 cm can be drained with aspiration whereas a larger collection will require placement of percutaneous drain. A pigtail catheter sized 8–10 Fr is usually sufficient for continuous drainage. Often hepatic abscess may have multiple septations or loculation, which may require placement of multiple catheters. Nature and volume of the drain should be evaluated following drainage. Repeat sonography should be performed to evaluate for reduction in the size of collection. Persistent drainage and change in nature can be due to biliary communication and should be evaluated using a sonogram. Drain removal should be based on resolution of collection, unproductive drainage, and clinical improvement of the patient. Residual loculations can be managed by catheter repositioning and aspiration.

Percutaneous cholecystostomy: In patients suffering from acute cholecystitis early cholecystectomy is the treatment modality of choice. The majority of cases in general population is due to stone, however, in ICU acalculous cholecystitis predominates. However, cholecystectomy for calculous or acalculous cholecystitis carries a high morbidity in critically ill patients with up to 30% mortality.[21] Ultrasound-guided percutaneous cholecystostomy offers a minimally invasive method for decompression of gallbladder in such patients and can be effectively done at bedside under local anesthesia. Placement of a percutaneous drain effectively decompresses the gallbladder and provides symptomatic relief as well as reduction in inflammatory response. Preprocedure preparation is similar to the other procedures and is discussed earlier. Prior to procedure ultrasound scanning should be performed using a 3–5 MHz

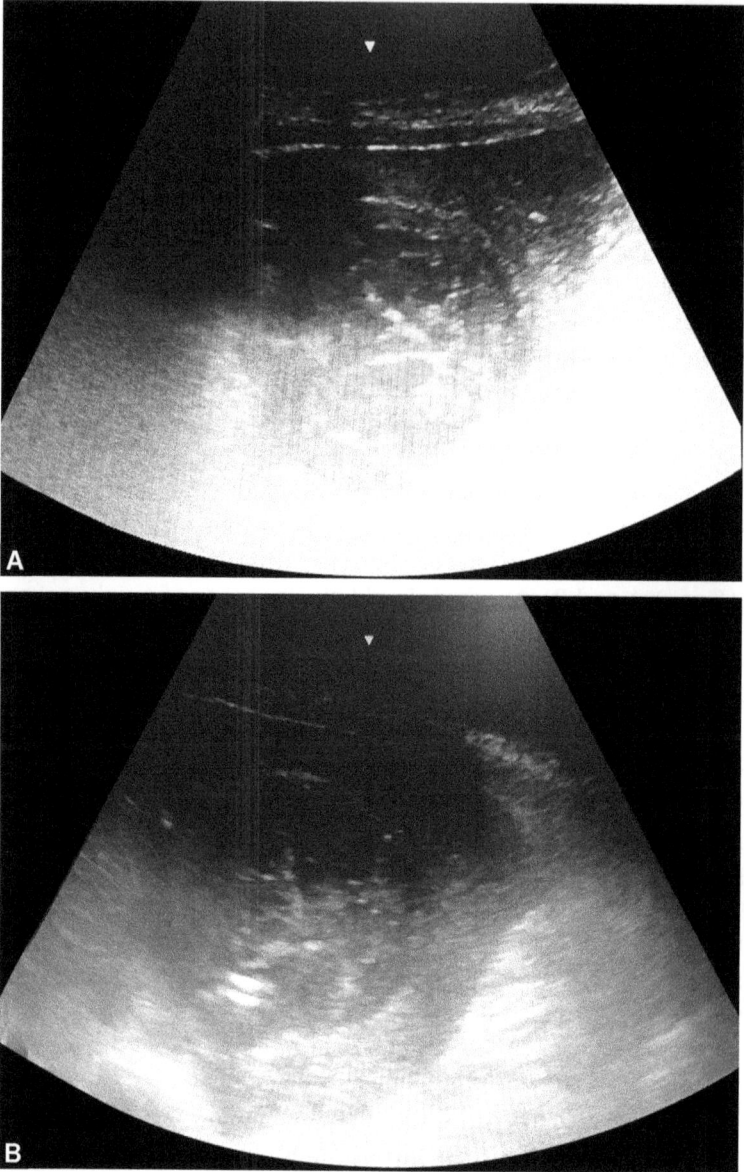

Figs. 20A and B: Ultrasound appearance of hepatic abscess with pigtail catheter in situ.

curvilinear probe and the safest access route is identified. A distended gallbladder more than 5 cm, gallbladder wall thickness more than 3 mm, transducer elicited tenderness over gallbladder, gallstone sludge, and pericholecystic fluid will confirm acute cholecystitis in up to 94% of patients (Fig. 21).[22] After skin asepsis subcutaneous local anesthetic is infiltrated. The procedure is performed in supine position. Two approaches are recommended: (1) Transhepatic, and (2) Transperitoneal. The transhepatic approach is an extraperitoneal approach and can be performed via both subcostal and intercostal approach. Transhepatic approach

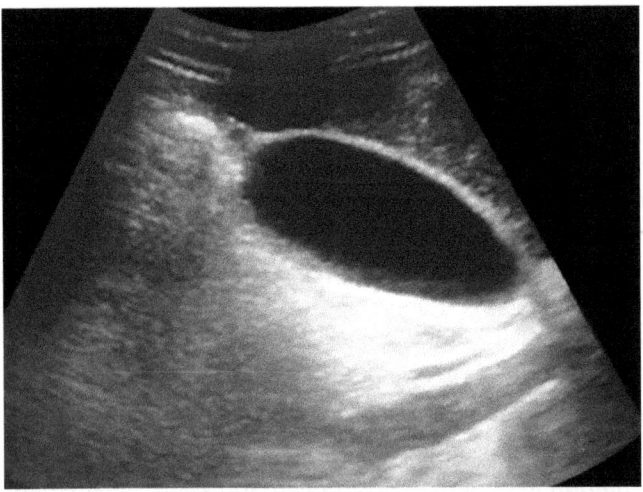

Fig. 21: Ultrasonographic appearance of acalculous cholecystitis. *Note:* The thick-walled distended gallbladder.

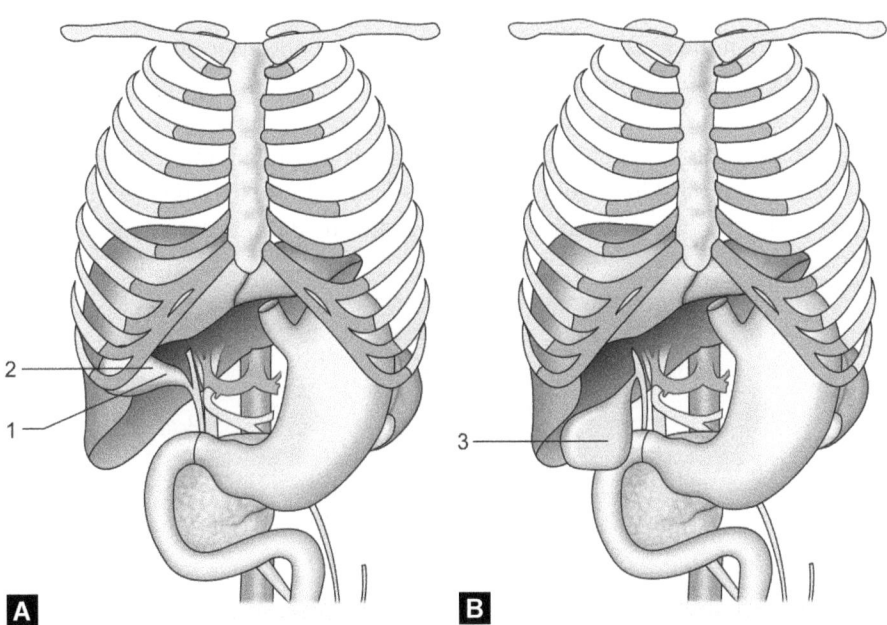

Figs. 22A and B: Diagrammatic representation of (1) subcostal, (2) intercostal, and (3) transperitoneal access for percutaneous cholecystostomy.

is associated with reduced risk of bile leak, colonic perforation, and provides greater catheter stability. During intercostal approach care must be taken to avoid injuring the pleura and needle should always be inferior to diaphragm. Also neurovascular bundle injury can be avoided by passing the needle inferior to the rib (Figs. 22A and B).

In patients with coagulopathy or in patients in whom there is anatomical difficulty transperitoneal approach can be considered. Both trocar and Seldinger technique are accepted methods of inserting the percutaneous drain. After initial puncture needle tip should be continuously visualized under ultrasound. Bile aspirated should be sent for bacteriologic evaluation. A pigtail catheter of 7–8 Fr is sufficient in most cases. Following insertion of catheter it should be connected to a drainage bag. Complications are similar to other procedures, however, there is a greater chance of pneumothorax in the intercostal approach whereas subcostal approach is associated with biliary leak which may cause biliary peritonitis. Postprocedure drain should be maintained for 3 weeks which allows for tract maturation. Catheter should be flushed twice daily to prevent clogging. After more than 3 weeks catheter can usually be safely removed. A cholecystogram should be performed to evaluate the cystic duct and presence of calculi.

Percutaneous nephrostomy: Obstructive uropathy occurs due to blockage of urinary flow which can result in increased intrarenal pressure and lead to renal parenchymal damage. Urinary obstruction can further lead to pain, pyonephrosis, and sepsis and can be a potentially life-threatening condition and often requires decompression. Percutaneous nephrostomy is commonly performed procedure which allows upper urinary tract diversion in urinary tract obstruction with signs of sepsis or rapidly rising creatinine levels. Preprocedure preparation remains same as discussed earlier. Prior to performing the procedure patient's kidney should be scanned in the longitudinal plane in anterior to posterior direction using 3–5 MHz curvilinear probe. Presence of renal pelvis dilation more than 20 mm is usually suggestive of hydronephrosis (Fig. 23). The procedure is performed in supine position. The area with best sonographic visualization of dilated renal pelvis is identified and marked. The depth of the renal pelvis from the skin should be measured and attempt should be made to choose the puncture point at shortest distance. Conventionally the point of entry is along the Brodel's

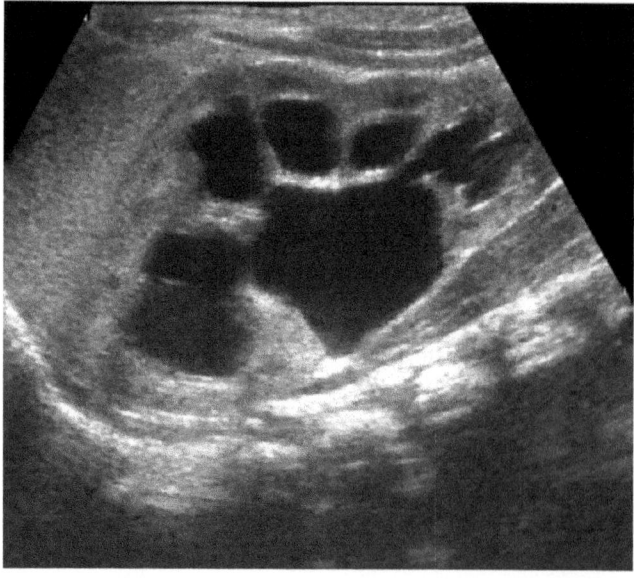

Fig. 23: Ultrasound appearance of severe hydronephrosis.

line near the posterior axillary line, approximately 3 cm below 12th rib. This approach is usually associated with least chances of arterial injury and also avoids colon-liver spleen, pleura, and erector spinae muscles.[23] After puncturing the skin with 18-G needle attached to syringe the needle is gradually advanced under USG guidance. Aspiration of urine or direct visualization of needle in the renal pelvis confirms the position of the needle. Seldinger technique is usually recommended for placement of a percutaneous drain. 8–14 Fr nephrostomy catheters is usually adequate for placement. Following placement of catheter its position can be confirmed by direct visualization or flow of urine/pus from the catheter. Catheter is attached to a drainage bag and is sutured securely to skin and dressing applied. The most common complications are bleeding and organ injury (bowel, spleen, and liver) and bacteremia which can be avoided by direct visualization under ultrasound.

Postprocedure Care

- Monitor vitals of the patient following the procedure.
- *Catheter management:* Up to the time of catheter removal daily monitoring should be done of catheter output. Drain output should gradually decrease over days, however, if drain output drops significantly it suggests clogging and should be flushed with 3-5 mL of saline. Often it will require exchanging or upsizing. Fistulous communications can result in persistent large volume drainage and may necessitate fluoroscopy to identify the fistula. Catheter removal may be considered, if drain output is less than 10 mL in 24 hours, clinical improvement, and reduction in leukocytosis.
- Prior to removal repeat sonographic assessment should be made to confirm resolution of the fluid cavity.

Complications[23,24]

- Pain.
- *Hemorrhage and hematoma formation:* This can occur due to injury to the overlying vessels and can lead to intra-abdominal bleed or relatively benign hematoma formation.
- *Sepsis:* Local skin site infection, bacteremia, or spread of abscess to adjacent structures.
- Fistula formation with bowel bladder or urinary bladder due to accidental injury of these structures. Diagnosed by presence of persistent high volume output from the fistula.
- Abdominal viscus perforation.

Clinical Pearls

- Ensure adequate skin incision prior to introduction of dilator since application of force may cause guidewire to kink.
- If there is resistance to passage of either dilator or catheter gently withdraw the guidewire and then readvance slowly. Buttress the tissue with free hand, hold the catheter close to skin, and advance the dilator or catheter in short firm strokes. This reduces radius of curvature and reduces buckling.

- *Choose drainable collections:* Nonliquefied contents and abscess less than 3 cm are poorly amenable to catheter drainage and can be managed with aspiration or antibiotic therapy.
- *Correct catheter size:* Locking pigtail catheters are often preferred over straight catheters which are more prone for dislodgment. Serous contents can be drained with catheters of smaller caliber (8–10 Fr). Larger catheters may be required to drain abscesses, 12–14 Fr for complex abscesses, and 24–30 Fr for drainage of pancreatic and peripancreatic fluid collections.

REFERENCES

1. Joint working party of the Association of Anaesthetists of Great Britain and Ireland, the Royal College of Anaesthetists, and the Intensive Care Society Ultrasound in Anaesthesia and Intensive Care. The Association of Anaesthetists of Great Britain and Ireland, The Royal College of Anaesthetists, and The Intensive Care Society: A Guide to Training. Anaesthesia. 2010.
2. Chapman GA, Johnson D, Bodenham AR. Visualisation of needle position using ultrasonography. Anaesthesia. 2006;61(2):148-58.
3. Marhofer P, Chan VW. Ultrasound-guided regional anesthesia: current concepts and future trends. Anesth Analg. 2007;104(5):1265-9.
4. Mattison LE, Coppage L, Alderman DF, et al. Pleural effusions in the medical ICU: prevalence, causes, and clinical implications. Chest. 1997;111(4):1018-23.
5. Lichtenstein D, Goldstein I, Mourgeon E, et al. Comparative diagnostic performances of auscultation, chest radiography, and lung ultrasonography in acute respiratory distress syndrome. Anesthesiology. 2004;100(1):9-15.
6. Rahman NM, Singanayagam A, Davies HE, et al. Diagnostic accuracy, safety and utilisation of respiratory physician-delivered thoracic ultrasound. Thorax. 2010;65(5):449-53.
7. Havelock T, Teoh R, Laws D, et al. Pleural procedures and thoracic ultrasound: British Thoracic Society pleural disease guideline 2010. Thorax. 2010;65(Suppl 2):i61-76.
8. Balik M, Plasil P, Waldauf P, et al. Ultrasound estimation of volume of pleural fluid in mechanically ventilated patients. Intensive Care Med. 2006;32(2):318.
9. Sikora K, Perera P, Mailhot T, et al. Ultrasound for detection of pleural effusions and guidance of thoracocentesis procedure. ISRN Em Med. 2012;2:1-10.
10. Sternbach G. Claude Beck: cardiac compression triads. J Emerg Med. 1988;6(5):417-9.
11. Harper RJ. Pericardiocentesis. In: Roberts JR, Hedges JR (Eds). Clinical procedures in emergency medicine, 5th edition. Philadelphia: Saunders Elsevier; 2010. pp. 287-307.
12. Salem K, Mulji A, Lonn E. Echocardiographically guided pericardiocentesis—the gold standard for the management of pericardial effusion and cardiac tamponade. Can J Cardiol. 1999;15(11):1251-5.
13. Kennedy UM, Mahony NJ. A cadaveric study of complications associated with the subxiphoid and transthoracic approaches to emergency pericardiocentesis. Eur J Emerg Med. 2006;13(5):254-9.
14. Goldberg BB, Goodman GA, Clearfield HR. Evaluation of ascites by ultrasound. Radiology. 1970;96(1):15-22.
15. Wiese S, Mortensen C, Bendtsen F. Few complications after paracentesis in patient with cirrhosis and refractory ascites. Dan Med Bull. 2011;58(1):A4212.
16. Mercaldi CJ, Lanes SF. Ultrasound guidance decreases complications and improves the cost of care among patients undergoing thoracentesis and paracentesis. Chest. 2013;143(2):532-8.
17. De Gottardi A, Thévenot T, Spahr L, et al. Risk of complications after abdominal paracentesis in cirrhotic patients: a prospective study. Clin Gastroenterol Hepatol. 2009;7(8):906-9.
18. Politano AD, Hranjec T, Rosenberger LH, et al. Differences in morbidity and mortality with percutaneous versus open surgical drainage of postoperative intra-abdominal infections: a review of 686 cases. Am Surg. 2011;77(7):862-7.

19. Venkatesan AM, Kundu S, Sacks D, et al. Practice guidelines for adult antibiotic prophylaxis during vascular and interventional radiology procedures. Written by the Standards of Practice Committee for the Society of Interventional Radiology and Endorsed by the Cardiovascular Interventional Radiological Society of Europe and Canadian Interventional Radiology Association [corrected]. J Vasc Interv Radiol. 2010;21(11):1611-30.
20. Huang CJ, Pitt HA, Lipsett PA, et al. Pyogenic hepatic abscess. Changing trends over 42 years. Ann Surg. 1996;223(5):600-9.
21. Venara A, Carretier V, Lebigot J, et al. Technique and indications of percutaneous cholecystostomy in the management of cholecystitis in 2014. J Visc Surg. 2014;151(6):435-9.
22. Ralls PW, Colletti PM, Lapin SA, et al. Real-time sonography in suspected acute cholecystitis. Prospective evaluation of primary and secondary signs. Radiology. 1985;155(3):767-71.
23. Kadir S. Teaching Atlas of Interventional Radiology: Nonvascular Interventional Procedures. New York: Thieme; 2005.
24. Lorenz J, Thomas JL. Complications of percutaneous fluid drainage. Semin Intervent Radiol. 2006;23(2):194-204.

Chapter 11

Ultrasonography in Neurocritical Care

Gyaninder Pal Singh, Niraj Kumar

INTRODUCTION

Ultrasonography has emerged as a noninvasive, low-cost, accurate and fast tool that contributes to the diagnosis and monitoring of various clinical conditions. Today, ultrasonography has an application in almost every field of medicine including emergency medicine, operating rooms and critical care units. However, its use in neurological and neurosurgical population is still limited though it is expected that the application of ultrasonography for central nervous system (CNS) examination will increase in the years to come. At present, the important use of ultrasonography in neuro patient includes measurement of intracranial pressure (ICP), cerebral blood flow (CBF) and velocities, diagnosis of the intracranial mass lesion and midline shifts, and examination of pupils. Measurement of optic nerve sheath diameter (ONSD) with ocular ultrasound helps to diagnose the raised ICP at a very early phase which is particularly useful in uncooperative trauma patient where using other modalities, such as computed tomography (CT) is not feasible. Transcranial ultrasound is used to identify various structures inside the brain tissue, detect intracranial space occupying lesions, real time monitoring of cerebral blood flow velocity (CBFV), measure midline shift (MLS), etc. The diagnosis of brain death is a clinical finding, but in some situations, few ancillary tests are also performed to confirm the diagnosis. Transcranial sonography (TCS) is a noninvasive, safe, low cost and bedside technique used for the diagnosis of the cerebro-circulatory arrest (CCA) that precedes brain death (BD). Reverberant pattern (to and fro oscillating motion), systolic spikes, and disappearance of flow in cerebral vessels which was previously present are the accepted pattern for CCA. Moreover, ultrasound is a fast and practical tool for pupillary light reflex (PLR) assessment when the direct visual inspection is not possible either due to soft tissue damage, corneal opacity or ocular trauma.

MEASUREMENT OF OPTIC NERVE SHEATH DIAMETER

Raised ICP is an emergency condition that requires early diagnosis and therapeutic intervention. Measurement of ICP can be achieved with either invasive or noninvasive method of monitoring. Among the noninvasive method CT of the brain or ocular ultrasound are popular options. However, CT may not always be available or feasible in many circumstances.

It has been proposed that the ocular ultrasound evaluation of the ONSD is a safe and noninvasive method to detect the raised ICP.[1] Optic nerve ultrasonography is more useful in emergency services where other devices to measure ICP are not available or is not feasible. Ultrasound takes less time to perform when compared with other neuroimaging studies. Early ultrasound detection allows for prompt therapeutic action. Additionally, serial measurements can be done assessing response to treatment.[2-4] It also contributes to assessing the need for further invasive monitoring and transfer of the patient to another specialized center.

The eyes with ocular ultrasonography can frequently suggest disease states somewhere else in the body outside of the eye. The optic nerve is an extension of the CNS and it is enclosed by cerebral spinal fluid (CSF) and dura matter known as optic nerve sheath. The optic nerve is more distensible near the eyeball because it is only loosely attached to the dural sheath. Due to direct communication of the subarachnoid space between the brain and optic nerve, the pressure changes in the brain will also be reflected in the optic nerve and its sheath. ONSD measurement with ultrasound can diagnose the raised ICP at an early stage.[5] This is an easy, safe, noninvasive method and is especially useful in uncooperative trauma patient where using other modalities, such as CT is not feasible or not available.

Method of Optic Nerve Sheath Diameter Measurement

- A *linear probe* having a high frequency (5–10 MHz) is used to perform an ocular ultrasound exam. Depth is adjusted to visualize structures up to 5–6 cm deep.
- The patient is placed and examined in the supine position, and cooperative patients can be directed to maintain the midline position of the eye.
- Apply a copious amount of ultrasound gel over the upper closed eyelid.
- Place the transducer softly over the superior aspect of the closed eye avoiding too much of pressure (Fig. 1).[6,7]

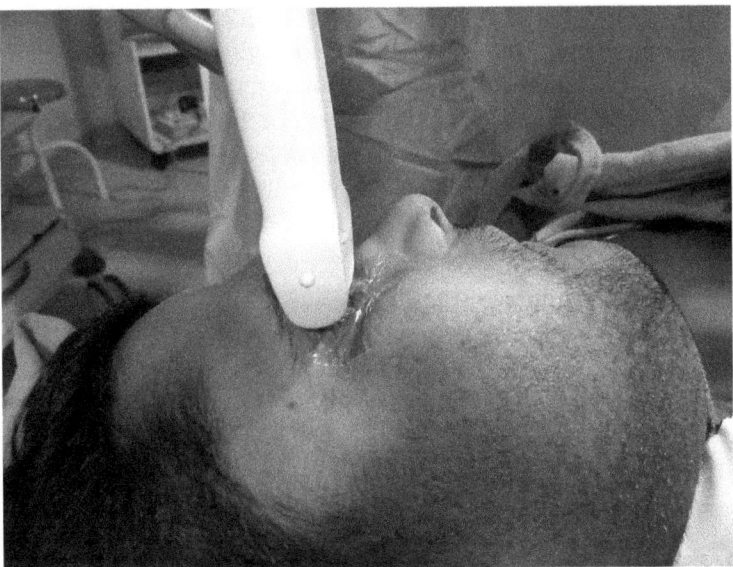

Fig. 1: Placement of probe over the upper eyelid.

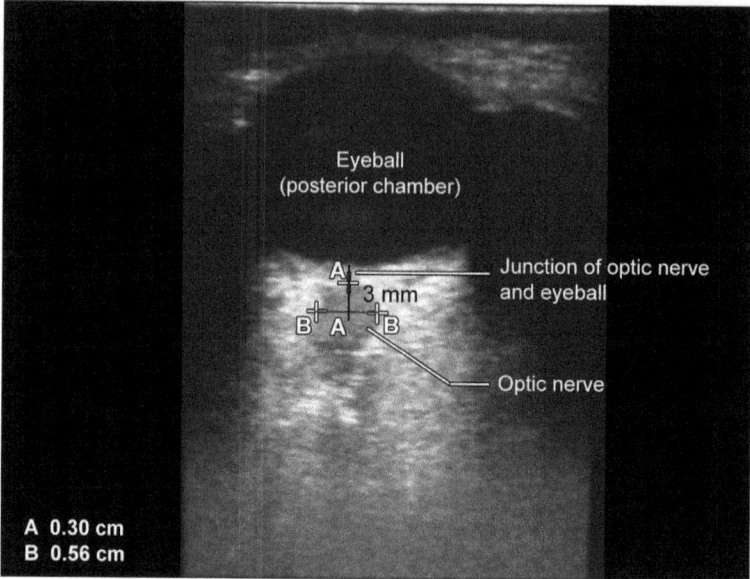

Fig. 2: Sonographic image of optic nerve and eyeball. For measurement of optic nerve sheath diameter (ONSD) a 3 mm line (A-A) is drawn from junction of eyeball and optic nerve. At this point another line is drawn perpendicular to the first line (B-B) to measure the ONSD.

- The probe is adjusted, so that cross-section of the globe, as well as optic nerve, can be seen.
- The optic nerve sheath visualized as a *tubular hypoechoic band*, shifting away from the eyeball.
- ONSD measurement is taken *3 mm* posterior to the globe.
- For the assessment of ONSD, two lines are drawn (Fig. 2):

1st line: From the junction of the optic nerve and the eyeball, a straight line is drawn 3 mm long (Fig. 2; A-A) that is the reference point for next measurement.

2nd line: At the 3 mm reference point, a second perpendicular line is drawn across the optic nerve, which provides ONSD measurement (Fig. 2; B-B).[3,4]

The normal value of ONSD in adults is 4.5 mm. There is a wide range of cut off values to denote raised ICP however, a value of more than 5.0 mm is usually considered to denote raised ICP.

Advantages of ONSD for monitoring ICP:
- Readily available
- Bedside
- Can be repeated
- Noninvasive
- Cost effective
- Results are easily reproducible
- High sensitivity and specificity
- Early indicator of raised ICP.

Contraindications for ONSD measurement:
- Orbital trauma
- Optic neuritis
- Optic nerve atrophy
- Orbital hematoma
- Glaucoma
- Hyperthyroidism with exophthalmos.

MEASUREMENT OF MIDLINE SHIFT OF BRAIN

Midline shift (MLS) of the cerebrum is frequently assessed in various neurological and neurosurgical conditions which indicate the severity of the condition and might indicate the need for surgical management in certain instances. With the help of TCS, it is possible to visualize the midline through the temporal bone window.[8]

Techniques for Measuring Midline Shift of Brain

There are two different methods for measuring MLS using ultrasonography:

1st technique: This technique was described by Seidel et al.[9]
- Initially, the third ventricle is identified and it is taken as a marker of midline for the brain.
- After that length of one side of the temporal bone and third ventricle is measured (A)
- On the contralateral side also, the same measurement is obtained (B)
- Finally, MLS can be calculated by the formula:

MLS = (A − B) ÷ 2

This technique has been widely accepted and there are reports which suggest a good correlation between CT and sonographic measurement of MLS.[10,11] However, in the patients in which decompressive craniectomy has already been done; measurement with this technique can be difficult and may induce bias.

2nd technique: This technique was described by Caricato et al.[12] This method is simple because it is measured by directly seeing the ultrasound image rather than applying any formulae to calculate the MLS. A curvilinear (convex) probe is used in the axial plane.
- First, the midline is identified, which is taken as the boundary between the two lateral ventricles (Fig. 3).
- Then localize the falx cerebri, from frontal to occipital area.
- Finally, MLS can be calculated by measuring the distance between the interventricular line and the falx cerebri. Measurement of MLS with this method has a good correlation with the measurement done by CT.

ASSESSING ADEQUACY OF CEREBRAL BLOOD FLOW WITH TRANSCRANIAL DOPPLER

Transcranial Doppler (TCD) technique is used to assess CBF and CBFV.

Traumatic brain injury (TBI): TCD can identify hemodynamic changes following TBI that helps in predicting the outcome with some degree of reliability. TBI usually leads to

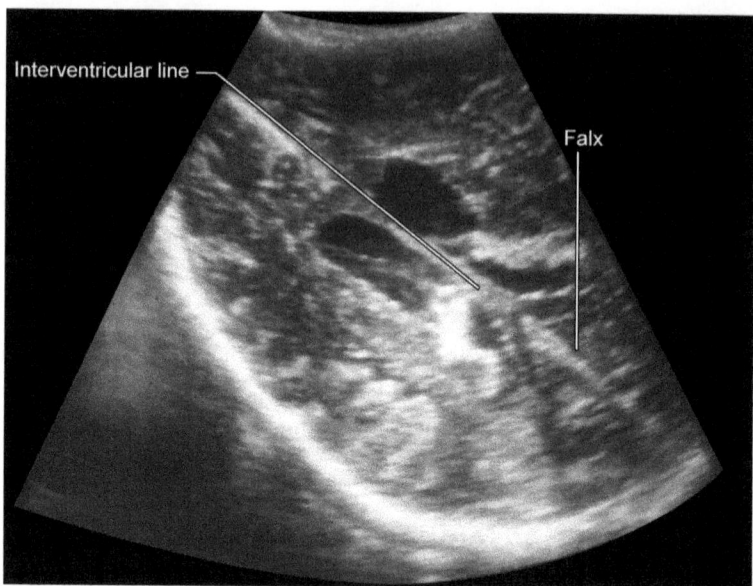

Fig. 3: Ultrasound image showing the interventricular line and falx cerebri.

variation in CBF, characterized by, hypoperfusion initially followed by hyperemia between day 1 and 3 and lastly vasospasm from day 4 to 15, ultimately resulting in increased ICP.[13]

Transcranial Doppler can identify these post traumatic cerebrovascular hemodynamic alterations in a noninvasive manner. There is also a direct correlation between cerebral hypoperfusion and the outcome after 6 months of TBI. TCD (a noninvasive technique) also provide similar information as provided by invasive means of cerebral blood flow measurement about the prognosis that may help in the management of TBI patients. [14-16]

A decreased velocity of blood flow in MCA is a predictor of poor outcome assessed within 72 hours of TBI.[17] In a similar manner, peak mean-CBFV is also an indicator of outcome after TBI.[18] The degree of vasospasm, especially that of basilar artery is also one aspect that is related to poor outcome in such patients.[19]

The sequence of event that occurs with the gradual rise in ICP are recognized by TCD, starting with the decrease in end diastolic velocity (EDV) to zeroing of EDV.[20] TCD is also useful in determining the cerebral perfusion pressure (CPP) in a noninvasive manner.[21] Despite the fact that TCD is used for the calculation of ICP and CPP, this method still need substantial evidence of support for its wide use.[14] TCD is more useful for assessing the dynamic changes in CPP rather than a single value.[22]

Subarachnoid hemorrhage (SAH): Intracranial vasospasm (VSP) is the vasoconstriction of intracranial arteries after the rupture of cerebral aneurysm leading to decrease in cerebral blood flow.[23] It is estimated that in case of SAH, nearly 70% of the patient develop vasospasm (seen on angiography). The vasospasm is usually absent during the first 3 days of SAH. It starts peaking between 6 days and 12 days and resolving between 15 days and 30 days of onset of hemorrhage.[24] Delayed cerebral ischemia (DCI) develops as a result of vasospasm, seen in about 25% of the patients with SAH.[25,26] Although angiography is considered to be

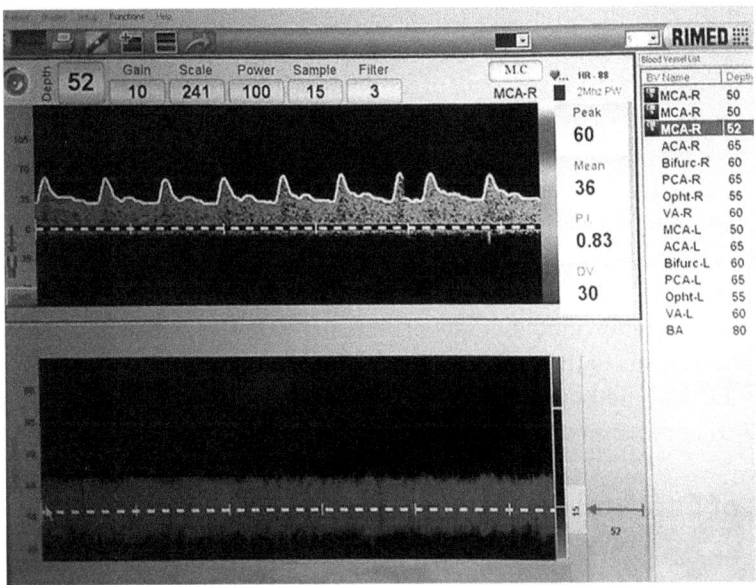

Fig. 4: Transcranial Doppler (TCD) with normal flow in middle cerebral artery (MCA).

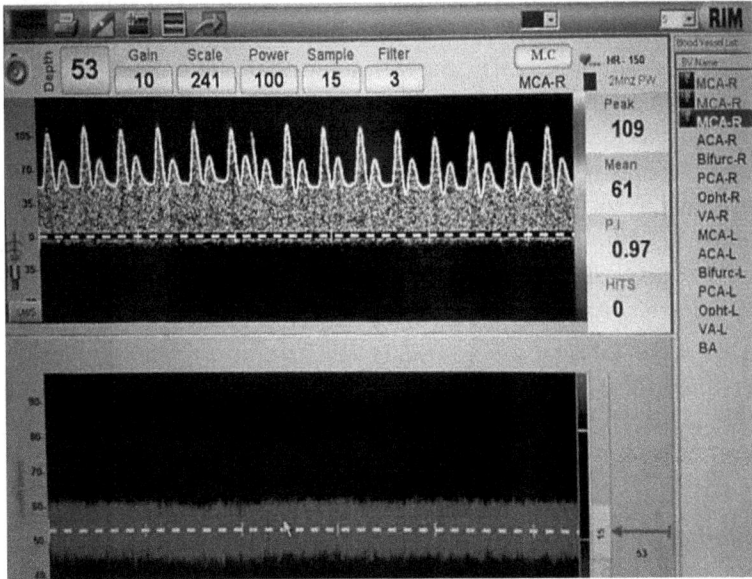

Fig. 5: Transcranial Doppler (TCD) showing vasospasm.

gold standard (excellent specificity) for the diagnosis of vasospasm following SAH it is an invasive procedure and cannot be used for day to day or continuous monitoring.[27]

Transcranial Doppler is an established method to diagnose and monitor vasospasm. It is noninvasive, relatively inexpensive and can be repeated several times for this purpose (Figs. 4 and 5).[28] TCD measurement should be done on a daily basis for the early diagnose of VSP so that prompt therapeutic strategies can be implemented in patients with SAH.[29] TCD can also monitor the effectiveness of an intervention in SAH patients. TCD has got very good sensitivity for identifying the vasospasm of the cerebral blood vessels especially of MCA.

In a systematic review by Lysakowski C et al., comparing the accuracy of TCD to that of angiography, it was found that TCD has 99% specificity, 67% sensitivity, and positive predictive value of 97% for the diagnosis of vasospasm in MCA.[30] Diagnostic accuracy of TCD is lower in a case of the basilar artery (BA) as compared to MCA to detect vasospasm.[31] For VSP detection of anterior cerebral artery (ACA) and posterior cerebral artery (PCA) sensitivity is further very low with TCD.

It is important to understand that rate of increase of mean CBFV can predict the development of DCI which leads to the poor outcome of the patient.[32] During the course of management, TCD is a useful method to detect and monitor the development of vasospasm in the patient with SAH as recommended by American Heart Association.[33]

Diagnosis of Cerebral Circulatory Arrest

The characteristic feature of BD is permanent loss of brain function. The diagnosis of BD is primarily done by clinical findings, though confirmatory tests are sometimes required.[34] The main pathophysiology involved for the occurrence of BD is raised intracranial pressure and a loss in cerebral autoregulation. As a result of this, cerebral blood flow is affected leading to cerebral circulatory arrest.[35]

The diagnosis of BD is confirmed by the various ancillary tests. The TCD is also an accepted method for this purpose and is a safe, noninvasive, low cost, and bedside technique.[36] TCD was first utilized in 1987 for confirmation of CCA.

Transcranial Doppler requires equipment having the pulsed-Doppler probe of 2 MHz. TCD and transcranial color Doppler (TCCD) both can be used to confirm of CCA. In TCCD, both the B-mode and the pulsed wave mode are combined to form the image of brain parenchyma as well as vascular structures of the brain. Here operators have the advantage to correct insonation angle also.

Prerequisites: To avoid false positives results of cerebral circulatory arrest, it is recommended that patient must be hemodynamically stable (systolic blood pressures more than 90 or mean arterial blood pressures above 60 mm Hg.) and there should be no signs of hypoxemia, hypercarbia, hypothermia or metabolic derangements.[37,38]

Interpretation: There are three patterns which are generally accepted for CCA—
1. *Reverberant flow or "To and Fro" oscillating flow or alternating flow:* It is characterized by equalization of antegrade systolic flow and retrograde diastolic flow in the cerebral blood vessels resulting in stoppage of cerebral perfusion. (Fig. 6).[39]
2. *Systolic spikes:* Short systolic peaks (duration less than 200 msec. with < 50 cm/s of systolic velocity), are seen when ICP equals systolic blood pressure (Fig. 7). Sharma et al. showed that when the clinical diagnosis is doubtful TCD can be a useful first line supplementary test to confirm BD.[40]
3. *Lack of signal in a previously detected blood flow signal:* When the analyzed vessels show no flow signal that previously showed an acceptable blood flow.

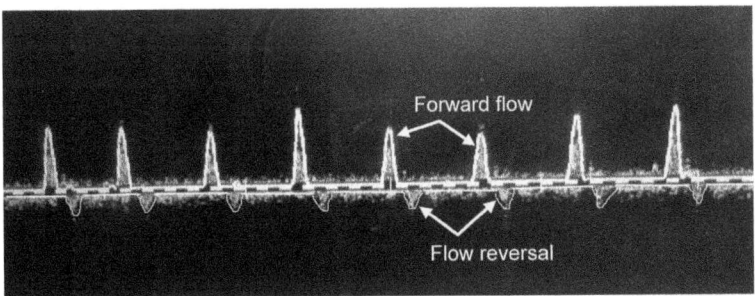

Fig. 6: Reverberating pattern.

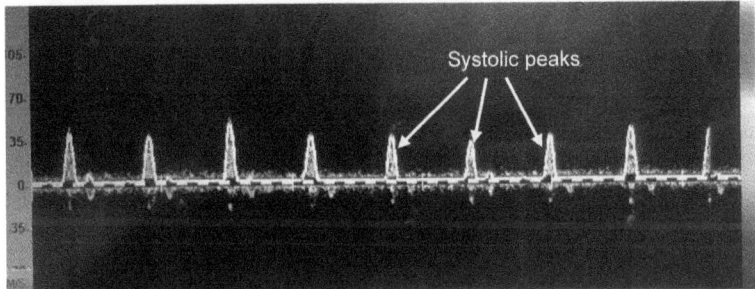

Fig. 7: Systolic spikes.

ASSESSING PUPILS IN PATIENTS

Assessing pupils is difficult in patients who have a grossly swollen eyelid or soft tissue damage and the eyelids cannot be opened. Ultrasonography can be used to assess the pupils in such patients.

Pupillary Light Reflex

Pupillary light reflex is a crucial neurological test with several clinical implications.[41,42] The consensual pupillary reflex is used to examine the integrity of the retina, optic nerve, portion of the midbrain, and the oculomotor nerve. There are several methods for assessment of PLR but most of them require expertise and availability especially in emergency settings.

Ultrasound imaging can be an alternative for assessment of PLR and is especially indicated when the visual assessment is not possible to perform, such as in case of soft tissue damage where eyelid cannot be opened. Current ultrasound machines along with high-frequency linear probes offer high-quality ophthalmic images.[43,44] Ultrasound assessment of PLR is a fast, safe, dynamic and practical method which may also act as a possible adjunct to physical examination.

Technique: According to Sargsyan et al. there are two approaches for obtaining PLR using ultrasound viz. superior and inferior.[45]

In the superior approach (Fig. 8):
- A high frequency (5–12 MHz) linear probe is used.
- In the supine position, the patient is directed to gaze downward towards the feet.

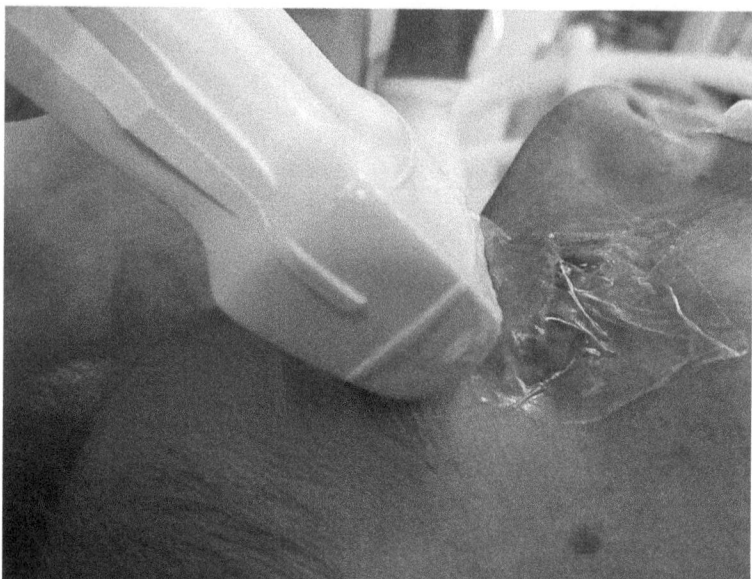

Fig. 8: Placement of probe over upper eyelid to obtain the image of pupil.

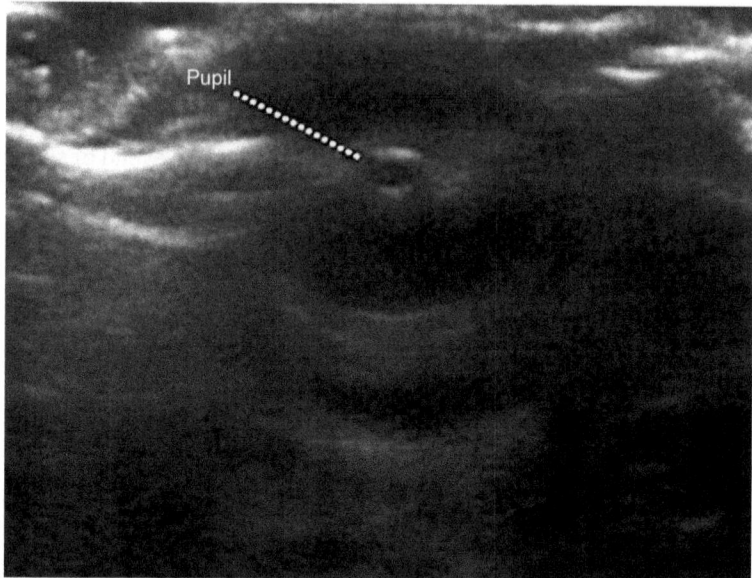

Fig. 9: B-mode image of ultrasonography showing image of pupil.

- A copious amount of ultrasound gel is placed over the upper eyelid.
- The probe is positioned transversely on the superior aspect of the orbit.
- The probe is adjusted and angled such that it is aligned with the plane of the iris.
- After obtaining the anechoic pupillary image (Fig. 9) consensual PLR is obtained by stimulation of light in the contra lateral eye.

The *inferior approach* is similar to superior approach except the fact that probe is placed on the lower eyelid and tilted upward to reach the plane of the iris.

KEY POINTS

- Ultrasound of the CNS nowadays is performed by nonradiologist for various applications like measurement of ONSD, assessment of MLS, CBF, diagnosis of CCA, and assessment of pupil because of its low-cost, easy availability, accuracy, bedside availability and easy reproducibility.
- Even today it is an underestimated imaging device that necessitates a more dispersion.
- Among the various ancillary test, TCS is also a method for the diagnosis of CCA and BD.
- The modern ultrasound system is a quick and practical tool for papillary assessment in cases where direct visual access in difficult.
- Transcranial ultrasound is extremely subjective technique and skill of the clinician is an important determinant for the correct assessment.

REFERENCES

1. Qayyum H, Ramlakhan S. Can ocular ultrasound predict intracranial hypertension? A pilot diagnostic accuracy evaluation in a UK emergency department. Eur J Emerg Med. 2013;20:910-7.
2. Strumwasser A, Kwan RO, Yeung L, et al. Sonographic optic nerve sheath diameter as an estimate of intracranial pressure in adult trauma. J Surg Res. 2011;170:265-71.
3. Dubourg J, Javouhey E, Geeraerts T, et al. Ultrasonography of optic nerve sheath diameter for detection of raised intracranial pressure: a systematic review and meta-analysis. Intensive Care Med. 2011;37:1059-68.
4. Dubost C, Geeraerts T. Possible pitfalls when measuring the optic nerve sheath with sonography. J Surg Res. 2012;173:e43-45.
5. Fledelius HC. Ultrasound in ophthalmology. Ultrasound Med Biol. 1997;23:365-75.
6. Rajajee V, Vanaman M, Fletcher JJ, et al. Optic nerve ultrasound for the detection of raised intracranial pressure. Neurocrit Care. 2011;15:506-15.
7. Moretti R, Pizzi B. Ultrasonography of the optic nerve in neurocritically ill patients. Acta Anaesthesiol Scand. 2011;55:644-52.
8. Rincon F. Bedside transcranial sonography: a promising tool for the neurointensivist. Crit Care Med. 2012;40:1969-70.
9. Seidel G, Kaps M, Gerriets T, et al. Evaluation of the ventricular system in adults by transcranial duplex sonography. J Neuroimaging. 1995;5:105-8.
10. Stolz E, Gerriets T, Fiss I, et al. Comparison of transcranial color-coded duplex sonography and cranial CT measurements for determining third ventricle midline shift in space-occupying stroke. Am J Neuroradiol. 1999;20:1567-71.
11. Llompart Pou JA, Abadal Centellas JM, Palmer Sans M, et al. Monitoring midline shift by transcranial color-coded sonography in traumatic brain injury. A comparison with cranial computerized tomography. Intensive Care Med. 2004;30:1672-5.
12. Caricato A, Mignani V, Bocci MG, et al. Usefulness of transcranial echography in patients with decompressive craniectomy: a comparison with computed tomography scan. Crit Care Med. 2012;40:1745-52.
13. Martin NA, Patwardhan RV, Alexander MJ, et al. Characterization of cerebral hemodynamic phases following severe head trauma: hypoperfusion, hyperemia, and vasospasm. J Neurosurg. 1997;87:9-19.

14. White H, Venkatesh B. Applications of transcranial Doppler in the ICU: a review. Intensive Care Med. 2006;32:981-94.
15. Sloan MA, Alexandrov AV, Tegeler CH, et al. Assessment: transcranial Doppler ultrasonography: report of the Therapeutics and Technology Assessment Subcommittee of the American Academy of Neurology. Neurology. 2004;62:1468-81.
16. Jaggi JL, Obrist WD, Gennarelli TA, et al. Relationship of early cerebral blood flow and metabolism to outcome in acute head injury. J Neurosurg. 1990;72:176-82.
17. van Santbrink H, Schouten JW, Steyerberg EW, et al. Serial transcranial Doppler measurements in traumatic brain injury with special focus on the early posttraumatic period. Acta Neurochir (Wien). 2002;144:1141-9.
18. Zurynski YA, Dorsch NW, Fearnside MR. Incidence and effects of increased cerebral blood flow velocity after severe head injury: a transcranial Doppler ultrasound study II. Effect of vasospasm and hyperemia on outcome. J Neurol Sci. 1995;134:41-6.
19. Soustiel JF, Shik V, Feinsod M. Basilar vasospasm following spontaneous and traumatic subarachnoid haemorrhage: clinical implications. Acta Neurochir (Wien). 2002;144:137-44.
20. Hassler W, Steinmetz H, Gawlowski J. Transcranial Doppler ultrasonography in raised intracranial pressure and in intracranial circulatory arrest. J Neurosurg. 1988;68:745-51.
21. Czosnyka M, Matta BF, Smielewski P, et al. Cerebral perfusion pressure in head-injured patients: a noninvasive assessment using transcranial Doppler ultrasonography. J Neurosurg. 1998;88:802-8.
22. Saqqur M, Zygun D, Demchuk A. Role of transcranial Doppler in neurocritical care. Crit Care Med. 2007;35(5 Suppl):S216-223.
23. Arnolds BJ, von Reutern GM. Transcranial Doppler sonography. Examination technique and normal reference values. Ultrasound Med Biol. 1986;12:115-23.
24. Biller J, Godersky JC, Adams HP. Management of aneurysmal subarachnoid hemorrhage. Stroke J Cereb Circ. 1988;19:1300-5.
25. Velat GJ, Kimball MM, Mocco JD, et al. Vasospasm after aneurysmal subarachnoid hemorrhage: review of randomized controlled trials and meta-analyses in the literature. World Neurosurg. 2011;76:446-54.
26. Dorsch N. A clinical review of cerebral vasospasm and delayed ischaemia following aneurysm rupture. Acta Neurochir Suppl. 2011;110(Pt 1):5-6.
27. Topcuoglu MA, Pryor JC, Ogilvy CS, et al. Cerebral vasospasm following subarachnoid hemorrhage. Curr Treat Options Cardiovasc Med. 2002;4:373-84.
28. Bederson JB, Connolly ES, Batjer HH, et al. Guidelines for the management of aneurysmal subarachnoid hemorrhage: a statement for healthcare professionals from a special writing group of the Stroke Council, American Heart Association. Stroke J Cereb Circ. 2009;40:994-1025.
29. McGirt MJ, Blessing RP, Goldstein LB. Transcranial Doppler monitoring and clinical decision-making after subarachnoid hemorrhage. J Stroke Cerebrovasc Dis Off J Natl Stroke Assoc. 2003;12:88-92.
30. Lysakowski C, Walder B, Costanza MC, et al. Transcranial Doppler versus angiography in patients with vasospasm due to a ruptured cerebral aneurysm: A systematic review. Stroke J Cereb Circ. 2001;32:2292-8.
31. Skjelland M, Krohg-Sørensen K, Tennøe B, et al. Cerebral microemboli and brain injury during carotid artery endarterectomy and stenting. Stroke J Cereb Circ. 2009;40:230-4.
32. Tsivgoulis G, Alexandrov AV, Sloan MA. Advances in transcranial Doppler ultrasonography. Curr Neurol Neurosci Rep. 2009;9:46-54.
33. Connolly ES, Rabinstein AA, Carhuapoma JR, et al. Guidelines for the management of aneurysmal subarachnoid hemorrhage: a guideline for healthcare professionals from the American Heart Association/American Stroke Association. Stroke J Cereb Circ. 2012;43:1711-37.
34. Wijdicks EFM. Brain death worldwide: accepted fact but no global consensus in diagnostic criteria. Neurology. 2002;58(1):20-5.

35. Günther A. Determination of Brain Death: An Overview with a Special Emphasis on New Ultrasound Techniques for Confirmatory Testing. Open Crit Care Med J. 20117;4:35-43.
36. Ropper AH, Kehne SM, Wechsler L. Transcranial Doppler in brain death. Neurology. 1987;37:1733-5.
37. Ducrocq X, Hassler W, Moritake K, et al. Consensus opinion on diagnosis of cerebral circulatory arrest using Doppler-sonography: Task Force Group on cerebral death of the Neurosonology Research Group of the World Federation of Neurology. J Neurol Sci. 1998;159:145-50.
38. Segura T, Calleja S, Irimia P, et al. Recommendations for the use of transcranial Doppler ultrasonography to determine the existence of cerebral circulatory arrest as diagnostic support for brain death. Rev Neurosci. 2009;20:251-9.
39. Hassler W, Steinmetz H, Pirschel J. Transcranial Doppler study of intracranial circulatory arrest. J Neurosurg. 1989;71:195-201.
40. Sharma D, Souter MJ, Moore AE, et al. Clinical experience with transcranial Doppler ultrasonography as a confirmatory test for brain death: a retrospective analysis. Neurocrit Care. 2011;14:370-6.
41. Kardon R. Pupillary light reflex. Curr Opin Ophthalmol. 1995;6:20-6.
42. Schreiber MA, Aoki N, Scott BG, et al. Determinants of mortality in patients with severe blunt head injury. Arch Surg Chic Ill 1960. 2002;137:285-90.
43. Blaivas M, Theodoro D, Sierzenski PR. A study of bedside ocular ultrasonography in the emergency department. Acad Emerg Med. 2002;9:791-9.
44. Bedi DG, Gombos DS, Ng CS, et al. Sonography of the eye. Am J Roentgenol. 2006;187:1061-72.
45. Sargsyan AE, Hamilton DR, Melton SL, et al. Ultrasonic evaluation of pupillary light reflex. Crit Ultrasound J. 2009;1:53-7.

Chapter 12

Focused Assessment Sonography of Trauma

Kapil Dev Soni

INTRODUCTION

The use of sonography as a modality in trauma began for assisting triage in hypotensive blunt abdominal or thoracic trauma patients. The purpose was to rapidly identify free fluid, either intraperitoneal or in pericardial cavity. Later on, it was incorporated in the regular examinations and is being used for normotensive blunt torso trauma as well.

Focused assessment sonography of trauma (FAST), formally, termed as focused abdominal sonography for trauma—initial application limited to assessment of abdominal fluid—later on, its use extended to identify fluid in pericardial cavity and recently to identify fluid in pleural cavity, pneumothorax, inferior vena cava (IVC) assessment for fluid status, known as extended focused assessment with sonography for trauma (EFAST).

Focused assessment sonography of trauma is an adjunct to primary survey in trauma resuscitation and used primarily for evaluation of circulation part. (C in ABCDE-ATLS). Patient can lose blood chiefly in five areas following an injury: (1) exterior, (2) pleural cavity, (3) abdominal cavity, (4) retroperitoneum, and (5) around long bones. Since, the signs of abdominal bleed may not be apparent on clinical examination and can often be missed even by an experience person, FAST serves as an excellent noninvasive modality to identify hemoperitoneum.

It has largely replaced use of diagnostic peritoneal lavage (DPL) in the primary survey of ATLS protocol for evaluation of intraperitoneal bleed.

The differences between the FAST and DPL have been shown in Box 1.

The FAST identifies only the fluid within the cavity. It cannot differentiate fluid from blood. Since, the modality is primarily used in trauma, the fluid is assumed to be blood unless

Box 1: Differences between the focused assessment sonography of trauma (FAST) and diagnostic peritoneal lavage (DPL).

- FAST is noninvasive
- It can be repeated multiple times
- Shorter learning curve
- It can be performed rapidly
- Only identifies fluid
- Do not locate the source

- DPL is invasive
- It can be performed only once
- Need some expertise
- Takes time
- Highly sensitive to detect
- Intraperitoneal bowl injury

proved otherwise. It is poor in localizing source of intraperitoneal bleed. The reported sensitivity and specificity and of FAST varies from 70–99%.[1-3]

Literature review: There has been a considerable debate centered around the use of FAST protocol in trauma care pathways. Does it improve outcomes? Is it better than CT scan? A recent Cochrane review found that US-based trauma pathways had no negative impact on overall survival or morbidity in comparison to non-US-based trauma pathways. However, there exists a significant heterogeneity in the trials and the results of review are best exploratory. It is unlikely that a randomized controlled trial (RCT) would ever be conducted between US-based trauma management and a non-US-based trauma care, since its use has become a standard of care worldwide.[4] Another controversy is about the operator. Is the use by non-radiologist is safe? In a study of 200 unstable patients, authors compared diagnostic accuracy of emergency medicine residents (EMRs) and radiology residents (RRs) in performing FAST. The EMRs can performed sonography on trauma patients as successfully as RRs with equivalent accuracy (94%) in both groups.[5] Similar findings were reported from Australian trauma center where they prospectively analyzed the accuracy of FAST performed be non-radiologist. They found high sensitivity, specificity, positive and negative predictive values for FAST—78%, 97%, 91%, and 93%.[1] Is the FAST exam reliable across different cohorts of patients? In a retrospective study, authors studied the accuracy in severely injured patients based on injury severity score (ISS) scoring. They found decreased performance.[6]

TECHNIQUE

Essentially, the FAST identifies fluid in four quadrants:
1. Right upper quadrant—Morrison's pouch, perihepatic, subphrenic space.
2. Left upper quadrant—splenorenal recess, subphrenic space.
3. Suprapubic—fluid around the bladder, rectovesicle pouch (male), rectouterine pouch (females).
4. Subxiphoid—around the heart-pericardial cavity.
    - *Selection of US probe*: Ideally the low frequency probes are warranted and convex probe is the desired one however, one may use phase array as well.
    - *The direction of the pointer*: The mark on the probe is toward patients right or toward head end (cephalad) when scanning in coronal plane, it corresponds to mark on the US screen which is on the right.
    - *Patient's position*: In severe blunt torso trauma with shock, it is desirable to keep patient supine and to avoid unwanted movements. In this position, intraperitoneal fluid tends to accumulate in dependent regions: hepatorenal pouch (Morrison's space) or between the liver and diaphragm (subphrenic space) in right upper quadrant view, between spleen and diaphragm (subphrenic space) or between spleen and kidney (splenorenal) in left upper quadrant view and around the bladder (rectovesicle, rectouterine) in suprapubic view. One can identify fluid around the heart (pericardium) in subxiphoid view even when the patient is supine.

- *Operator*: Although the modality is operator dependent, it is associated with steep learning curve. The key is to obtain good images. If the images are not ideal or perfect interpretation would be faulty. One who is performing scanning he should be well-versed with the basics of ultrasound, image acquisition, and probes.
- *The landmarks*:
  - For right upper quadrant is the liver.
  - For left upper quadrant is the spleen.
  - For suprapubic view it is the urinary bladder.
  - For pericardial fluid it is the heart.
- *Images*: Ultrasound produces gray and black images depending upon the tissue characteristics and surrounding. Fluid on ultrasound appears hypoechoic or an echoic whereas surrounding structures may vary depending upon the tissue density and fat.
  - *For right upper quadrant view*: One sees liver as the probe is placed in coronal plane at midaxillary line between the 8th and 11th ribs, pointing toward posterior axillary line. Normally, there is clear hyperechoic interface between liver and kidney. The Morrison's space looks obliterated. If there is any fluid (blood) one sees anechoic or hypoechoic shadow as rim between liver and kidney. It is utmost important to scans above the liver and kidney as much as possible so that even minimal fluid could be picked up in subphrenic space or in right paracolic gutter.

    A common artifact which confounds the view could be rib shadow. To differentiate it from true bleed, it is important to remember that rib shadow move with respiration whereas fluid remains stationary (Figs. 1 and 2).
  - *Left upper quadrant view*: The left kidney is more posterior and superior. The probe is placed in the posterior axillary line between the 8th and 11th ribs, pointing toward

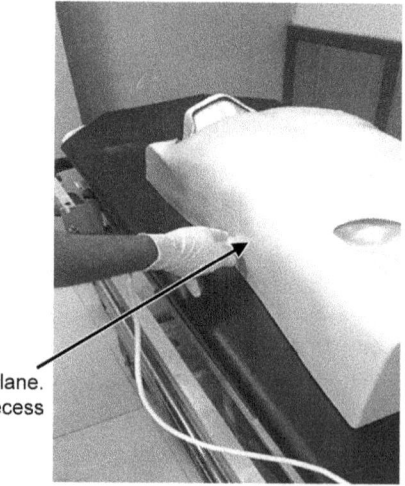

Probe is placed in coronal plane.
Visualizing hepatorenal recess

Fig. 1: Right upper quadrant view.

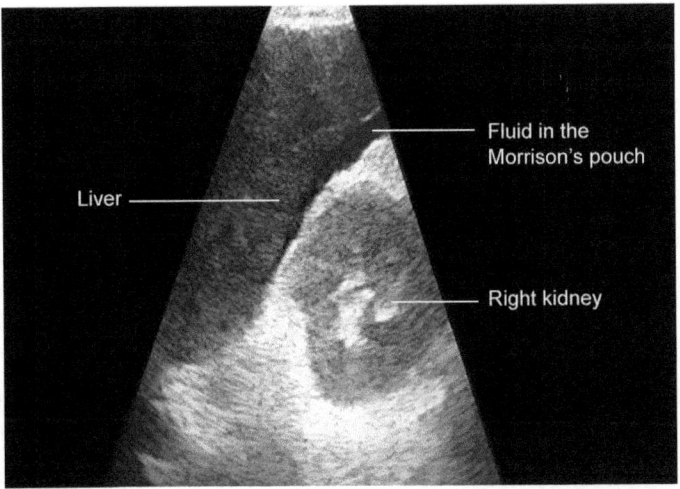

Fig. 2: Fast positive in right upper quadrant view.

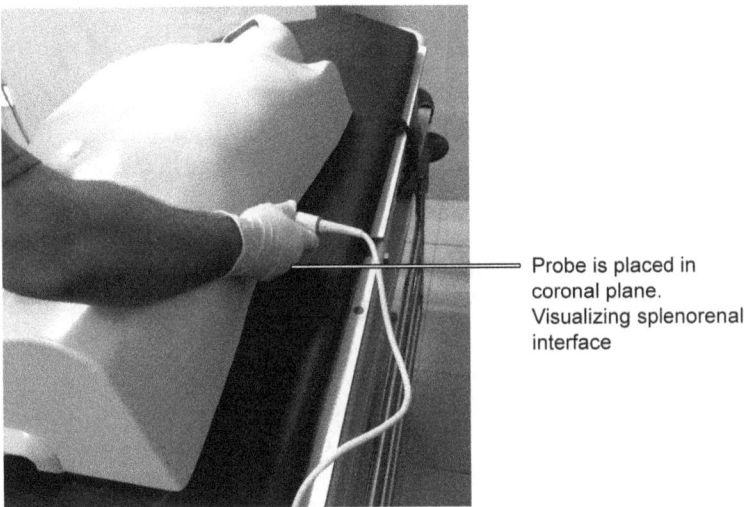

Fig. 3: Left upper quadrant view.

mid-axillary line. The mark on the pointer points cephalad in coronal plane. Often, the fluid gets accumulates between spleen and diaphragm (subphrenic space) rather than between spleen and kidney (splenorenal recess). Therefore, one must scan the subphrenic space thoroughly. Fluid will be anechoic or hypoechoic on imaging. One needs to constantly adjust gain to obtain good contrast and images (Figs. 3 and 4).
- *Suprapubic view:* The fluid around the bladder can be scan both in transverse plane and in sagittal plane. The probe is placed above pubic symphysis pointing

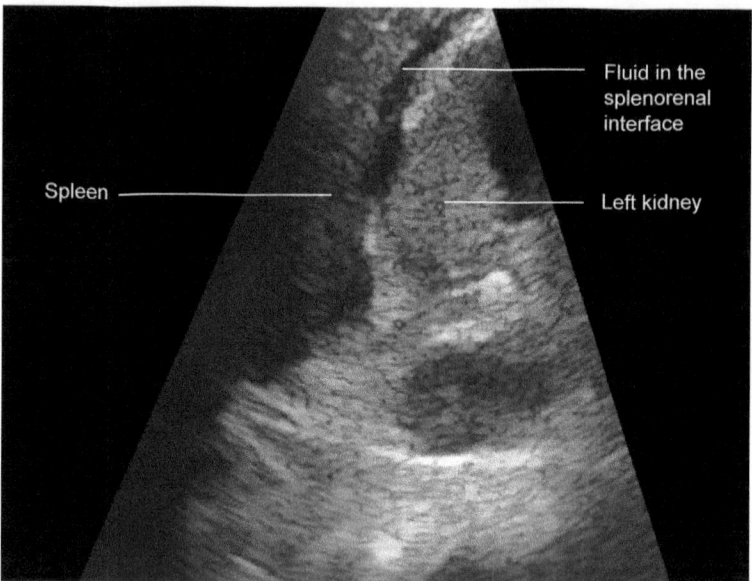

Fig. 4: Fast positive in left upper quadrant view.

posteriorly. The filled bladder appears rectangular in transverse view and triangular in sagittal. One must differentiate fluid in the bladder and around it. A filled urinary bladder is used as an acoustic window to identify fluid in rectovesical space in males and in rectouterine space in females. It is desirable to obtain this view before the Foley's is placed in an injured victim. Tilting the transducer side-to-side provides complete assessment of pelvic cavity. The fluid within the bladder and around appears hypoechoic to an echoic in consistency and signifies hemoperitoneum (Figs. 5 and 6).
- *Subxiphoid view:* The purpose of this view is to rule out cardiac tamponade as a possible cause of hemodynamic instability. The probe is placed below the xiphoid pointing toward left shoulder. One sees the four chambers of heart. Since the right ventricle is anterior, the chamber close to probe is right chamber whereas the chamber faraway would be the left ventricle. The liver is used as an acoustic window. It is desirable one must be able to differentiate fluid between pericardium and epicardium and fluid between pericardium and pleura. The fluid between pericardium and ventricle is the pericardial fluid whereas fluid between pericardium and pleura is pleural effusion. Fluid appears as anechoic and hypoechoic similar to fluid present within the ventricles. One must be cautious not to misinterpret pericardial fat from pericardial effusion. The fluids tends to accumulate first in most dependent portion that is posterior to left ventricle and with increasing volume it encircles the heart. However, it is the rapidity of fluid collection which determines the severity of cardiac tamponade rather than the volume. A collapsing right ventricle and right atrium signifies impending shock (Figs. 7 and 8).

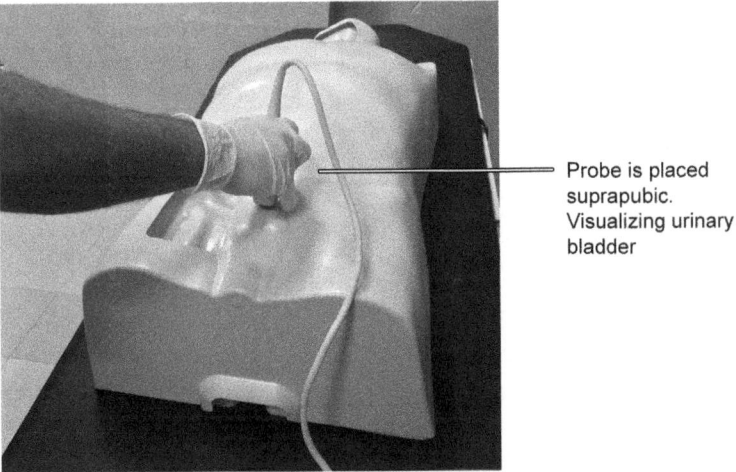

Fig. 5: Suprapubic view.

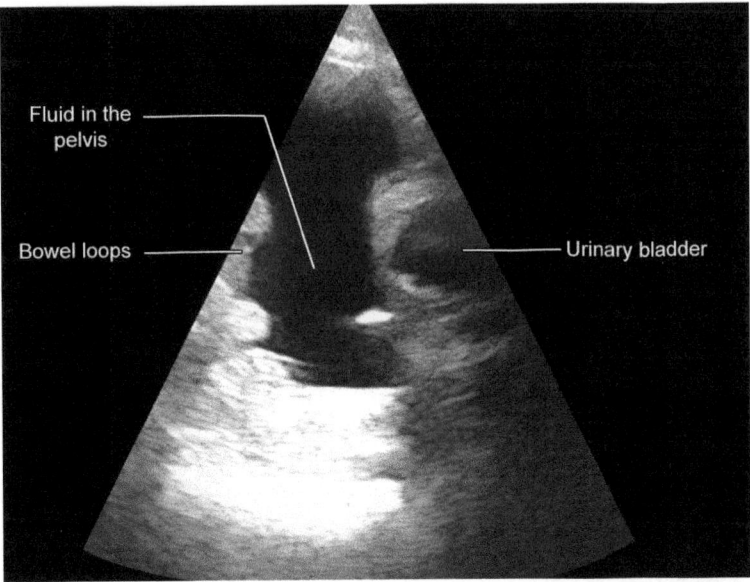

Fig. 6: Fast positive in suprapubic view.

CLINICAL APPLICATION

Focused assessment sonography of trauma plays a crucial role in management of trauma patients coming with hemodynamic instability. However, FAST positive does not mean that patient requires exploratory laparotomy. The decision of laparotomy depends upon the

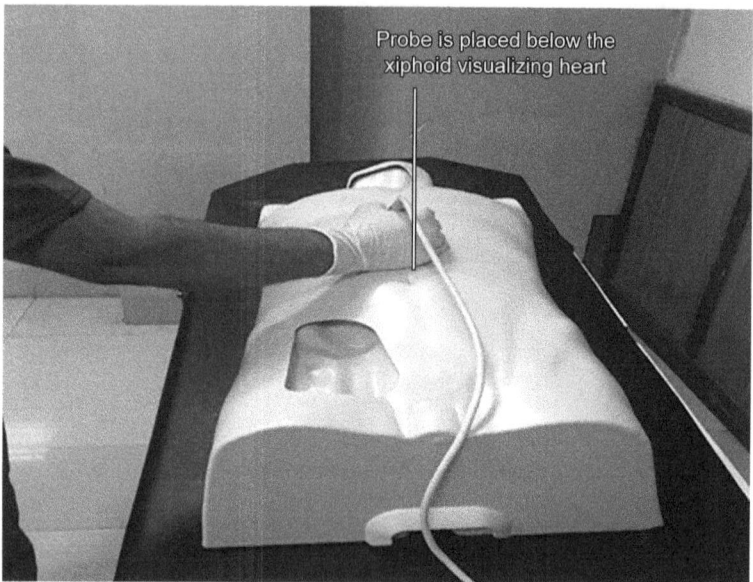

Fig. 7: Subxiphoid view.

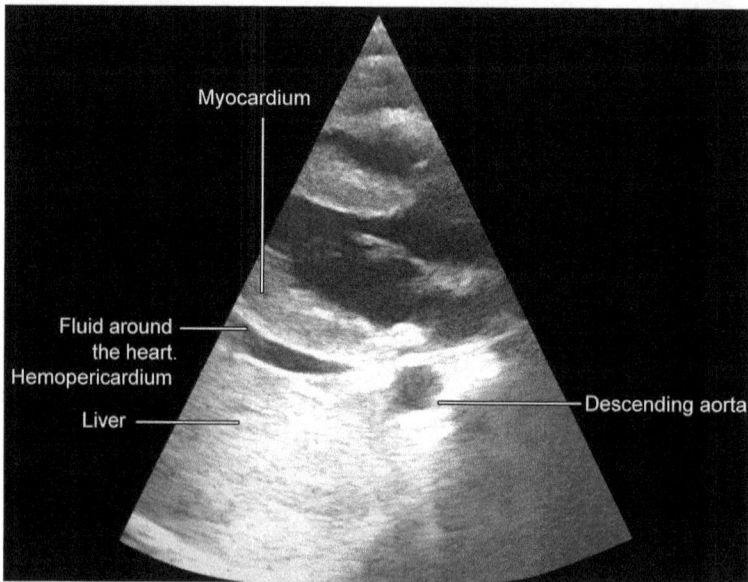

Fig. 8: Fast positive in subxiphoid view.

hemodynamic instability. If the patient is FAST positive and responder, one can wait and complete primary survey and secondary survey. If the patient is FAST positive and either transient responder or nonresponder after fluid challenge, an urgent laparotomy is warranted without the need for further investigations.

EFAST: The utility of ultrasound in trauma continues to evolve. With increasing use, the potential exist for wider application in identifying cause of hemodynamic instability. This rapidity and ease has enabled to use it for evaluation of fluid in pleural cavity, pneumothorax and IVC assessment for fluid status. This extended use is termed as EFAST or extended FAST.[7]

LIMITATIONS

As the modality is operator dependent, it needs some experience to convincingly identify fluid. Small amount of fluid may be missed by novices therefore it is advisable to perform initial scans with supervision from experts and once some positive scans are identified correctly one may conduct FAST scan independently.

As we know that air is the enemy of ultrasound waves and the sound waves do not traverse in air, any air column present between the skin and subcutaneous tissue (surgical emphysema) will make the examination futile.

Also, if bowl loops are filled with air as in case of bowel distension one may find difficulty in obtaining good quality images and interpretation.

Since, fluid can be present at any dependent areas in hemoperitoneum, it is often impossible to identify source of bleeding with certainty. One may find fluid in right upper quadrant even in case of splenic laceration.

The amount of bleeding is also impossible to quantify on FAST scan.

As previously discussed, the USG is not the ideal modality to identify bowl injury. DPL is more sensitive for this.

REFERENCES

1. Hsu JM, Joseph AP, Tarlinton LJ, et al. The accuracy of focused assessment with sonography in trauma (FAST) in blunt trauma patients: experience of an Australian major trauma service. Injury. 2007;38(1):71-5.
2. Schnüriger B, Kilz J, Inderbitzin D, et al. The accuracy of FAST in relation to grade of solid organ injuries: A retrospective analysis of 226 trauma patients with liver or splenic lesion. BMC Med Imaging. 2009;9:3.
3. Fleming S, Bird R, Ratnasingham K, et al. Accuracy of FAST scan in blunt abdominal trauma in a major London trauma centre. 2012;10(9):470-4.
4. Stengel D, Bauwens K, Rademacher G, et al. Emergency ultrasound-based algorithms for diagnosing blunt abdominal trauma. Cochrane Database Syst Rev. 2013;(7):CD004446.
5. Arhami Dolatabadi A, Amini A, Hatamabadi H, et al. Comparison of the accuracy and reproducibility of focused abdominal sonography for trauma performed by emergency medicine and radiology residents. Ultrasound Med Biol. 2014;40(7):1476-82.
6. Becker A, Lin G, McKenney MG, et al. Is the FAST exam reliable in severely injured patients? Injury. 2010;41(5):479-83.
7. Flato UA, Guimarães HP, Lopes RD, et al. Usefulness of Extended-FAST (EFAST-Extended Focused Assessment with Sonography for Trauma) in critical care setting. Rev Bras Ter Intensiva. 2010;22(3):291-9.

Chapter 13

Abdominal Ultrasonography in the Intensive Care Unit

Riddhi Kundu, Puneet Khanna, Manisha Jana

INTRODUCTION

Point-of-care bedside ultrasonography (USG) increasingly available to the intensivists provides useful information that can facilitate diagnosis, guide important therapeutic decisions, and alter management. The practical applications of ultrasonography are no longer the sole domain of the radiologist. While the radiologist can always be consulted for a more detailed evaluation of structures, basic appreciation of common anomalies by the intensivist can alter management, save undue delay in logistics, and impact outcome. Therefore, it becomes imperative that the modern day intensivist is well-versed in ultrasonographic imaging of the heart and great vessels, lungs, diaphragm, abdomen, and extremities. The following review aims to focus on the practical aspects of abdominal USG that should be a part of armamentarium of every physician working in the intensive care unit (ICU).

The intensivist uses abdominal sonography to detect intra-abdominal free-fluid collections, detect pathologies of the gallbladder, kidneys, and urinary bladder, image major vascular structures, and detect intra-abdominal emergencies. In addition, USG can help in guiding safe performance of therapeutic procedures, such as abdominal paracentesis and bladder catheterization. Imaging of the abdomen requires a low frequency curvilinear probe or a phased array "cardiac" probe (Figs. 1A and B). The sensitivity and specificity is operator dependent. Therefore, one tends to get better with more imaging.

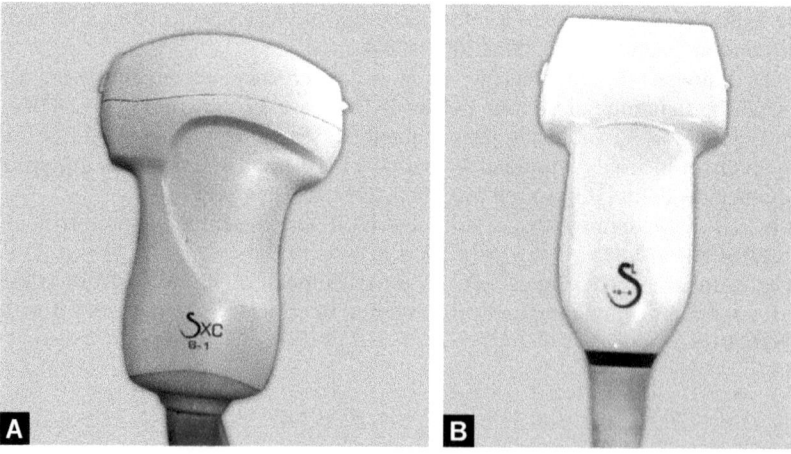

Figs. 1A and B: (A) Curvilinear probe of 1–6 MHz; (B) Linear phased array transducer of 6–10 MHz.

DETECTION OF FREE-FLUID ACCUMULATION (FOCUSED ASSESSMENT WITH SONOGRAPHY FOR TRAUMA AND EXTENDED FOCUSED ASSESSMENT WITH SONOGRAPHY FOR TRAUMA)

The most common application of USG in intensive care is for the focused assessment with sonography for trauma (FAST) examination. FAST provides a framework for basic abdominal image acquisition. Traditionally FAST has been used to detect bleeding in the pericardial and peritoneal spaces in the setting of trauma. It is an integral part of American Trauma Life Support (ATLS) algorithm in detection of concealed intra-abdominal and myocardial injuries that may require emergent surgical exploration. The sensitivity and specificity of FAST for detection of free intraperitoneal fluid were 64–98% and 86–100%, respectively.[1] However, its use is no longer limited to trauma victims. FAST has been used as part of a protocolized examination to evaluate causes of unrecognized hypotension in the emergency department and ICU. It can efficiently detect spontaneous hemoperitoneum, bleeding from ectopic pregnancy, aneurysmal rupture, and virtually any condition that leads to accumulation of free fluid in the abdomen. In the medical ICU, FAST can be applied for detection of intraperitoneal fluid which may be ascitic fluid or peritoneal dialysate fluid. However, abdominal sonography cannot differentiate between the nature of the fluids, whether it is ascites or blood. Neither can it identify the origin of free-fluid extravasation. Besides, the FAST examination requires a minimum of 150–250 mL of fluid accumulation for detection.

The FAST examination involves scanning the abdomen in the subxiphoid cardiac, right upper quadrant, left upper quadrant, and suprapubic views.

- *Right upper quadrant view*: The right upper quadrant view is used to image the hepatorenal recess which is considered to be the most dependent area in the upper abdomen. The external landmark for probe placement is the 10th/11th intercostal place in the posterior/midaxillary line along the long axis with the orientation marker directed toward the patient's head (Fig. 2). A proper view provides with coronal images of the liver, right kidney, Morrison's pouch, and right dome of the diaphragm.

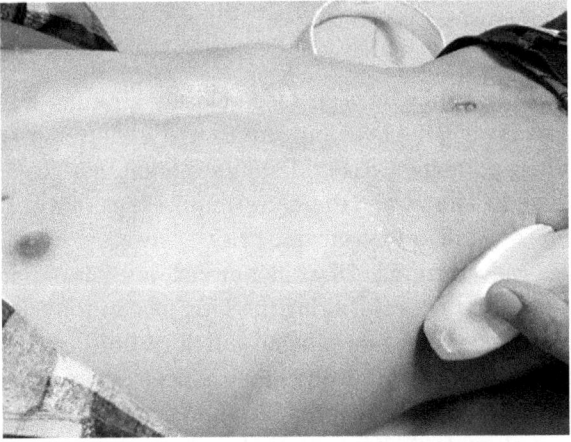

Fig. 2: Ultrasonography (USG) transducer position in right upper quadrant view.

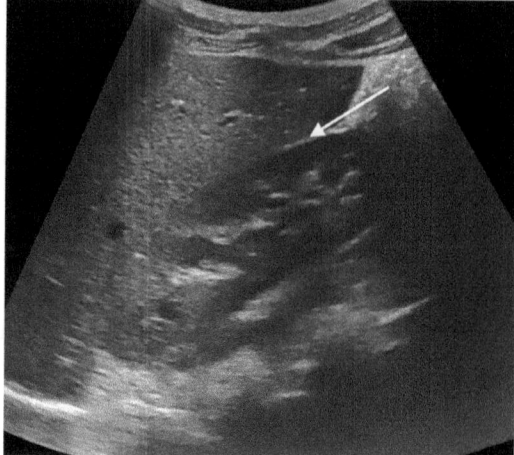

Fig. 3: Normal ultrasound image of the hepatorenal pouch of Morrison.

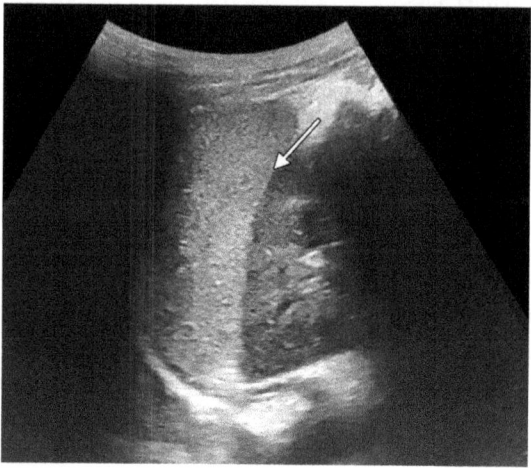

Fig. 4: Normal ultrasound image of the splenorenal recess.

The fascial plane of the kidney is seen to be closely abutting the liver. Therefore, any anechoic collection between the liver and kidney in the Morrison's pouch represents free fluid and is potentially pathological (Fig. 3). Fluid is often seen to collect initially adjacent to the inferior pole of the kidney and therefore, requires scanning of the kidney in entirety by moving the probe upward and downward.

- *Left upper quadrant view*: Imaging of the splenorenal recess is performed in the left upper quadrant view by placing the probe along the long axis in the 6th–9th intercostal space with the orientation marker directed cephalad and the probe directed posteriorly. The view provides with a coronal image of the left kidney, spleen, splenorenal recess, left dome of diaphragm, and the area above the diaphragm. Fluid in the left upper quadrant can appear as an anechoic space above (subphrenic) or below the spleen and left kidney (Fig. 4) unlike on the right side where fluid collects between the liver and kidney.

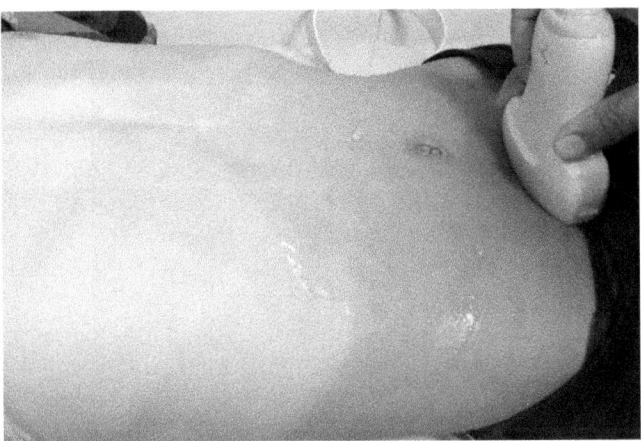

Fig. 5: Ultrasonography (USG) transducer position in suprapubic short-axis view.

- *Suprapubic view*: The suprapubic view looks for any fluid collection in the pelvis which is anatomically the most dependent part of the abdomen. It is considered to be the most sensitive view for intra-abdominal fluid collection. However, scanning the pelvis is technically more challenging in comparison to the upper quadrant views. The pelvis can be imaged in both long and short axis with the probe orientation marker pointing toward the right and cephalad, respectively (Fig. 5). The transverse/short axis views of the bladder are easy to obtain and helps in delineating whether we are dealing with an empty/full bladder. The long axis/longitudinal view, however, is more useful in detecting any fluid collection posterior to the bladder in the rectovesical pouch and is therefore advocated as the initial image to obtain as part of FAST protocol. Fluid outside the bladder wall signifies intraperitoneal fluid collection in which the hyperechoic bladder wall is seen separating the two anechoic spaces on either side.
  The uterus in females is seen as a hypoechoic structure posterior to the bladder in the long-axis view. Any anechoic area surrounding the uterus separating it from the bladder wall in an appropriate clinical setting can be a pointer toward a ruptured ectopic pregnancy with bleeding in the pouch of Douglas (Fig. 6). Indeed USG performed by emergency physicians decreased the time to appropriate diagnosis in suspected ectopic pregnancy.[2]
- *Subxiphoid cardiac view*: The USG probe is placed transversely below the xiphoid process to obtain a coronal section of the heart. The probe orientation marker is directed to the right and the probe directed toward the patient's head or the left shoulder to obtain an optimal image (Fig. 7). The heart is imaged through the left lobe of the liver in this view and is therefore one of the easiest views to obtain even in patients with hyperinflated lung fields. The free wall of the right ventricle is seen to be in apposed to the liver lobe in this view. Pericardium is recognized by its bright echogenicity next to the less echogenic myocardium. Intrapericardial fluid accumulation is visible as anechoic/hypoechoic shadow in between the pericardium and myocardium (Fig. 8). Mere accumulation of fluid in the pericardial space does not signify tamponade unless there is ventricular collapse during diastole.

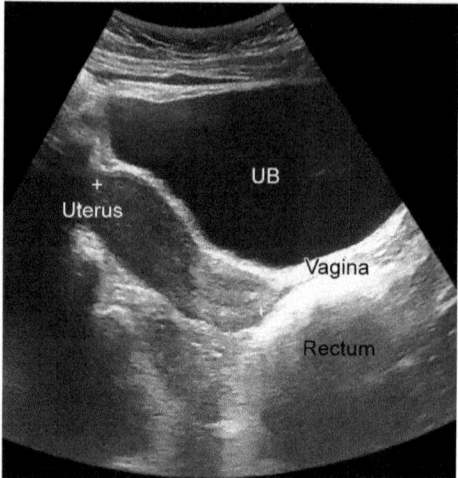

Fig. 6: Suprapubic long-axis visualization of bladder and uterus.

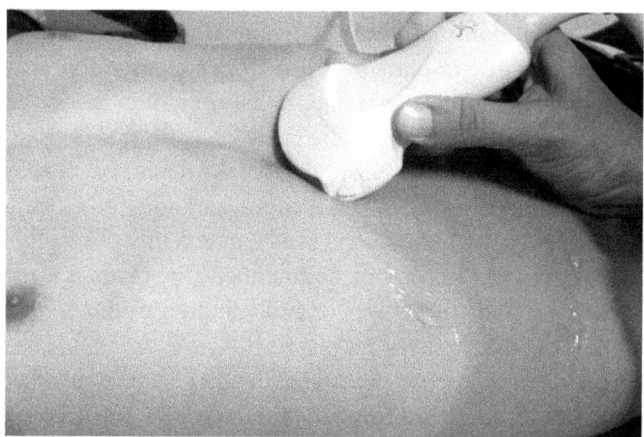

Fig. 7: Ultrasonography (USG) transducer position in subxiphoid cardiac view.

- *Extended focused assessment with sonography for trauma*: The extended FAST (eFAST) additionally includes scanning of the thoracic cavities to detect pleural fluid (hemothorax) or air (pneumothorax) collections. Detection of potential bleeding in the thoracic cavity entails sliding the USG probe one or two interspaces higher in the right and left upper quadrants while scanning the hepatorenal and splenorenal recess for collections in the pleural cavity above the diaphragm. USG can detect as little as 20–50 mL of pleural fluid against 150–200 mL fluid collection required for detection by a chest X-ray (CXR).

  Pneumothorax detection requires image acquisition in midclavicular line in the 3rd–5th intercostal spaces bilaterally to look for absence of lung sliding and "stratosphere/bar code" signs in motion mode (Fig. 9).

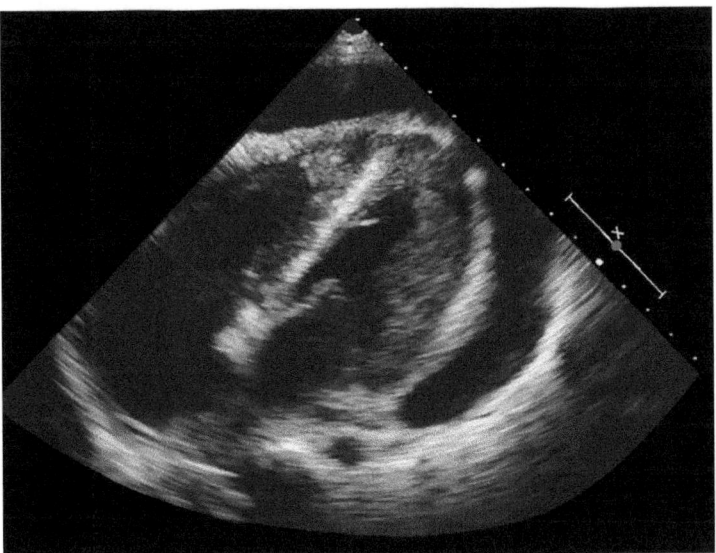

Fig. 8: Echocardiogram showing pericardial effusion only with "NO" collapse of RV; therefore no tamponade.

Fig. 9: Bar code/stratosphere sign seen in pneumothorax in motion mode.

ULTRASOUND OF THE KIDNEYS AND URINARY BLADDER

Detection of pathologies in the kidneys and urinary bladder require scanning of the respective organs in both transverse and longitudinal views. In the setting of decreased output in the critically ill USG can easily differentiate between a full bladder (problem with drainage) and empty bladder (problem with urine production). A distended bladder is identified in the transverse suprapubic view as a large anechoic space in between the echogenic bladder walls. A collapsed bladder on the other hand may be difficult to appreciate recognized only by the fluid-filled anechoic Foley's catheter balloon lying inside (Fig. 10). Bedside estimation of urine volume in the bladder can be performed using standard mathematical formula.

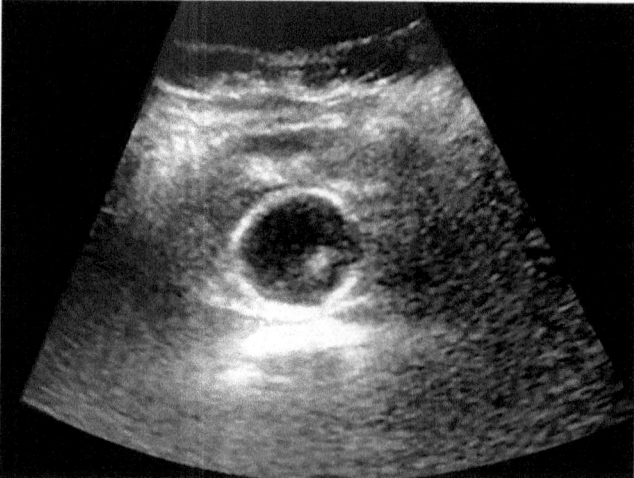

Fig. 10: Collapsed bladder with fluid-filled anechoic Foley's catheter balloon.

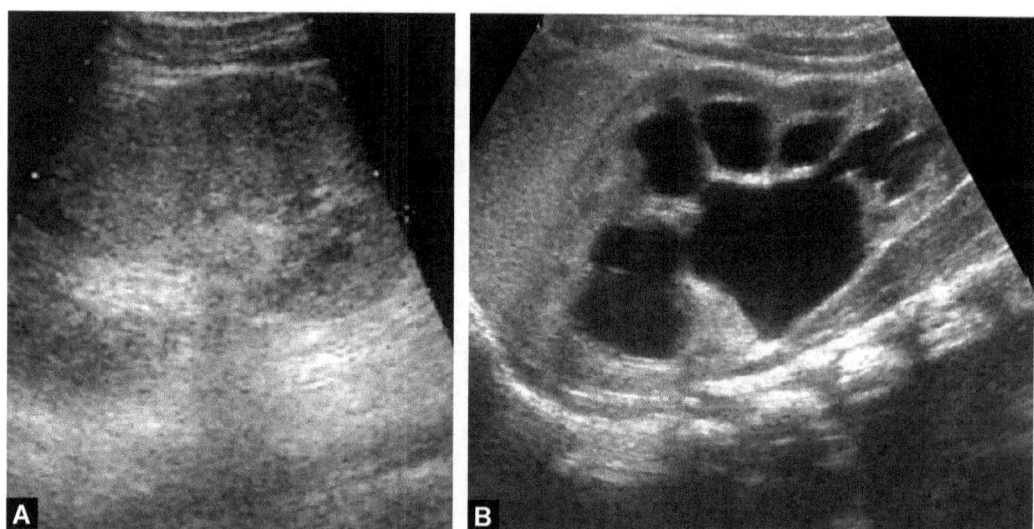

Fig. 11: (A) Loss of corticomedullary differentiation; (B) Hydronephrosis of kidney with dilated renal pelvis.

Renal ultrasound provides the intensivist in assessing renal perfusion, identifying hydronephrosis, and postobstructive changes in kidneys secondary to obstruction. The liver and the spleen provide the acoustic windows for renal ultrasound. Probe placement is similar to right and left subcostal views in FAST examination. The kidneys normally appear as hyperechoic cortex outside surrounding an inner hypoechoic medulla. Loss of corticomedullary differentiation is a characteristic feature of renal failure (Fig. 11A). Postobstructive hydronephrosis leads to dilatation of the collecting duct system leading a widened medulla with a thinned out cortex (Fig. 11B).

Assessment of renal perfusion requires application of Doppler USG to the renal arteries. Renal resistive index (RI) is calculated using the following formula: (peak systolic

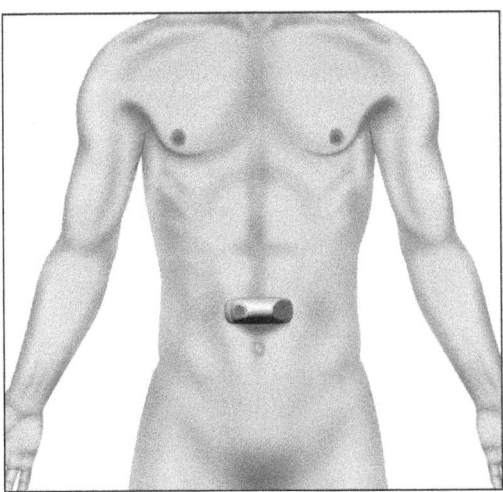

Fig. 12: Ultrasonography (USG) transducer position for short-axis view of abdominal aorta.

velocity− end-diastolic velocity)/peak systolic velocity in an interlobar or arcuate artery, with a normal value of 0.58 ± 0.10. Values more than 0.70 are generally considered to be indicative of renal hypoperfusion and a predictor of acute kidney injury.[3]

IMAGING THE MAJOR VESSELS

Abdominal Aorta

Ruptured abdominal aortic aneurysm is a surgical emergency requiring immediate laparotomy. Delay in identification can have grave consequences. Bedside abdominal USG can be a useful tool to exclude aortic aneurysm in a patient with unexplained hypotension. Indeed USG performed by the physician in the emergency department has shown to have high sensitivity and specificity in diagnosing aneurysmal dilatation of the abdominal aorta.[4] However, it is difficult to identify the retroperitoneal bleed that accompanies a ruptured aneurysm.

Imaging of the aorta requires interrogation of the vessel along its entire length from the origin of the celiac trunk till its bifurcation into the right and left common iliac arteries. The aortic diameter measures less than 3 cm normally. The risk of rupture increases when the diameter exceeds 5.5 cm. The probe is placed below the xiphisternum perpendicular to the abdominal wall with the orientation marker pointing toward the patient's right (Fig. 12). The aorta is imaged in the short axis and diameter estimated by measuring the distance between outer wall and outer wall. The aorta may be difficult to image owing to excessive bowel gas or excessive abdominal obesity. Firm downward pressure can be used to displace the bowel loops. One may also use a low frequency probe to increase penetration (Fig. 13). The initial landmark in the ultrasound image is the lumbar spinous process. Just overlying the spine the inferior vena cava (IVC) is localized with its characteristic respiratory phasic variations toward the patient's right, and the aorta is seen toward the left (Fig. 13). Application of pulse wave Doppler can be used to identify the characteristic pulsations of the aorta. The origin of the

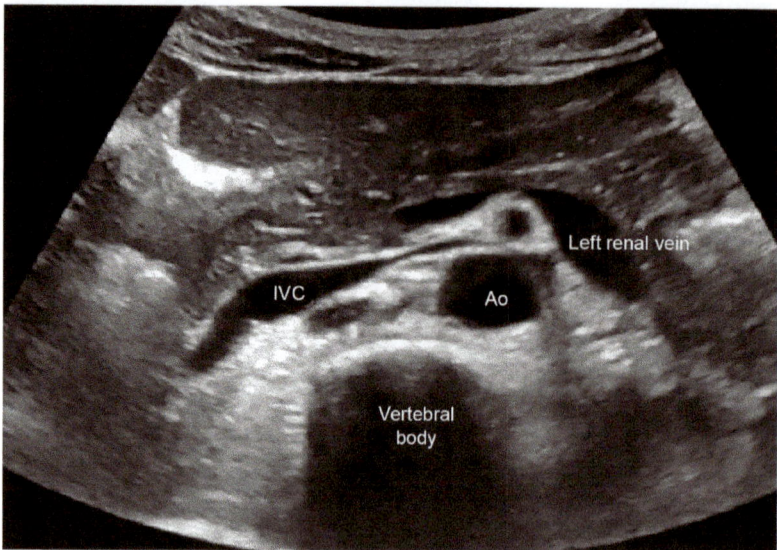

Fig. 13: Short-axis view of abdominal aorta and surrounding structures.
Abbreviations: Ao: Aorta; IVC: Inferior vena cava.

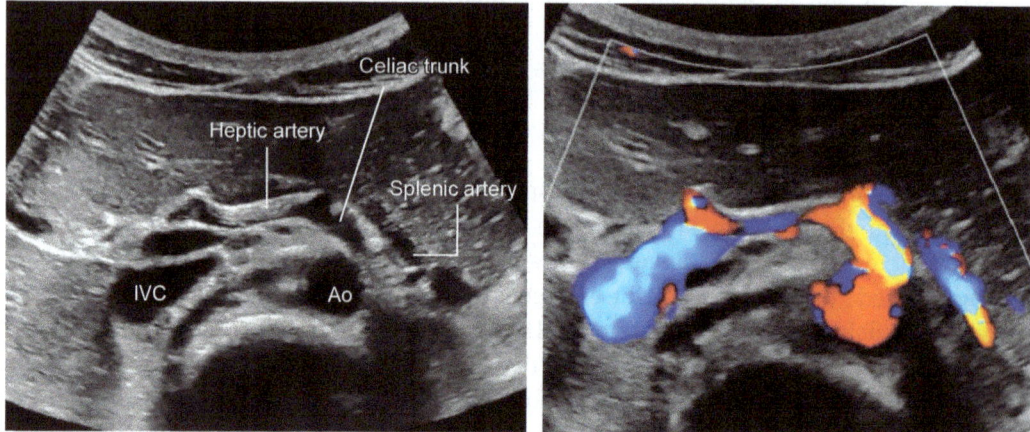

Fig. 14: Origin of celiac trunk from abdominal aorta—"seagull sign".
Abbreviations: Ao: Aorta; IVC: Inferior vena cava.

celiac trunk can be identified at this level with its eventual bifurcation into hepatic and splenic arteries. The celiac trunk is the largest branch of the abdominal aorta and is recognized by the characteristic "seagull sign". The right wing of the seagull corresponds to the hepatic artery coursing toward the patient's right and the left wing corresponds to the splenic artery coursing toward the patient's left (Fig. 14). The probe should be gently slide downward to identify the renal arteries and further downward till distal aortic bifurcation into common iliac arteries. Most of the aortic aneurysms are infrarenal in location. The aorta should similarly be probed in the long axis to locate any aneurysms missed in the short-axis views.

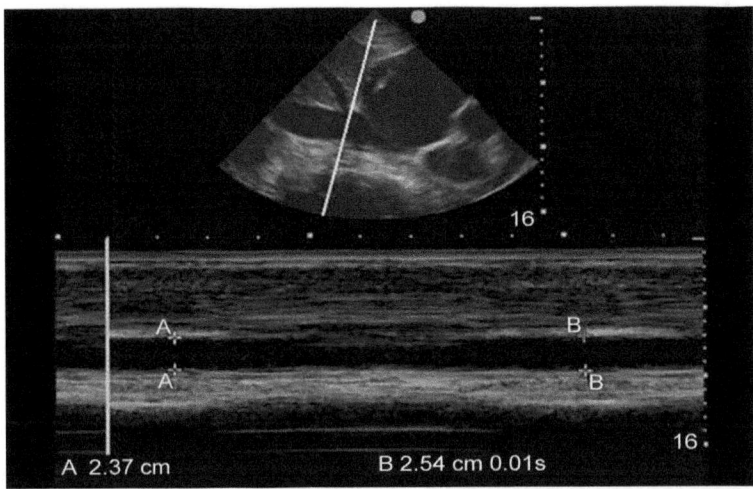

Fig. 15: Measurement of phasic variations of inferior vena cava (IVC) diameter in motion mode.

Inferior Vena Cava

The IVC is imaged frequently in the ICU to assess fluid responsiveness. The IVC is imaged using a transthoracic subcostal approach with the probe orientation marker directed cephalad. The transducer is placed below the xiphisternum, a few centimeters to the right of the midline. The IVC is visualized draining into the right atrium. Once an appropriate image is obtained, motion mode is applied to the IVC 1 cm caudad to the joining of hepatic vein (Fig. 15). Phasic respiratory variations of the IVC are used to measure IVC collapsibility and distensibility indices. An IVC collapsibility more than 12% in mechanically ventilated patients is an indicator of fluid responsiveness.[5]

IMAGING THE GALLBLADDER

Gallbladder pathologies are frequently encountered silent causes of abdominal sepsis in the critically ill. Evaluation of the gallbladder is required for detection of stones, identifying wall edema, and pericholecystic fluid accumulation. USG performed by the emergency physician (EP) performed US for the detection of acute cholecystitis had a sensitivity of 87% [95% confidence interval (CI), 66-97%], specificity of 82% (95% CI, 74-88%).[6] Sonographic Murphy's sign elicited by pain on compression of the gallbladder under the USG probe can be useful aid to diagnose acute cholecystitis.

The gallbladder can be imaged in both transverse and longitudinal axis. The longitudinal axis approach involves placing the transducer below the subcostal margin to the right of the xiphisternum with the orientation marker directed cephalad. The transducer is then fanned laterally to locate the gallbladder. Gallbladder appears as pear-shaped hypoechoic structure with hyperechoic walls (Fig. 16). The gallbladder can also be imaged in its short/transverse axis by rotating the probe at 90° with the orientation marker toward the patient's right.

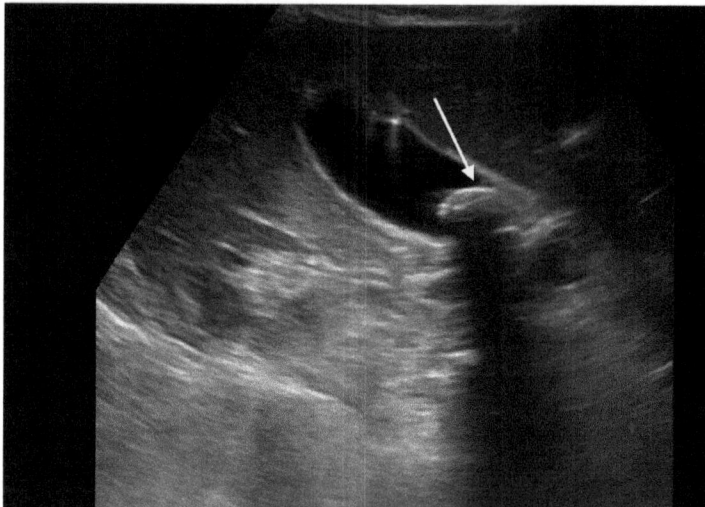

Courtesy by: Devasenathipathy Kandasamy, Ananya Panda

Fig. 16: Ultrasonography (USG) image of gallbladder showing calculus with characteristic acoustic shadow.

The gallbladder appears spherical in short-axis view. Careful fanning of the gallbladder will help to identify any pathology.

Gallstones are identified as hyperechoic structures within the gallbladder with its characteristic posterior acoustic shadow. Gallbladder edema is assessed by measuring anterior wall thickness of the gallbladder in the short axis with the probe placed perpendicular to the wall. A gallbladder wall thickness more than 3 mm is considered to be abnormal indicative of inflammation of the gallbladder. Pericholecystic fluid appears as anechoic area surrounding the gallbladder and is potentially pathogenic. Dilatation of the common bile duct can indicate obstruction in the biliary system.

DETECTION OF PNEUMOPERITONEUM

Pneumoperitoneum following hollow viscus perforation has been classically detected using abdominal X-ray and computed tomography scans. USG may be a useful investigation modality in the detection of intraperitoneal air.[7] Presence of air is a hindrance to ultrasound imaging. Air inside the peritoneal cavity tends to accumulate superficially below the peritoneal layer. In the ultrasound image, such air beneath the peritoneum gives rise to characteristic reverberation artifacts with enhancement of echogenicity of the peritoneal layer (Fig. 17).

GUIDING THERAPEUTIC PROCEDURES

Ultrasonography can be used to guide commonly performed therapeutic interventions, such as abdominal paracentesis, urethral catheterization, and nasogastric (NG) tube placement.

Abdominal paracentesis has been performed traditionally by a blind landmark-based technique. However, USG assistance has made the procedure safer avoiding complications. USG helps in delineating the optimal needle insertion site where the ascitic fluid pocket is at its

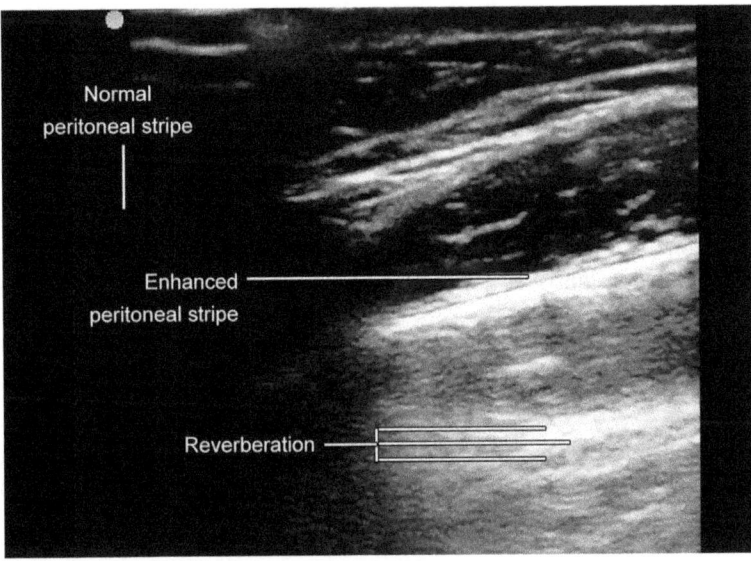

Fig. 17: Ultrasonography (USG) image of pneumoperitoneum showing enhancement of the peritoneal stripe and reverberation artifact.

maximum depth. It allows real-time visualization of needle puncture and avoids accidental puncture of epigastric vessels, collateral veins, bladder, and bowel perforations. When compared to landmark-based technique, USG guidance improved the success rate of paracentesis markedly.[8]

Nasogastric (NG) tube placement confirmation clinically requires epigastric auscultation follow-ing a push of air through the NG tube. One can also aspirate and inspect the gastric contents. Alternatively a CXR is used to confirm position of NG which exposes the patient to additional radiation hazard. USG, as an adjunct to clinical examination can confirm NG tube placement by direct visualization of the tube in the stomach or observing fogging in response to manual instillation of saline mixed with air.[9]

Urethral catheterization is a procedure routinely performed in majority of patients admitted to the ICU. Cases of difficult catheterization may benefit from USG guidance. Transabdominal USG can help to locate the tip of the catheter in the posterior and bulbar urethra. US-guided catheterization with transrectal pressure was found to be useful in situations where standard techniques have failed.[8]

The application of bedside USG has revolutionized clinical practice for the intensivist. It compares reliably to the more traditional diagnostic tools and provides the ICU physician with an excellent tool for point-of-care goal-directed therapy. The clinical applications of abdominal USG seek to answer a wide range of diagnostic dilemmas encompassing gastrointestinal, respiratory, cardiovascular, and genitourinary systems. The quality of imaging only gets better with practice and it is essential that all ICUs actively promote bedside abdominal USG by the intensivist rather than delaying intervention pending a length radiology consultation.

REFERENCES

1. Körner M, Krötz MM, Degenhart C, et al. Current role of emergency US in patients with major trauma. Radiographics. 2008;28:225-42.
2. Moore C, Todd WM, O'Brien E, et al. Free fluid in Morison's pouch on bedside ultrasound predicts need for operative intervention in suspected ectopic pregnancy. Acad Emerg Med. 2007;14:755-8.
3. Lerolle N, Guérot E, Faisy C, et al. Renal failure in septic shock: predictive value of Doppler-based renal arterial resistive index. Intensive Care Med. 2006;32:1553-9.
4. Rubano E, Mehta N, Caputo W, et al. Systematic review: emergency department bedside ultrasonography for diagnosing suspected abdominal aortic aneurysm. Acad Emerg Med. 2013;20: 128-38.
5. Barbier C, Loubières Y, Schmit C, et al. Respiratory changes in inferior vena cava diameter are helpful in predicting fluid responsiveness in ventilated septic patients. Intensive Care Med. 2004; 30(9):1740-6.
6. Summers SM, Scruggs W, Menchine MD, et al. A prospective evaluation of emergency department bedside ultrasonography for the detection of acute cholecystitis. Ann Emerg Med. 2010;56:114-22.
7. Chen SC, Yen ZS, Wang HP, et al. Ultrasonography is superior to plain radiography in the diagnosis of pneumoperitoneum. Br J Surg. 2002;89:351-4.
8. Kameda T, Murata Y, Fujita M, et al. Transabdominal ultrasound-guided urethral catheterization with transrectal pressure. J Emerg Med. 2014;46:215-9.
9. Brun PM, Chenaitia H, Lablanche C, et al. 2-point ultrasonography to confirm correct position of the gastric tube in prehospital setting. Mil Med. 2014;179:959-63.

Chapter 14

Blue, Falls, and Sesame Protocol

Dalim Kumar Baidya, Anil Malik

INTRODUCTION

Bedside ultrasonography is now considered an integral part of the routine assessment of patients admitted to the intensive care units (ICUs). It has gained a unique place in the quick systemic evaluation of critically ill-patients as it is noninvasive, easy and quick to perform. Ultrasonography has been widely used for venous and arterial cannulation as it decreases the risks of procedural complications. It is also used for bedside screening of lung pathologies like pleural effusion, hemothorax, pneumothorax, and pulmonary edema.[1] Ultrasonography of lung involves two protocols as described by Lichtenstein et al.; (1) the BLUE protocol and (2) the FALLS protocol.[2] BLUE protocol is used for the bedside screening of causes of acute hypoxemia while FALLS protocol is used to rule out the causes of shock and hypoperfusion. These protocols involve easy screening of pleural and lung interfaces, pleural effusions, interstitial edema, consolidations and pneumothorax. All of these conditions are associated with few predefined signs, which are accurate in diagnosing them. The BLUE protocol will be able to quickly diagnose acute conditions like pneumothorax, embolism, pulmonary edema and infective conditions, such as pneumonia and empyema. It will also rule out chest conditions associated with heart failure as well as few chronic lung diseases. At the same time FALLS protocol will be useful bedside tool for the screening of acute conditions causing shock and hypoperfusion like hypovolemia, cardiac pump failure due to myocardial infarction, pericardial tamponade, valvular heart diseases and also massive pulmonary embolism causing right ventricular failure.

THE BLUE PROTOCOL (BEDSIDE LUNG ULTRASOUND IN EMERGENCY)

This protocol is designed to rapidly and accurately diagnose various chest conditions associated with acute hypoxemia in seriously ill-patients, so that appropriate therapeutic measure can be initiated to correct it. There are predefined thoracic points, as described by Lichtenstein et al. for the easy identification and reproducibility of the imaging findings called as *BLUE points*.[3] The *BLUE points* are located just below the clavicle and can be easily identified by placing two hands over the anterior chest wall with the lower hand overlapping the upper thumb, while the little finger of the upper hand will be positioned below the clavicle. The *upper BLUE point* will be located at the center of the upper palm while the *lower BLUE point* will be situated at the center of the lower hand. These points on both sides of the chest wall will

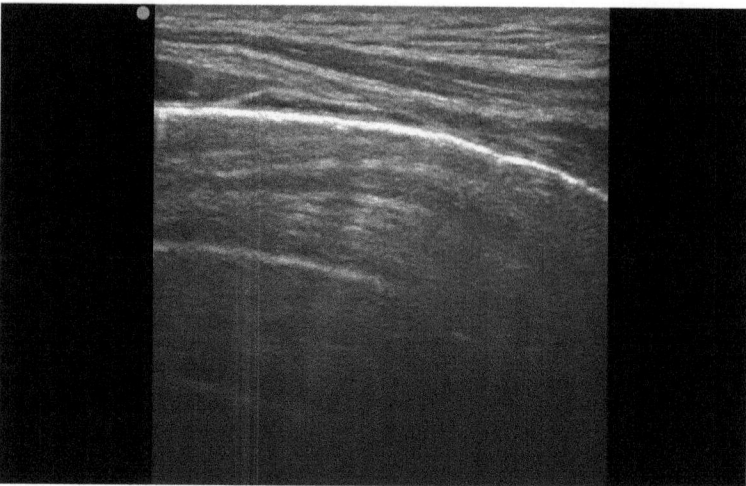

Fig. 1: A lines (A profile).

provide the best images of the lung pathologies and will help in establishing the diagnosis. There is two more points called as the *PLAPS point* (posterolateral alveolar and/or pleural syndrome). These are located at the posterior axillary line. If a horizontal line is drawn from the lower BLUE point, then the point where it meets with the posterior axillary line is known as the *PLAPS point*.[4] These points are useful in the identification of the lung consolidations.

The analysis of the images in lung ultrasonography is based on various basic physical principles. Usually free gas will be located more toward the upper fields while fluids will be generally located in the lower fields. Other important features of lung ultrasonography are the artifacts. These are images produced by the ultrasonography which does not correspond to any true anatomical structure. One example of the artifact is the A-line (Fig. 1). It is an ultrasonographic image artifact due to the presence of gas below the pleural surface. This line can be easily identified by a sign called the *bat sign*. The shadows of the adjacent ribs along with equally spaced repeated hyperechoic white line will form the figure of a bat, hence the bat sign. The pleural line in lung ultrasound always corresponds to the parietal pleura. Another normal finding in lung ultrasonography is the *lung sliding* or the *seashore sign*. Lung sliding can be visualized using the 2-D mode of the ultrasonogram as well as with the M-mode. In the M-mode, the image will have two layers; the soft tissue above the pleura will be still without any movement, while below it due to lung excursion the pattern will be sandy in appearance. This typical ultrasonography image of a normal lung is referred as *seashore sign*. Lung ultrasonography showing A-lines as well as normal lung sliding will indicate a normal lung surface and exclude the presence of any pneumothorax.[4]

Lung ultrasonography is also relying on another simple physical principle that most pathological conditions which cause severe hypoxemia are superficial in natures, hence are easy to visualize. These superficial pathological conditions include pleural effusion, empyema, lung consolidations, lung collapse and pneumothorax. Pleural effusion (Fig. 2) which is a very common finding in the ICU patients can be easily and quickly diagnosed by using different signs.

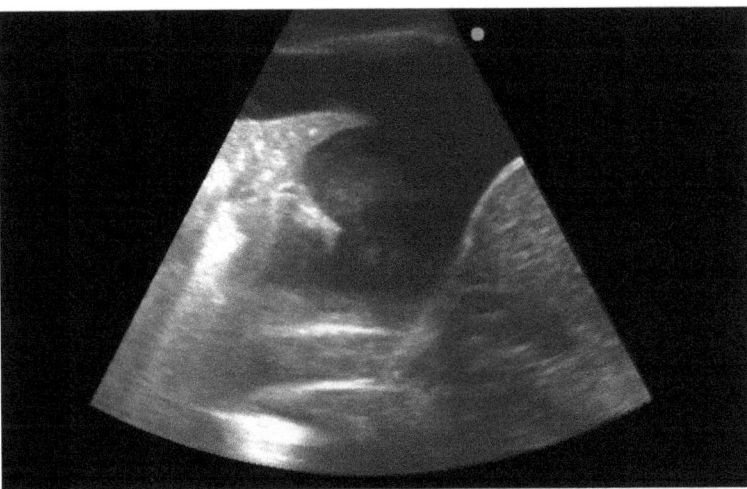

Fig. 2: Pleural effusion.

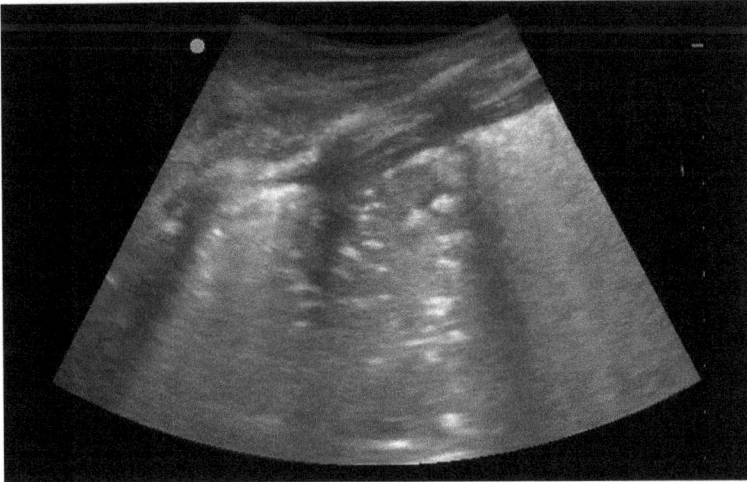

Fig. 3: Fractal (or shred) sign.

The *quad sign* is an image whose boundaries are formed laterally by the shadows of the ribs, superiorly by the chest wall and inferiorly by the lung parenchyma. An hyperechoic line may be visualized above the lung parenchyma. And the hypoechoic content of the quadrangular space indicates the pleural effusion. When visualized with M-mode through the effusion, it will form the sinusoidal image pattern called as the *sinusoid sign*; as the lung moves toward and away from the probe with each respiration. This sign will be absent with very viscous effusions.[5]

Similarly lung consolidation can be easily visualized using bedside lung ultrasonography. Partial lobar consolidations of lungs, limited to single lobes will form the *fractal (or shred) sign* (Fig. 3) on ultrasonography imaging. There will be visible underlying aerated lung tissues in between lobar consolidation. Also the margins of the tissues will be irregular and shredded.

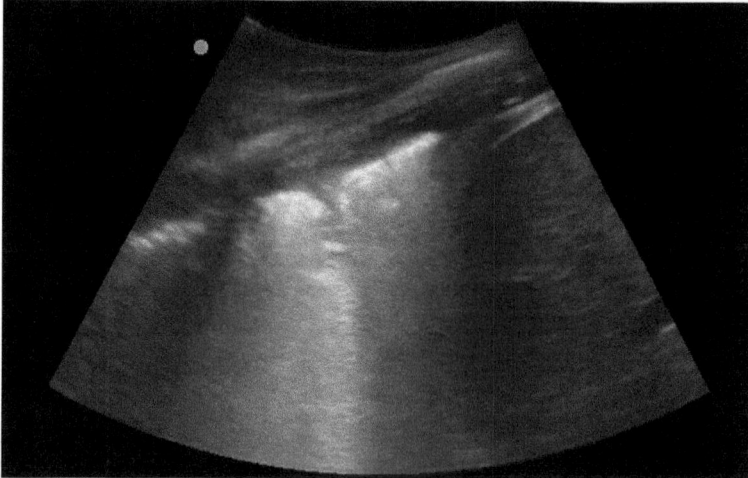

Fig. 4: Tissue-like sign.

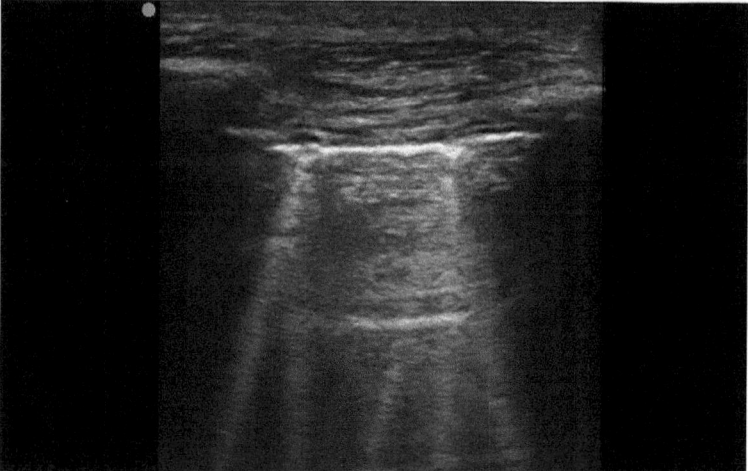

Fig. 5: B-profile with multiple B-lines.

In case of massive lobar consolidation there will be no underlying aerated lung tissue and there will be dense hyperechoic images. The consolidation mostly involves multiple lobes of the lung and forms ultrasonographic image similar to the image formed by the splenic tissue. Hence, it is called as the tissue-like sign (Fig. 4) and indicates massive lobar consolidation.[6] Both these are very common findings in patients of pneumonia and indicates infection as the cause. On visualization with 2-D mode, there will be absent lung sliding.

Pulmonary edema (interstitial edema) will form artifacts on ultrasonography known as *B-line* (Fig. 5).[7] They are also termed as comet tail artifact. These B-lines are also found in normal lungs. They will indicate interstitial edema, if there is confluence of more than three B-lines within two rib margins (*lung rockets*). They always arise from the pleural surface and extend till the end of the screen. They are discrete artifacts which move with the respiration and are hyperechoic in nature. They should not be confused with E-lines, which are similar artifacts like B-lines but they arises from the skin and terminates at the pleural surface. Presence of E-lines in ultrasonography indicates subcutaneous emphysema.

Pneumothorax is a life-threatening condition and requires immediate diagnosis. Its identification on lung ultrasonography is easy and requires only few steps. It begins with the identification of the A'-profile on lung ultrasonography. A'-profile is defined as the abolition of normal lung-sliding with respiration. Also there will be no visible B-lines on the ultrasonography screen. On M-mode, the image will show the *stratosphere sign* or the *bar code sign*, which are homogeneous patterns similar to the patterns in a bar code. The finding of A'-profile on ultrasonography is highly sensitive for pneumothorax. The next step is to confirm the presence of pneumothorax, which requires the identification of *lung point*. It is a point adjacent to the A'-profile on the chest wall. It will be marked by the presence of normal lung movement B-lines. Identification of lung point is a pathognomonic sign for pneumothorax.[8]

Lung Profiles

There are established profiles associated with signs and their location on the chest wall as described by Lichtenstein et al. These profiles are defined based on their pathophysiological correlations with the disease process. The A-profile (*see* Fig. 1) is identified at the frontal chest wall. Ultrasonography will show normal lungs movements with respiration with A-lines. Its presence indicates a normal anterior lung surface. In the A'-profile there will be absent lung movement with respiration along with A-lines. Typically there will be no visible B-lines present in the screen. It is a specific finding in the presence of pneumothorax.

The *B-profile* is (*see* Fig. 5) described as the presence of confluent B-lines associated with normal lung movements with respiration. It is usually seen in the presence of pulmonary edema due to fluid overload or heart failure. The impedance gradient generated between gas and fluid interspaces in pulmonary edema create these B-lines. *B'-profile* is described as the presence of confluent B-lines with the absence of normal lung movements during respiration. It may occur in pneumonia due to the presence of exudates. The *A/B profile* is a mixed profile combining both A-profile and B-profile on the both sides of the chest. This profile indicates interstitial inflammation which corresponds to pneumonia.

The C-profile is defined as the ultrasonographic presence of consolidation patches over the anterior chest wall (*see* Figs. 3 and 4). They can be multiple in numbers with different sizes. The C-profile is always associated with pneumonia, as consolidations due to pulmonary edema or pulmonary embolism are usually posteriorly located.[9] Even a small anteriorly located lung consolidations will suggest pneumonia rather than pulmonary embolism as the cause of hypoxemia.[2]

The *A-V-PLAPS-profile* is described as the combination of the A-profile, presence of consolidations and no ultrasonographic evidence of deep vein thrombosis (DVT). This profile is typically present in patients with pneumonia. Another profile which shows the presence of normal lung sliding but without any evidence of DVT or consolidation is called as *nude profile*. It is seen in patients with bronchial diseases, such as asthma and chronic obstructive pulmonary disease (COPD).

Approach

The BLUE protocol should be done by the physician after performing a thorough physical examination. The standardized point should be evaluated using bedside lung ultrasound to rule out the cause of the respiratory failure. Initially the physician should look for normal lung movement with respiration. If normal lung movement with respiration is present along with A-lines than it is termed as A-profile. The next step is to look for any evidence of DVT with a venous scan.[9] Presence of ultrasonographic evidence of venous thrombosis will be highly suggestive of pulmonary embolism. If DVT scan is negative than PLAPS points should be evaluated to look for the A-V-PLAPS-profile, which will indicates pneumonia as the cause. If no abnormalities are detected then it will suggest the nude profile indicating COPD or asthma, which will require further investigations. In the presence of A'-profile, the physician should be alert and should search for the appearance of adjacent normal lung movements, as it is associated with pneumothorax. If the patient has a predominant B-profile then it will clearly indicates pulmonary edema. In patients with pneumonia the lung profile will show varying degree of anterior consolidations. acute respiratory distress syndrome (ARDS) can show features of any of the profile suggesting either interstitial edema or consolidation. BLUE protocol provides 90% accuracy in the prediction of the cause of hypoxemia.[2]

THE FALLS PROTOCOL (FLUID ADMINISTRATION LIMITED BY LUNG SONOGRAPHY)

Echocardiography which uses the ultrasonographic probe is one of the most popular equipment to evaluate the cardiovascular status of a sick patient.[10,11] The FALLS protocol is based on a simple physical principle; pulmonary edema causes thickening of the interlobular septa of the lung tissue, due to which their subpleural end are visible on lung ultrasound.[12,13] In critically ill-patients, B-lines will appear at a pulmonary artery occlusion pressure of 18 mm Hg.[14] Another principle is that B-line appear with hemodynamic edema and disappear once it is cleared, making it an useful objective parameter.

The FALLS protocol is guided by pathophysiology of shock as described by the Weil classification of shock,[15] such as obstructive, cardiogenic, hypovolemic, and distributive shock. The first step is to look for the cardiac window to rule out the obstructive causes of shock like tamponade due to effusion, pulmonary embolism and massive pneumothorax. Initially the physician will look for any substantial pericardial effusion to rule out tamponade. Then the size of the right ventricle is evaluated to rule out right ventricular (RV) dilation, which occurs in pulmonary embolism. If the ultrasound shows an A'-profile then tension pneumothorax should be ruled out as the cause of the shock. If the physician is not able to visualize properly all the cardiac chambers, then the steps of BLUE protocol can be followed along with venous thrombosis scan. These entire three steps will rule out any causes of obstructive shock.

The next step in the FALLS protocol is to look for the causes of cardiogenic shock. The presence of B-profile, on lung ultrasound will suggest hemodynamic pulmonary edema, most commonly due to left ventricular failure. Presence of B-profile has 97% sensitivity and 95% specificity for accurately predicting the presence of acute pulmonary edema.[2]

Once above two profiles are excluded, then the remaining cause of shock includes hypovolemic or distributive shock. Initially the physician will look for A-profile. If the patient shows an A-profile then it is called *FALLS responder* and the physician can consider for therapeutic fluid administration. If after adequate fluid administration the patients hemodynamic parameters improved and patients' lung ultrasound continue to show A-profile, then the diagnosis will be of hypovolemic shock.

On the contrary, if the patient's hemodynamic condition does not improve and the lung ultrasound changes to show B-lines and lung rockets as seen in B-profile, then a diagnosis of distributive shock became obvious as all other causes has been ruled out. At this step the fluid administration should be discontinued as it is called *FALLS-endpoint*. Here the FALLS protocol will end, and it will prevents the development of interstitial edema due to fluid overload, as B-lines appears much earlier than the clinical detection of alveolar edema. The most common and devastating cause of distributive shock is septic shock and its diagnosis will be the end of the protocol.

Therefore, FALLS protocol provides an initial guidance for the starting as well as discontinuation of fluid therapy in patient presenting with undiagnosed shock based on their pathophysiology. It can be done with simple equipments and also without the knowledge of echocardiography, which requires greater expertise. Although it has limited use when patient initially presented with B-profile and also it did not measure any cardiovascular parameters to guide fluid therapy. In those cases caval vein assessment as well as other hemodynamic parameters should be monitored to guide fluid therapy.

THE SESAME PROTOCOL

This stands for the mnemonic "Sequential Echographic Scanning Assessing Mechanism or Origin of Severe Shock of Indistinct Cause". It is a technical tool used to evaluate the causes of acute cardiac arrest, by using the ultrasound in a step-by-step approach. As cardiac arrest gives little time for the appropriate evaluation of patient to rule out the reversible causes; ultrasound has gained immense popularity among emergency physicians and intensivists. The focus is on complete examination of the sick and hemodynamically unstable patients. It involves a step-by-step approach to evaluate for the probable reversible causes of cardiac arrest.

The first step is to look for the presence of pneumothorax as discussed above in the BLUE protocol. Although it is not a very common condition, but missing it will be met with devastating complication. And secondly large pneumothorax causing hemodynamic instability will take only few seconds to diagnose. The location of the lung ultrasonography probe will be positioned at the lower BLUE point on the anterior chest wall. Cardiopulmonary resuscitation (CPR) should be continued and chest compression must not be interrupted. The lungs should be scanned as fast as possible. Presence of the A'-profile will strongly suggest pneumothorax.

The second life-threatening condition is pulmonary embolism which needs immediate diagnosis. This step is one of the difficult scan to expertise as it requires a good cardiac window. Combination of lung ultrasound and venous thrombosis scan has very high specificity for its rapid diagnosis. However it is time consuming as it is described in the BLUE protocol. Therefore SESAME protocol proposed a point called V point on the lower femoral vein, for the rapid diagnosis of DVT. If DVT scan is negative then physician should proceed to the next step.

The next step is to look for free fluid collection in the abdominal cavity and chest cavity. This scan should be done as fast as possible without wasting valuable time. If no free fluid collection is detected then the next step is to rule out pericardial tamponade as the cause of cardiac arrest.

The fourth step is to check for pericardial effusion causing tamponade. As pericardium is more superficial and diagnosing free fluid encircling heart requires less time, it is the logical next step. A subcostal approach is advised as it will cause less interruption in the chest compression. Another advantage is an immediate ultrasound guided pericardiocentesis can be performed to relieve the obstructive shock. If pericardial effusion is absent than evaluation of the heart is the final step of this protocol.

The fifth step involves transthoracic echocardiography (TTE) by the physician. It has various technical difficulties, as it requires a great deal of expertise to master it. Also it requires the interruption of the cardiac compression. And lastly it will require a cardiac probe to perform the TTE. Different findings are expected on TTE. Asystole is the simplest and easiest pattern to diagnose. A dilated RV will suggest the diagnosis of pulmonary embolism as the cause. A collapsed right side of the heart will suggest pericardial tamponade. Even ventricular fibrillation can be diagnosed sometimes.

Although it is a useful protocol, it requires a great deal of expertise to master the technique. Secondly it requires a sophisticated ultrasound machine to perform all the scans. Lastly it demands fine coordination between the team performing chest compression as well as the physician doing simultaneous ultrasound. Even the operator has to switch between different probes rapidly during the scanning, so as to proceed through all the steps. If executed properly it will be of great use in saving many lives of critically ill-patients having cardiac arrest due to reversible cause.

REFERENCES

1. Lichtenstein D, Axler O. Intensive use of general ultrasound in the intensive care unit. Prospective study of 150 consecutive patients. Intensive Care Med. 1993;19(6):353-5.
2. Lichtenstein DA, Mezière GA. Relevance of lung ultrasound in the diagnosis of acute respiratory failure: the BLUE protocol. Chest. 2008;134(1):117-25.
3. Lichtenstein DA. Whole Body Ultrasonography in the Critically Ill. Berlin, Germany: Springer-Verlag; 2010.
4. Lichtenstein D, Meziere G, Biderman P, et al. The comet-tail artifact: an ultrasound sign ruling out pneumothorax. Intensive Care Med. 1999;25(4):383-8.
5. Lichtenstein D, Hulot JS, Rabiller A, et al. Feasibility and safety of ultrasound-aided thoracocentesis in mechanically ventilated patients. Intensive Care Med. 1999;25(9):955-8.

6. Lichtenstein DA, Lascols N, Meziere G, et al. Ultrasound diagnosis of alveolar consolidation in the critically ill. Intensive Care Med. 2004;30(2):276-81.
7. Lichtenstein D, Meziere G, Biderman P, et al. The comet-tail artifact. An ultrasound sign of alveolar-interstitial syndrome. Am J Respir Crit Care Med. 1997;156(5):1640-6.
8. Lichtenstein D, Meziere G, Biderman P, et al. The lung point: an ultrasound sign specific to pneumothorax. Intensive Care Med. 2000;26:1434-40.
9. Lichtenstein DA. The BLUE-protocol, venous part: deep venous thrombosis in the critically ill. Technique and results for the diagnosis of acute pulmonary embolism. In: Lichtenstein DA (Ed). Lung Ultrasound in the Critically Ill. Berlin, Germany: Springer-Verlag; 2015. pp. 123-42.
10. Jardin F, Farcot JC, Boisante L, et al. Influence of positive end-expiratory pressure on left ventricle performance. New Engl J Med. 1981;304(7):387-92.
11. Vieillard-Baron A, Slama M, Cholley B, et al. Echocardiography in the intensive care unit: from evolution to revolution? Intensive Care Med. 2008;34(2):243-9.
12. Kerley P. Radiology in heart disease. BMJ. 1933;2(3795):594-7.
13. Lichtenstein D. Diagnostic échographique de l'oedèmepulmonaire. Rev Im Med. 1994;6:561-2.
14. Lichtenstein DA, Mezière GA, Lagoueyte JF, et al. A-lines and B-lines: lung ultrasound as a bedside tool for predicting pulmonary artery occlusion pressure in the critically ill. Chest. 2009;136(4):1014-20.
15. Weil MH, Shubin H. Proposed reclassification of shock states with special reference to distributive defects. Adv Exp Med Biol. 1971;23(0):13-23.

Chapter 15A

Ultrasound of Diaphragm

Sandeep Sahu, Alka Verma, Rashmi Soori

INTRODUCTION

"Diaphragma" is a Greek word; means partition. Diaphragm is a musculotendinous dome-shaped structure made up of muscles and tendon.[1] It separates thoracic cavity from abdominal cavity.[2] Being the important muscle of both inspiration and exhalation, it plays a pivotal role in respiration. It has thin, skeletal muscle that can contract voluntarily.

APPLIED ANATOMY OF DIAPHRAGM

It is located in the inferior aspect of the rib cage and forms the floor of the thoracic cavity and the roof of the abdominal cavity. It has three peripheral attachments (Fig. 1).[3]
1. Lumbar vertebrae and arcuate ligament. (Lumbar part has medial and lateral parts of origin. Medial part originates from L1–L3 vertebral bodies, intervertebral disks and anterior longitudinal ligament forming the right and left crus of the diaphragm, respectively; and the lateral part arises from medial and lateral arcuate ligaments).[3]

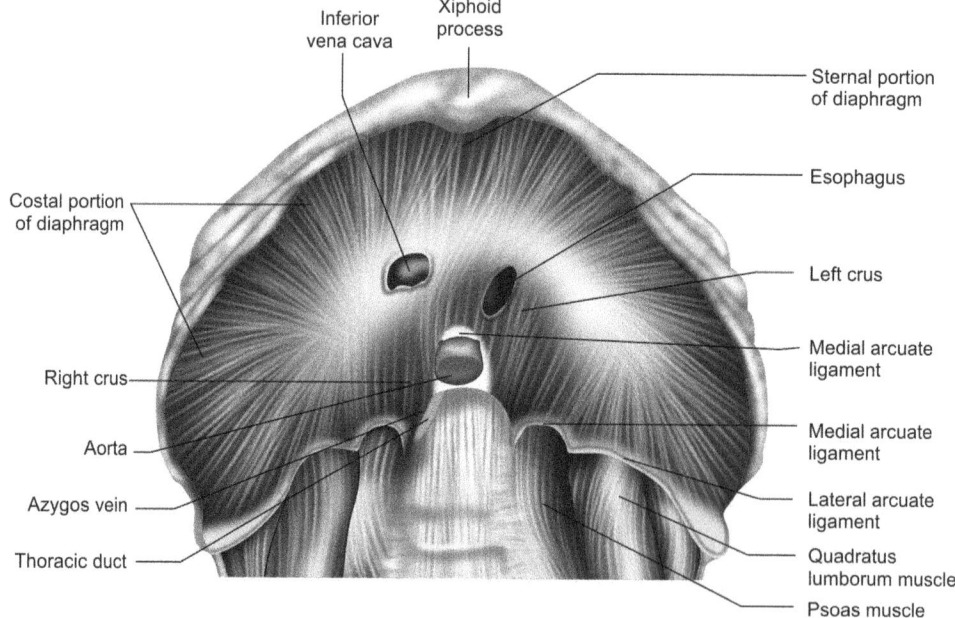

Fig. 1: Anatomy of diaphragm showing its parts and attachments.
Source: (Adapted from: Sugarbaker DJ, et al. Adult Chest Surgery, 2nd edition. New York: McGraw Hill Education; 2009).

2. Costal cartilages of rib 7-12 (costal part originates from inner surface of 7th-12th ribs at the lower margin of costal arch).
3. Xiphoid process of sternum (sternal part arises from posterior surface of xiphoid process).

The tendinous vertebral attachment on the right called as the right crus arises from L1-L3 and the left crus from L1-L2 and their intervertebral discs, respectively. Some fibers from the right crus surround the esophageal opening, and acts as a physiological sphincter preventing reflux of gastric contents into the esophagus.[4]

The muscle fibers fan at the periphery while the tendons coalesce forming the aponeurosis at the center called the *"central tendon"*. The central tendon fuses with the inferior surface of the fibrous pericardium[5] and has three openings:

1. *Esophageal opening (T10 level):* Transmits the esophagus, vagus nerves, and esophageal branches of the left gastric vessels.
2. *Aortic opening (T12 level):* Transmits the aorta, thoracic duct (a large lymphatic vessel) and azygos vein.
3. *Caval opening (T8 level):* Transmits the inferior vena cava.[2]

On either side of the pericardium, the diaphragm has a concavity towards the abdominal cavity with convexity towards the thorax; forming the left and right *"domes"*. At rest, the right dome lies slightly higher than the left probably due to the presence of the liver.

The arterial supply to the diaphragm is primarily by the *"inferior phrenic arteries"*, arising from the abdominal aorta. The other arteries supplying include the superior phrenic, pericardiophrenic and musculophrenic arteries. Motor and sensory innervation to the diaphragm is by the right and left phrenic nerves. The peripheral part of diaphragm also receives sensory supply by the lower six or seven intercostal nerves.[6]

VARIANT ANATOMY

- *Eventration:* A rare congenital anomaly characterized by partial or total replacement of the diaphragmatic muscle fibers by fibrous tissue, manifested as an abnormal bulge in the hemidiaphragm.
  Partial eventration leads to a focal upward bulge, often affecting the right hemidiaphragm and is clinically insignificant.
  Total eventration usually affects the left hemidiaphragm and must be distinguished from phrenic nerve paralysis, atelectasis and subpulmonic effusion.

  Differential diagnosis: Elevation of hemidiaphragm mnemonic is *"PEAS"*.
   P: Phrenic nerve palsy or paralysis
   E: Eventration (total)
   A: Atelectasis
   S: Subpulmonic effusion.
- *Scalloping:* This is like that of eventration that is an age-related process besides being congenital. Multiple focal bulges may present commonly within the right hemidiaphragm. It is clinically unimportant.
- *Diaphragmatic slip*
- *Hypertrophic crus*
- *Inversion.*

FUNCTION

Diaphragm is the primary muscle used for quiet breathing. As a person breathes in and out, the diaphragm contracts and relaxes, forcing air to rush into and exit the lungs, respectively. The lungs are enclosed within the thorax by the rib cage on the front, back, and sides with the diaphragm forming the floor of the cavity.[1]

During inspiration, when the diaphragm contracts, the dome flattens, and the external intercostals between the ribs contract, thereby elevating the anterior part of the rib cage like the handle of a bucket. This increases the volume of the chest cavity and creates a suction that draws breath into the lungs. Upon relaxation, the diaphragm re-assumes the dome shape, thereby decreasing the volume of the chest cavity which pushes the breath out of the lungs.

Diaphragm along with the anterolateral abdominal muscles help to increase the intra-abdominal pressure in processes like defecation, urination, emesis and childbirth.[7] Diaphragm acts as a sphincter to prevent gastroesophageal reflux. Diaphragm supports the vertebral column and prevents flexion weight lifting with Valsalva maneuver.[8]

The diaphragm may occasionally contract involuntarily due to direct irritation by the acid in the stomach or following a carbonated drink. Hiccups are produced, if the air is inhaled at the times of diaphragm contraction, because the space between the vocal cords at the back of the throat closes suddenly.

DYSFUNCTION

- Dysfunction can be caused by conditions that directly damage the diaphragm, like:
  ○ Trauma
  ○ Cardiothoracic surgery
  ○ Adjacent thoracic pathology
  ○ Abdominal pathology
  ○ Muscular dystrophies.
- Diaphragm movement can also be affected by:
  ○ Central nervous system diseases
  ○ Phrenic nerve involvement (during cardiothoracic or neck surgery or due to compression caused by bronchogenic or mediastinal tumors)[9]
  ○ Motor neuron disease, infections (e.g. herpes zoster and Lyme disease) and inflammatory disorders[10]
  ○ Diseases of the neuromuscular junction.
- Diaphragm dysfunction can be caused by hypothermia, traction, cauterizing or severing of the phrenic nerve.

CLINICAL PRESENTATION OF DIAPHRAGM DYSFUNCTION

Unilateral (U/L) diaphragmatic palsy: Unexplained dyspnea especially in the supine position, difficulty in weaning from oxygen or mechanical ventilation, recurrent unilateral pneumonia and recurrent unilateral lung collapse.

Bilateral (B/L) diaphragmatic palsy: Abdominal paradox with the maximum transdiaphragmatic pressure generated against closed airway of less than 30 cm of H_2O.[11] Abdominal paradox is the abnormal pattern of breathing where the accessory inspiratory muscles of the rib cage and neck contract and lower pleural pressure causing the weakened or flaccid diaphragm to move in cephalad direction and inward movement of the abdominal wall during inspiration.[12] The critical-illness myopathy and polyneuropathy are some of the common causes of diaphragmatic weakness and ventilator dependency in patients in the intensive care unit.[13]

DIFFERENT TECHNIQUES FOR EVALUATING THE DIAPHRAGM

Different techniques are available for evaluating the diaphragm (Table 1).

ULTRASONOGRAPHY

The word "*ultrasound*", refers to sound waves with a frequency higher than that which is audible to human ear. Ultrasound imaging is popular due to its noninvasive nature and frequencies between 2 MHz and 18 MHz are commonly used.[21] Higher frequencies provide better quality

Table 1: Techniques for evaluating the diaphragm.

Techniques	Uses	Drawbacks
Chest radiography[14]	Diaphragm weakness	Cannot predict normal motion of diaphragm
Fluoroscopy[15]	Assess excursion of domes Shift of diaphragm via sniff test of Hitzenberger Unilateral or bilateral paralysis of diaphragm	Radiation exposure Need spontaneous breathing Not ideal for critically ill-patient
Computed tomography[16]	Assess diaphragm structure	Cannot assess dynamic imaging
Magnetic resonance[17]	Evaluation of excursion, synchronicity and velocity of diaphragmatic motion	Operator dependence, limited availability, expensive and the need for patient transport
Pulmonary function test[11]	Not an imaging technique To diagnose diaphragm weakness and lung pathology	Dependence on patient effort and confounding factor could be chest wall and lung diseases
Transdiaphragmatic pressure via esophageal or gastric transducer[18]	Diagnosis of bilateral diaphragm paralysis	Invasive, time consuming, and is not useful in diagnosing unilateral weakness
Electromyography[19]	Can detect denervation and differentiate between neuropathic and myopathic causes of paralysis	Uncomfortable Technically challenging to perform Risk of pneumothorax
Ultrasonography[19,20]	Portable, no risk of ionizing radiation, to assess both the structural and functional components of the diaphragm	Operator dependent

images but are more readily absorbed by the skin and other superficial tissues, causing dampening. Lower frequencies can penetrate deeper, with poor image quality (Table 2).

In 1969, ultrasonography (USG) of the diaphragm, in 1989 diaphragm thickness, in 2011 excursion in ICU patient, in 2012 validated thickness in healthy patients were assessed. Four different modes of ultrasound are used in imaging. These are (Table 3):

- *A-mode*: A single transducer scans a line through the body with the echoes plotted on screen as a function of depth.
- *B-mode*: In B-mode ultrasound, a linear array of transducers simultaneously scans a plane through the body that can be viewed as a two-dimensional image on screen.
- *M-mode*: M stands for motion. In M-mode, a rapid sequence of B-mode scans whose images follow each other in sequence on screen enables doctors to see and measure range of motion, as the organ boundaries that produce reflections move relative to the probe.

Ultrasonographical Appearance of Diaphragm

Curved, muscular echotexture, deep located, mixed echogenic. Composed of four components:
- Transverse septum (which is anterior and becomes the central tendon of the diaphragm)
- Pleuroperitoneal folds
- Esophageal mesentery
- Muscular body wall.

Ultrasound Probes (Tables 2 and 4)

Patient Position

Supine spontaneously breathing patient position is considered the best during sonographic examination of the diaphragm. The reason being supine positioning causes maximal

Table 2: Types of probes and their uses.

Curvilinear low frequency probe	Linear high frequency probe
1–3 MHz	5–18 MHz
To evaluate deeper structures	To examine superficial structures
To assess diaphragmatic excursion	To measure diaphragmatic thickness
Limitation is less spatial resolution. M-mode is used to assess diaphragmatic excursion	B-mode is used for examining the echotexture and measure the thickness

Table 3: Comparison of two ultrasound views.

Longitudinal view (mixed echogenic appearance)	Transverse view
Hypoechoic muscle fibers	Hypoechoic muscle fibers
Hyperechoic fibroadipose septae (perimysium)	Two echogenic layers of pleura and peritoneum Starry night appearance[22]

Table 4: Comparison of B- and M-modes of ultrasound.

B-mode ultrasound	M-mode ultrasound
Real-time imaging	A single beam of image on the Y-axis as it changes over time on the X-axis
Thickness and echogenicity	Excursion: velocity and response to phrenic nerve stimuli can be measured
It also allows identification of a good window for the use of M-mode	It uses a single beam of image and records the successive positions of the structures

diaphragmatic excursion for the same volume of air being inspired in sitting or standing position. Also, this position has shown to have less side-to-side variability and greater reproducibility.

Relationship between inspired volume of air and diaphragm movement has been shown to correlate better in the supine than in sitting position. Also, supine positioning causes exaggerated paradoxical motion of diaphragm, if present and uncovers any active exhalation caused by anterior abdominal muscle action. Spontaneous respiration with deep breathing or sniff maneuver helps to identify the moving diaphragm easily.

Modes

Ultrasound helps to evaluate mainly the posterior and lateral part of the diaphragm. The diaphragm is usually higher in children, young adults, and obese individuals. The position and motion of the diaphragm also depends on the posture of the subjects.[23,24]

Windows

On the right side, liver window helps in visualization of right diaphragm. Left diaphragm visualization is possible through the small spleen window. The examination of left diaphragm can therefore be easier in cases with splenomegaly, left pleural effusion, left upper quadrant mass, hepatomegaly with large left lobe.[24]

Views

USG description of different views of diaphragm and its appearance is described in Table 5.

THICKNESS OF DIAPHRAGM

The following are of importance to access thickness:
- High frequency transducer (7–18 MHz) should be used
    ○ Probe is placed at the anterior axillary line
    ○ 7th/8th or 8th/9th intercostal space
    ○ Sagittal plane view

Table 5: Ultrasonographic description of different views of diaphragm.

View	Transducer	Placement	Observation
Intercostal	High frequency linear array transducer	At the anterior axillary line, with the transducer positioned at the intercostal space between the 7th and 8th or 8th and 9th ribs. An image spanning two ribs, with the intercostal space between the ribs, can be obtained	In this view (Fig. 2), visualization of the diaphragm is limited to the zone of apposition where diaphragm thickness and echogenicity are assessed. In this view, the image may be obscured with deep inspiration when the lung is displaced downwards
Anterior subcostal	Low frequency curvilinear transducer	Subcostally between the midclavicular and anterior axillary lines, with the transducer directed medially, cranially, and dorsally. The ultrasound beam is directed to the posterior third of the right diaphragm approximately 5 cm lateral to the inferior vena caval foramen	B mode (Fig. 3) is used to visualize the diaphragm moving towards or away from the transducer. M-mode with the line perpendicular to the plane of the diaphragm is used for measuring the diaphragmatic excursion. Amplitude and velocity of excursion of the diaphragm can be calculated in M-mode[17,25,26]
Posterior subcostal	Low frequency curvilinear transducer	In a sitting position, probe placed in sagittal planes of posterior subcostal region on either side[27]	In measuring diaphragm excursion (Fig. 4)
Subxiphoid	Low frequency curvilinear transducer	Below the xiphoid in a transverse orientation, angled upwards towards the posterior leaflets of the diaphragm, especially in children[28]	B-mode (Fig. 5) is used to visualize portions of both domes together on subxiphoid view obtained a real-time comparison of excursion at the midline, (quantitative assessment). M-mode applied to each dome separately helps to assess excursion (side to side variability) amplitude.[17] (quantitative assessment)

- Visualization of both the pleural and peritoneal membranes at all times while imaging the diaphragm for thickness measurements
- Zone of apposition is easily visualized

Normal average thickness of diaphragm is 0.22–0.28 cm. Paralyzed diaphragm thickness is 0.13–0.19 cm. Atrophy of diaphragm thickness < 0.2 cm (at the end of expiration).[29] Serial measurement of diaphragm thickness at same intercostal space should be done. Ultrasonography of the diaphragm in mechanically ventilated patients shows that thinning of diaphragm begins as early as 48 hrs of initiation of mechanical ventilation and occurs at the rate of 6% per day of mechanical ventilation. High tidal volume has got a deleterious effect, with enhancing the thinning of diaphragm; while, high PEEP and respiratory rate are shown to be protective.[30]

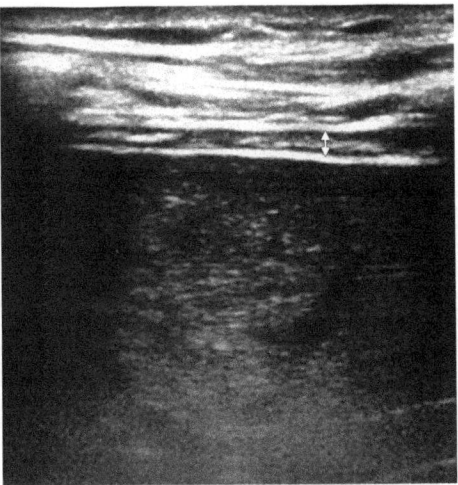

Fig. 2: B-mode sonography showing diaphragm with intercostal approach. A high frequency (10–12 MHz) probe was used.

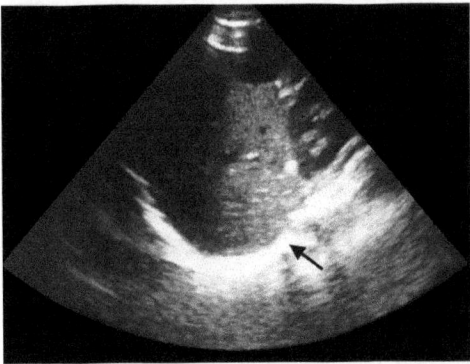

Fig. 3: B-mode sonography showing diaphragm in the zone of apposition during normal breathing. A curvilinear 4 MHz frequency probe was used with anterior subcostal approach.

Change in Thickness of Diaphragm

Diaphragm thickens on inspiration and thins out on exhalation (Fig. 6).
Change in thickness of diaphragm is measured as:
- T end inspiration—T end expiration
- T end expiration

(Where T end inspiration: thickness of diaphragm in end inspiration, T end expiration: thickness of diaphragm in end expiration.)

Diaphragm thickening less than 20% on maximal inspiration is consistent with paralysis. A chronically paralyzed diaphragm is thin, atrophic, and does not thicken on inspiration.[30] Diaphragm thickness correlates with muscularity and nutritional state of the person but is

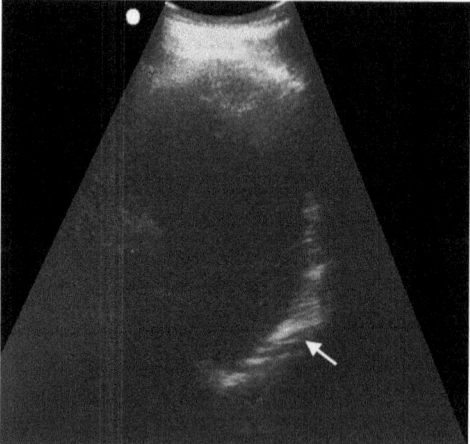

Fig. 4: B-mode sonography showing the diaphragm with posterior subcostal approach. A high frequency linear (10–12 MHz) probe was used.

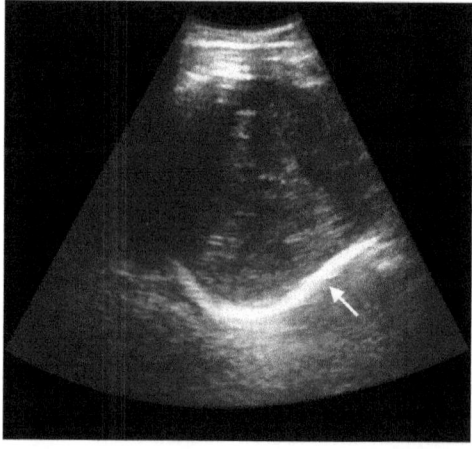

Fig. 5: B-mode sonography showing diaphragm with subxiphoid approach. A curvilinear 4 MHz frequency probe was used.

not affected by age, gender, body habitus or smoking. A side to side difference in diaphragm thickness at end expiration of more than 0.33 cm is abnormal. Normal range of motion of the diaphragm is greater posteriorly and laterally than anteriorly and medially.

Normal diaphragmatic excursion ranges from 1.9–9 cm during tidal respiration or deep breathing (Fig. 7).[31] Diaphragm excursion has been shown to be greater in men than in women and varies with age, weight and height.[32] Excursion of the diaphragm with maximum inspiration in healthy, standing patients is usually greater on the left side (Fig. 8). Normal difference in excursion between the right and left hemidiaphragms should be less than 50%. Normal range of side-to-side variability as defined by the right-to-left ratio of maximal excursion is 0.5–2.5 in the quiet respiration and 0.5–1.6 during deep breathing.[33,34]

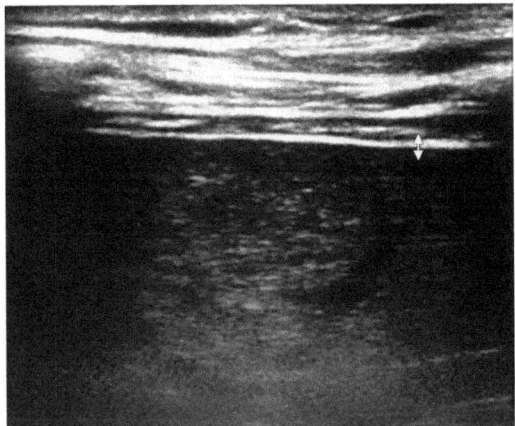

Fig. 6: B-mode sonography in the zone of apposition showing diaphragm thickness measured with an intercostal approach at the anterior axillary line, between 7th and 9th intercostal space with high frequency probe (10–12 MHz).

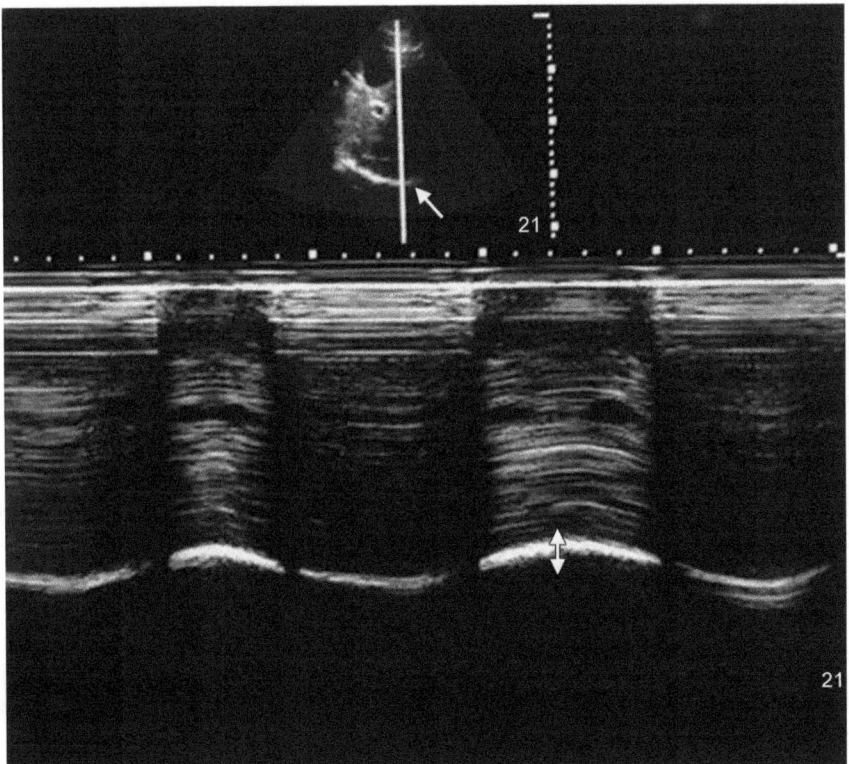

Fig. 7: Ultrasound view of diaphragmatic excursion in the zone of apposition during normal breathing in M-mode. A curvilinear 4 MHz frequency probe was used with anterior subcostal approach.

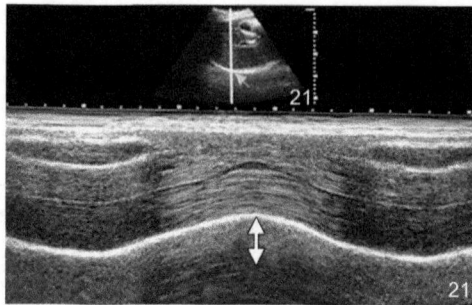

Fig. 8: The diaphragmatic excursion with deep breathing pattern in M-mode. A curvilinear 4 MHZ frequency probe was used with anterior subcostal approach.

The normal movement of diaphragm is towards the transducer during inspiration and away from the probe during exhalation. The reversal of this motion is defined as paradoxical motion of diaphragm. In adults during sniff maneuver excursion less than 2.5 cm precludes severe diaphragm dysfunction which denotes 50% decrease in vital capacity. Paradoxical motion during sniff is also one of the indicators of diaphragmatic dysfunction. Dysfunctional diaphragm with decreased excursion or paradoxical motion of diaphragm in USG related to prolong ventilation and delayed weaning.

DIAPHRAGMATIC VELOCITY

The sniff maneuver involves a quick, short, sharp breath which is maximal inspiratory effort through the nose. The velocity of diaphragm movement assessed in M mode; during the sniff maneuver, gives a reproducible and quantitative assessment of diaphragm strength (Fig. 7).[34] Increase in diaphragm velocity is almost 7-fold from 1.52 cm/s during quiet breathing to 10.4 cm/s during sniff.[32]

THE ECHOGENICITY OF DIAPHRAGM IN INTENSIVE CARE UNIT

A significant increase in the mean gray scale value (tibialis anterior only) indicates an increased muscle echogenicity, and a decrease in the grayscale standard deviation indicates that the muscle is more homogeneous. A homogenous muscle echotexture changes have been reported in muscular dystrophies and chronic inflammatory myopathies. Also, an increase in thickness of subcutaneous tissue over the diaphragm and rectus femoris has been documented during hospitalization.[35]

CLINICAL APPLICATIONS

Neuromuscular ultrasonography of diaphragm in both health and disease is evolving. The technique is used for various clinical applications like:
- Diagnosing diaphragmatic paralysis
- To detect the cause of diaphragmatic palsy (central nervous system etiology or lower motor neuron disease)

- Recovery following diaphragm palsy (improvement in diaphragmatic excursion on ultrasonography)
- Detect asynchrony between breathing efforts and the ventilator and optimize the same
- Patient who have paradoxical motion of diaphragm with mediastinal shift causing dyspnea, consider for surgical plication
- Diaphragmatic pacemakers in patients with persistent diaphragm palsy, allow them to become independent from mechanical ventilation.[31] Ultrasound is an excellent tool for providing a quantitative evaluation of diaphragm excursion while the pacemaker output is adjusted in real-time
- Respiratory dysfunction following acute stroke is attributed to decrease in unilateral or bilateral diaphragm motion during deep breathing but not seen on quiet breathing. It may help in decision on weaning from prolong mechanical ventilation[36]
- During spontaneous breathing trials, measured diaphragm excursion on M-mode ultrasound can predict weaning failure which is equal to the rapid shallow breathing index
- The cutoff of diaphragm excursion for predicting weaning failure is 1.4 cm for the right hemidiaphragm and 1.2 cm for the left hemidiaphragm, and less excursion is consistent with a greater chance of weaning failure[19]
- Electromyographic (EMG) examination of diaphragm carries risk of vital organ injury, which can be avoided by imaging. EMG needle can be guided into diaphragm via direct (needle in real-time) or indirect (measure depth prior to puncture) ultrasound guidance.[22]

LIMITATIONS OF ULTRASONOGRAPHY

- Operator dependent
- Intra- or inter-observer variability previously but in recent studies reliability has increased
- High correlation coefficient between and within observer
- Difficult to visualize in patient with large pleural effusion in standing position
- Paradoxical movement of an unparalyzed diaphragm found in respiratory diseases that cause restrict lung expansion, i.e. hydrothorax, negative pressure pneumothorax, lung fibrosis, atelectasis, and subphrenic abscess
- Difficult to measure diaphragmatic excursion as depends upon maximal voluntary inspiratory effort
- There are few studies on this topic to evaluated diaphragmatic parameter using ultrasound in patients with lung disease
- Both thickness and excursion of diaphragm can vary depending on the initial point of measurement either end of expiration or beginning of inspiration.[29,37]

REFERENCES

1. Downey R. Anatomy of the normal diaphragm. Thorac Surg Clin. 2011;21(2):273-9, ix.
2. Bordoni B, Zanier E. Anatomic connections of the diaphragm: influence of respiration on the body system. J Multidiscip Healthc. 2013;25(6):281-91.
3. Rives JD, Baker DD. Anatomy of the attachments of the diaphragm: their relation to the problems of the surgery of diaphragm hernia. Ann Surg. 1942;115(5):745-55.

4. Collis JL, Kelly TD, Wiley AM. Anatomy of the crura of the diaphragm and the surgery of hiatus hernia. Thorax. 1954;9(3):175-89.
5. Central Tendon of the Diaphragm in Primates. [online]. PubMed Journals. Available from https://ncbi.nlm.nih.gov/labs/articles/4968289/. [Accessed May 2017].
6. Pickering M, Jones JF. The diaphragm: two physiological muscles in one. J Anat. 2002;201(4):305-12.
7. Nason LK, Walker CM, McNeeley MF, et al. Imaging of the diaphragm: anatomy and function. Radiogr Rev Publ Radiol Soc N Am Inc. 2012;32(2):E51-70.
8. Harrison GR. The Anatomy and Physiology of the Diaphragm. In: Fielding J, Hallissey M (Eds). Upper Gastrointestinal Surgery. London: Springer; 2005. pp. 45-58.
9. Mehta Y, Vats M, Singh A, et al. Incidence and management of diaphragmatic palsy in patients after cardiac surgery. Indian J Crit Care Med. 2008;12(3):91-5.
10. Nicaise C, Hala TJ, Frank DM, et al. Phrenic motor neuron degeneration compromises phrenic axonal circuitry and diaphragm activity in a unilateral cervical contusion model of spinal cord injury. Exp Neurol. 2012;235(2):539-52.
11. McCool FD, Tzelepis GE. Dysfunction of the diaphragm. N Engl J Med. 2012;366(10):932-42.
12. Ahmed R, McNamara S, Gandevia S, et al. Paradoxical abdominal wall movement in bilateral diaphragmatic paralysis. Pract Neurol. 2012;12(3):184-6.
13. Chawla J, Gruener G. Management of critical illness polyneuropathy and myopathy. Neurol Clin. 2010;28(4):961-77.
14. Fretzayas A, Moustaki M, Nicolaidou P, et al. Transient elevation of the ipsilateral hemidiaphragm associated with pneumonia. Can Respir J J Can Thorac Soc. 2011;18(4):e66-7.
15. Mageras GS, Yorke E, Rosenzweig K, et al. Fluoroscopic evaluation of diaphragmatic motion reduction with a respiratory gated radiotherapy system. J Appl Clin Med Phys. 2001;2(4):191-200.
16. Sarwal A, Walker FO, Cartwright MS. Neuromuscular ultrasound for evaluation of the diaphragm. Muscle Nerve. 2013;47(3):319-29.
17. Chavhan GB, Babyn PS, Cohen RA, et al. Multimodality imaging of the pediatric diaphragm: anatomy and pathologic conditions. Radiogr Rev Publ Radiol Soc N Am Inc. 2010;30(7):1797-817.
18. Lerolle N, Guérot E, Dimassi S, et al. Ultrasonographic diagnostic criterion for severe diaphragmatic dysfunction after cardiac surgery. Chest. 2009;135(2):401-7.
19. Kim WY, Suh HJ, Hong SB, et al. Diaphragm dysfunction assessed by ultrasonography: influence on weaning from mechanical ventilation. Crit Care Med. 2011;39(12):2627-30.
20. Houston JG, Fleet M, Cowan MD, et al. Comparison of ultrasound with fluoroscopy in the assessment of suspected hemidiaphragmatic movement abnormality. Clin Radiol. 1995;50(2):95-8.
21. Ihnatsenka B, Boezaart AP. Ultrasound: Basic understanding and learning the language. Int J Shoulder Surg. 2010;4(3):55-62.
22. Boon AJ, Alsharif KI, Harper CM, et al. Ultrasound-guided needle EMG of the diaphragm: technique description and case report. Muscle Nerve. 2008;38(6):1623-6.
23. Openshaw P, Edwards S, Helms P. Changes in rib cage geometry during childhood. Thorax. 1984;39(8):624-7.
24. Gerscovich EO, Cronan M, McGahan JP, et al. Ultrasonographic evaluation of diaphragmatic motion. J Ultrasound Med. 2001;20(6):597-604.
25. Urvoas E, Pariente D, Fausser C, et al. Diaphragmatic paralysis in children: diagnosis by TM-mode ultrasound. Pediatr Radiol. 1994;24(8):564-8.
26. Ayoub J, Metge L, Dauzat M, et al. Diaphragm kinetics coupled with spirometry. M-mode ultrasonographic and fluoroscopic study; preliminary results. J Radiol. 1997;78(8):563-8.
27. Fedullo AJ, Lerner RM, Gibson J, et al. Sonographic measurement of diaphragmatic motion after coronary artery bypass surgery. Chest. 1992;102(6):1683-6.

28. Diament MJ, Boechat MI, Kangarloo H. Real-time sector ultrasound in the evaluation of suspected abnormalities of diaphragmatic motion. J Clin Ultrasound. 1985;13(8):539-43.
29. Wait JL, Nahormek PA, Yost WT, et al. Diaphragmatic thickness-lung volume relationship in vivo. J Appl Physiol (1985). 1989;67(4):1560-8.
30. Gottesman E, McCool FD. Ultrasound evaluation of the paralyzed diaphragm. Am J Respir Crit Care Med. 1997;155(5):1570-4.
31. Ayoub J, Cohendy R, Dauzat M, et al. Non-invasive quantification of diaphragm kinetics using M-mode sonography. Can J Anaesth. 1997;44(7):739-44.
32. Harris RS, Giovannetti M, Kim BK. Normal ventilatory movement of the right hemidiaphragm studied by ultrasonography and pneumotachography. Radiology. 1983;146(1):141-4.
33. Houston JG, Morris AD, Howie CA, et al. Technical report: quantitative assessment of diaphragmatic movement—a reproducible method using ultrasound. Clin Radiol. 1992;46(6):405-7.
34. Kantarci F, Mihmanli I, Demirel MK, et al. Normal diaphragmatic motion and the effects of body composition: determination with M-mode sonography. J Ultrasound Med. 2004;23(2):255-60.
35. Kumar S, Reddy R, Prabhakar S. Contralateral diaphragmatic palsy in acute stroke: An interesting observation. Indian J Crit Care Med. 2009;13(1):28-30.
36. Summerhill EM, El-Sameed YA, Glidden TJ, et al. Monitoring recovery from diaphragm paralysis with ultrasound. Chest. 2008;133(3):737-43.
37. Cartwright MS, Kwayisi G, Griffin LP, et al. Quantitative neuromuscular ultrasound in the intensive care unit. Muscle Nerve. 2013;47(2):255-9.

15B Ultrasonography for Assessment of Muscles Dysfunction in ICU

Vijay Hadda, Rohit Kumar

INTRODUCTION

Muscle dysfunction is commonly observed in critically ill patients admitted in intensive care unit (ICU). It is characterized by decreased muscle mass (sarcopenia) and manifests clinically as weakness. It is an independent risk factor associated with in-hospital mortality, morbidity such as difficult weaning, nosocomial infection, poor quality of life and increased cost of care. Also, the physical functioning and health-related quality of life may remain significantly impaired even after as long as 2 years following discharge from the ICU. The prevalence of weakness in ICU varies with the diagnostic criteria used, the timing at which the assessment was made and type of population studied. A systemic review reported the incidence of an ICU-acquired weakness syndrome to be 40% (95% confidence interval 38–42%).[1]

ASSESSMENT OF MUSCLE FUNCTION

Considering the impact of muscle dysfunction among critically ill patients, accurate evaluation of muscle function is important. The evaluation of muscle functions is challenging in patients admitted to ICU. It is done traditionally by using a combination of anthropometric measures and laboratory investigations.

Conventional Methods of Assessment of Muscle

These include clinical examination [Medical Research Council (MRC) Score], anthropometry [body mass index (BMI), body weight and mid-arm and mid-thigh circumference], electrophysiological studies [nerve conduction velocity (NCV), electromyography (EMG)], and biopsy of the muscle. Clinical examination for calculation of MRC score for assessment of muscle function in the ICU is limited by multiple factors, such as sedation, altered sensorium, use of neuromuscular blockers and noncooperativeness. Anthropometry measures including mid-arm and mid-thigh circumference examination are easy to perform but rely on normal hydration status which is not the case among patients in ICU. Electrophysiological studies require special equipment and shifting of patients from ICU to the electrophysiology laboratory. Also, expertise is required for interpretation of the results. Muscle biopsy may provide accurate account of the muscle changes in this setting; however, it is an invasive test. Due to its invasive nature, it is not a preferred test and repeat testing is not practical; it also requires expertise in analyzing the results.

Imaging for Assessment of Muscle Function

Muscle functions are determined by muscle mass, therefore, tools which can measure muscle mass may potentially be used in ICU also. Muscle mass can be measured as either muscle thickness, cross-sectional area or volume of the muscle. Among these, muscle thickness assessment seems easy to perform and may be used a good surrogate of muscle function. Various imaging tools which have been used to measure muscle thickness include computer tomography (CT) scan, magnetic resonance imaging (MRI) and dual energy X-ray absorptiometry (DEXA). These can measure muscle thickness with great accuracy. However, the equipment required for these measurements are bulky and expensive. Specific expertise may be required to interpret the images. However, one should note that both CT scan and DEXA are associated with significant exposure to ionizing radiation. Also, for performing these, most of the time patients need to be shifted to a dedicated laboratory outside the ICU, which may not be possible every time. Further, the utility of these imaging modalities as an outcome measure, where repeat testing is required, is limited.

Ultrasonography for Assessment of Muscle Function

Currently, most of the ICUs are equipped with ultrasound machines and ultrasonography (USG) has been integrated in common protocols for management of critically ill patients in ICU. USG due to its inherent property gives different echogenic shadows to subcutaneous tissue, muscle and bone. Further, due to its noninvasive nature, bedside availability, and repeatability with no radiation exposure, USG appears to be an excellent tool for the assessment of muscle functions. It can be used for assessment of both peripheral (limb) muscles as well as diaphragm.

Ultrasonography for Peripheral Muscles

The peripheral muscles of both upper as well as lower limbs are a convenient site to use USG for assessment of the muscle thickness. The peripheral muscle can be easily visualized with USG which can measure the thickness, cross-sectional area and echogenicity of the muscle. Among all, muscle thickness is widely used for the purpose of assessment of muscle functions.

Technique and equipment: Muscles which have been studied on USG include muscles of arm, forearm and thigh. For measurement of thickness of arm, the protocol used by Campbell[2] and recently by few others may be used.[3] We also demonstrated that this method was highly reliable and reproducible.[4] The measurements are done on the B-mode of USG using 5.0–13.0 MHz linear array probe (VF 13-5).

- *Muscles of arm:* For measurements of arm muscles, the patient should be lying supine, the elbow extended and forearm supinated with palm facing the ceiling. A point should be identified and marked for measurement and recording of the muscle thickness. The mid-arm or junction of upper 2/3 and lower 1/3 are preferred sites. This is important for the uniformity during the subsequent measurements. We prefer to measure the arm muscle thickness at the mid-arm level. For the measurements, first a circumferential mark is made midway between the tip of the greater tuberosity and tip of the olecranon process of humerus. Then, the linear probe is placed on this circumferential line, perpendicular to the skin and the probe positioned along the line drawn till a suitable image is obtained. Then a point on the circumferential mark which was corresponding to the center of the probe is marked with a vertical line. This point is used as the reference point for all subsequent measurements. This ensures that all measurements are made at the same level. For the measurement of muscle thickness, the probe is held perpendicular to the skin to negate any variation in the recording of muscle thickness due to angulation of the probe. During measurement of the muscle thickness, the force applied to the probe should be just sufficient to make adequate contact between skin and probe without compressing the underlying muscle. The images may be centered in the screen by little manipulation of the probe.

The thickness of the flexor compartment is measured between the superficial fat muscle interface and the humerus (Fig. 1). This includes both the biceps brachii and the coracobrachialis muscles.

- *Muscles of thigh:* For measurement of quadriceps muscle thickness, the patient should be lying supine with the knee extended and toe facing the ceiling. Same posture needs to be maintained whenever the measurements are being recorded. A circumferential mark is made at the point where one wants to measure the thickness of quadriceps muscle. We prefer the site midway between the tip of the greater trochanter and the lateral joint line of the knee.[5] The linear probe of USG is placed on this circumferential line, perpendicular to the skin and the probe is moved along the line till a suitable image is obtained. Then the point corresponding to the center of the probe is marked with a vertical line. This point

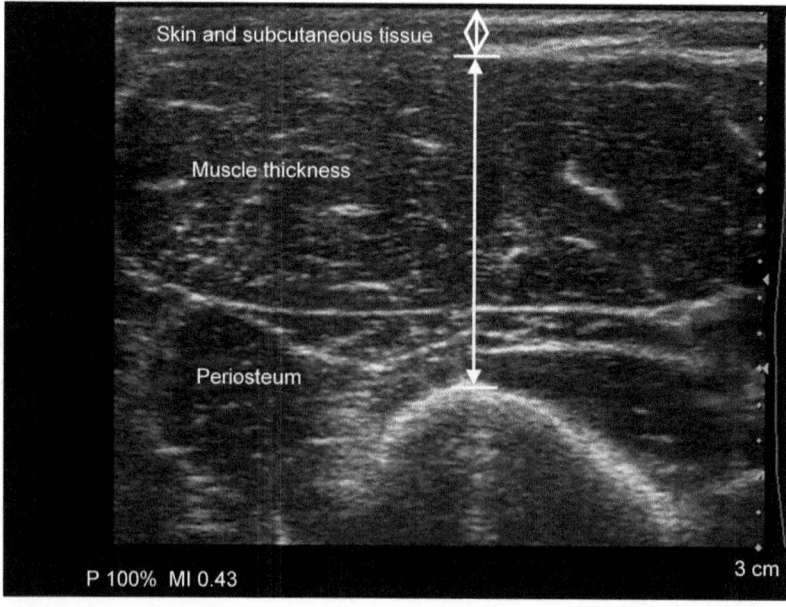

Fig. 1: Ultrasound image of the anterior arm. The muscle thickness is measured from the superficial fat-muscle interface to the periosteum.

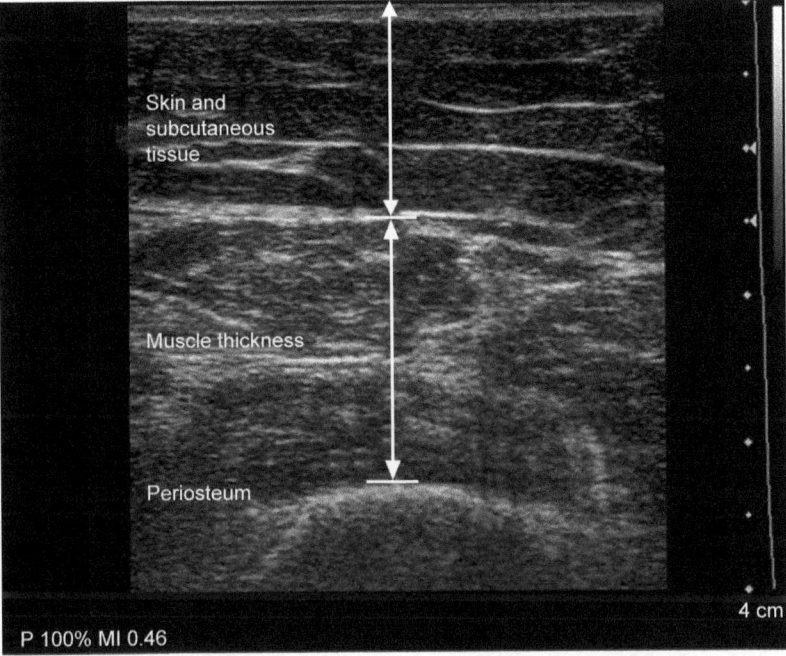

Fig. 2: Ultrasound image of the anterior thigh muscles (quadriceps). The muscle thickness is measured from the superficial fat-muscle interface to the periosteum.

should be marked as the reference point for all subsequent measurements. The skeletal muscle thickness of the quadriceps muscle group between the superficial fat-muscle interface and the periosteum of the femur may be measured as shown in Figure 2.

Ultrasonography for Diaphragm

- *Diaphragm:* USG can be used for both, assessment of thickness as well as the diaphragmatic excursion. Diaphragm thickness is usually measured using B-mode in the 8–10 intercostal space (zone of apposition). USG examination at this place allows for the direct visualization of the diaphragm thickness which can be easily measured in both phases of respiration. The diaphragm will be thickest at end inspiration and thinnest at end expiration. The thickening of the diaphragm (that is the diaphragm thickening fraction) during active breathing which may reflect the magnitude of diaphragmatic effort, similarly to an ejection fraction of the heart, can also be calculated using USG. The thickness fraction of diaphragm is calculated by formula: thickness at end inspiration—thickness at end expiration/thickness at end expiration.

For assessment of excursion of the diaphragm, the best place is subcostal region. For this ultrasound probe is applied at the subcostal region, with the liver acting as an acoustic widow, the diaphragm can be easily visualized (Fig. 3). The movement of the

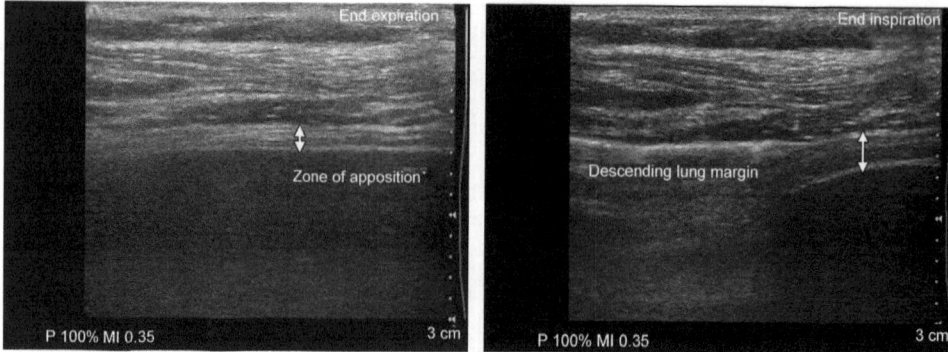

Fig. 3: Ultrasound images of the diaphragm at the zone of apposition (using the linear probe). The diaphragm was identified as the last set of parallel lines, the pleural and peritoneal membranes overlying the less echogenic muscle. The muscle thickness can be measured both in inspiration and the expiration.

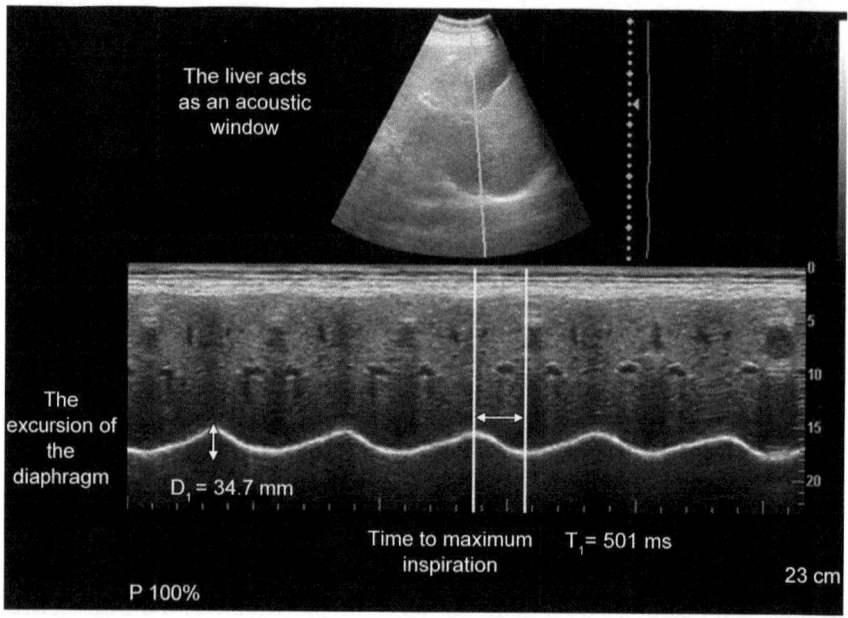

Fig. 4: Ultrasound image of the diaphragm (subcostal view, the liver as an acoustic window) using M-mode with curvilinear probe. The diaphragmatic excursions (D_1) and the time to maximum inspiration (T_1) can be seen. The speed of contraction was calculated as 69.2 mm/sec.

diaphragm may be appreciated as undulating line with increasing amplitude during inspiration as diaphragm comes towards probe and decreasing amplitude during expiration when diaphragm moves away from the USG probe. By using the M mode (motion mode), we can easily measure the thickening of the diaphragm (Fig. 4) and also determine the diaphragmatic excursion and the rapidity (velocity) of the diaphragm contraction (Fig. 5).

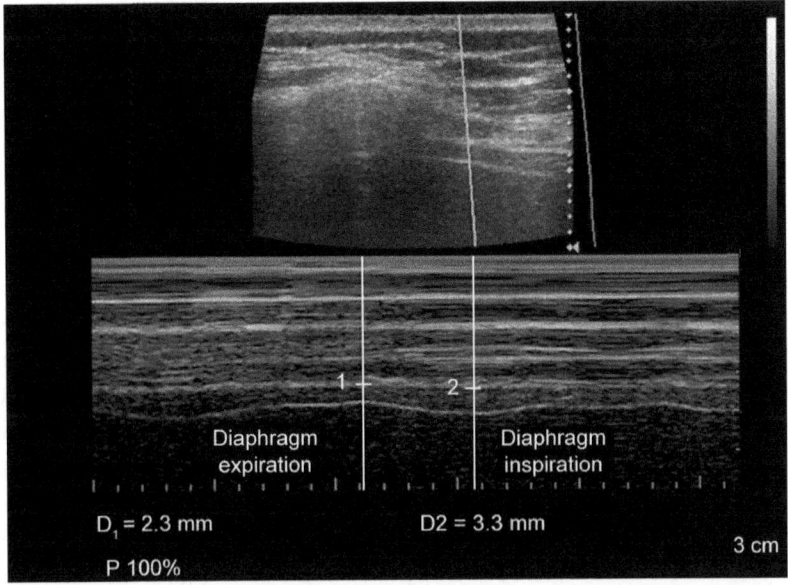

Fig. 5: Ultrasound image of the diaphragm (M-mode) using the linear probe. The muscle thickness is measured between the two intense lines. The thickening of the diaphragm during inspiration can be measured.

RELIABILITY OF ULTRASONOGRAPHY FOR ASSESSMENT OF MUSCLE FUNCTIONS

There is some concern regarding reliability and reproducibility of USG results. To investigate the accuracy of an USG for measuring the change in leg muscle volume, against the gold standard magnetic resonance imaging (MRI), a study was conducted.[6] USG results were compared with MRI data on 24 subjects undergoing 60 days of bed rest. Authors reported a good correlation between the muscle volume change measured with the USG and MRI.[6] This confirmed that the USG is sufficiently accurate for assessment of muscles. Furthermore, USG measurements of the peripheral muscles have good inter- and intraobserver reliability as shown by Tillquist and colleagues.[3] Similar results have been observed when reliability was assessed for ultrasound of the diaphragm. In a large study measuring diaphragmatic excursion, it was seen that there was excellent intraobserver and interobserver reproducibility.[7] Concerning the reproducibility of diaphragmatic thickness measurements, Vivier et al.[8] assessed analyzer reproducibility, intra-analyzer reproducibility (same settings analyzed repeatedly) and inter-analyzer reproducibility (same recordings obtained separately by two different ultrasonographers). The values reported for repeatability (intra-class correlation coefficients) were all above 0.97. In our experience, USG is an excellent tool for assessment of muscle thickness of anterior arm,[4] quadriceps[5] and diaphragm.

LITERATURE SUPPORTING THE USE OF ULTRASONOGRAPHY FOR ASSESSMENT OF MUSCLE FUNCTION IN INTENSIVE CARE UNIT

For Peripheral Muscle

There are few studies which have shown the utility of USG in ICU settings. Campbell et al.[2] in their study performed serial ultrasound measurements in nine critically ill-patients to demonstrate muscle loss. They performed a minimum of five measurements between day 1 and day 5 on each patient at the mid-arm, mid-forearm and mid-thigh level. They had shown a median rate of decline of 6% per day (ranging from 2–9.2% per day). Later on, Reid and colleagues[9] also used USG for the same purpose. They also showed that there was a decrease in muscle thickness on USG but the rate was lower (median 1.6% per day) than Campbell's study. Researchers have also used USG for serial examination of the rectus femoris cross-sectional area among patients admitted in ICU and demonstrated that cross-sectional area decreases during stay in ICU.[10] Recently, there has been an interest in the change of USG echogenic characteristics of muscle during ICU. In a study by Grimm et al.,[11] the echogenicity of muscles was graded semi-quantitatively in muscles and it was seen that patients showed a significant difference in mean muscle echotexture between patients and controls. In addition, from day 4 to day 14, the mean grades of muscle echotexture increased in the patient group, although the values did not reach significance levels. These results suggest that further studies are required to know the importance of USG echogenic characteristics of the muscle among critically ill-patients.

For Diaphragm

Ultrasonography has been used to demonstrate changes in diaphragm during ICU stay. Initially, Grosu and colleagues used USG for measurement of the diaphragm thickness, daily in seven mechanically ventilated patients, and reported a mean decline of 6% per day in the diaphragm thickness.[12] Similarly, other investigators also used USG and reported a significant decline in diaphragm thickness.[13,14] Francis et al.[13] reported USG thickness of diaphragm in four patients on control mode and they showed a diaphragm thickness decline of 4.7% per day.[13] Till date, largest published data on USG for diaphragm thickness on mechanical ventilated patients is by Schepens et al. They did USG on 54 mechanically ventilated patients and showed that the diaphragm thickness decreased by 9% after one day of mechanical ventilation, and a total decrease of 21% and 26% was observed after 2 days and 3 days of mechanical ventilation, respectively.[14]

The potential role of diaphragm USG as a weaning index is an area of active research.[15] In a study to assess the role of diaphragm thickening fraction (DTF) assessed by ultrasound as a weaning index, it was seen that a cut-off value of a DTF more than 36% was associated with a successful spontaneous breathing trial with a high sensitivity (0.82), specificity (0.88), positive predictive value (0.92), and negative predictive value (0.75).[16]

Muscle ultrasound is also being considered as a potential biomarker of muscle mass and assessment of nutritional status. It could be used for serial assessment of the muscle mass and for tracking changes in the muscle mass during the course of different diseases. It could also be a useful tool to study the impact of pharmacological and nonpharmacological interventions on the muscle mass of patients.

CONCLUSION

Ultrasonography is a useful tool in the hands of critical care physicians in ICU for assessment of muscle functions of peripheral muscles as well as diaphragm. Its bedside availability, non-invasive nature, easy to learn and no any additional risk make it practical tool for this purpose. Additionally, it can be used for serial assessment.

REFERENCES

1. Appleton RTD, Kinsella J, Quasim T. The incidence of intensive care unit-acquired weakness syndromes: A systematic review. J Intensive Care Soc. 2014;16(2):126-36.
2. Campbell IT, Watt T, Withers D, et al. Muscle thickness, measured with ultrasound, may be an indicator of lean tissue wasting in multiple organ failure in the presence of edema. Am J Clin Nutr. 1995;62(3):533-9.
3. Tillquist M, Kutsogiannis DJ, Wischmeyer PE, et al. Bedside ultrasound is a practical and reliable measurement tool for assessing quadriceps muscle layer thickness. JPEN J Parenter Enteral Nutr. 2014;38(7):886-90.
4. Hadda V, Kumar R, Dhungana A, et al. Inter- and intra-observer variability of ultrasonographic arm muscle thickness measurement by critical care physicians. J Postgrad Med. 2017;63(3):157-61.
5. Hadda V, Khilnani GC, Kumar R, et al. Intra- and inter-observer reliability of quadriceps muscle thickness measured with bedside ultrasonographic by critical care physicians. Indian J Crit Care Med. 2017;21(7):448-52.
6. Arbeille P, Kerbeci P, Capri A, et al. Quantification of muscle volume by echography: comparison with MRI data on subjects in long-term bed rest. Ultrasound Med Biol. 2009;35(7):1092-7.
7. Boussuges A, Gole Y, Blanc P. Diaphragmatic motion studied by M-mode ultrasonography: methods, reproducibility, and normal values. Chest. 2009;135(2):391-400.
8. Vivier E, Mekontso Dessap A, Dimassi S, et al. Diaphragm ultrasonography to estimate the work of breathing during non-invasive ventilation. Intensive Care Med. 2012;38(5):796-803.
9. Reid CL, Campbell IT, Little RA. Muscle wasting and energy balance in critical illness. Clin Nutr. 2004;23(2):273-80.
10. Puthucheary ZA, Rawal J, McPhail M, et al. Acute skeletal muscle wasting in critical illness. JAMA. 2013;310(15):1591-600.
11. Grimm A, Teschner U, Porzelius C, et al. Muscle ultrasound for early assessment of critical illness neuromyopathy in severe sepsis. Crit Care. 2013;17(5):R227.
12. Grosu HB, Lee YI, Lee J, et al. Diaphragm muscle thinning in patients who are mechanically ventilated. Chest. 2012;142(6):1455-60.
13. Francis CA, Hoffer JA, Reynolds S. Ultrasonographic evaluation of diaphragm thickness during mechanical ventilation in intensive care patients. Am J Crit Care. 2016;25(1):e1-8.
14. Schepens T, Verbrugghe W, Dams K, et al. The course of diaphragm atrophy in ventilated patients assessed with ultrasound: a longitudinal cohort study. Critical Care (London, England). 2015;19:422.
15. DiNino E, Gartman EJ, Sethi JM, et al. Diaphragm ultrasound as a predictor of successful extubation from mechanical ventilation. Thorax. 2014;69(5):423-7.
16. Ferrari G, De Filippi G, Elia F, et al. Diaphragm ultrasound as a new index of discontinuation from mechanical ventilation. Crit Ultrasound J. 2014;6(1):8.

Chapter
15C

Ultrasonography in Deep Vein Thrombosis

Akhil Kant Singh

INTRODUCTION

Chronic venous disease (CVD) describes a spectrum of signs and symptoms secondary to inefficient venous drainage from the lower limbs secondary to dysfunction in the superficial and/or deep venous systems. Although CVD encompasses diseases attributable to dysfunction in both superficial and deep venous systems, we will be limiting ourselves to thrombosis affecting the deep venous system and their consequences.

Venous thromboembolism (VTE) comprising of deep vein thrombosis (DVT), and its dreaded consequences, pulmonary thromboembolism (PTE) remains a major source of cardiovascular morbidity and mortality the world over, with 9–10% of all hospital deaths attributable to pulmonary embolism (PE) alone. Hospitalization contributes to more than half of the incidence of VTE, with almost around 24% of such cases attributable to admission for surgery.[1]

Most of the epidemiological data concerning VTE comes from the developed world. In the Unites States, the overall annual incidence of VTE has been described as being between 1 and 2 per thousand of the population, or 300,000 to 600,000 cases. The incidence of VTE increases with advanced age. Men have higher overall incidence, while women have a slightly increased risk during the reproductive years.[2]

The national prevalence of VTE in the United States was calculated for the period from 2002 to 2006 using a database of health insurance claims filed, of 12.7 million patients.[3] The prevalence of VTE was found to be 3.2 per 1000 in 2002, and 4.2 per 1000 in 2006. Cohen and colleagues estimated the total number of symptomatic VTE episodes in six European countries (France, Italy, Spain, Germany, Sweden and United Kingdom) to be 765,000, with 370,000 VTE related deaths.[4]

The mortality related to VTE is high. It is estimated that 10–30% of all patients of VTE have mortality with pulmonary embolism being the most common cause of death. Up to half of patients presenting with an initial episode of DVT of the lower limbs end up having chronic venous insufficiency and post-thrombotic syndrome, conditions which grossly impair quality of life.

The prospective registry on venous thromboembolic events (PROVE) study is a multinational study which enrolled 3,526 patients in 19 countries, out of which 667 patients were from India.[5] Compared with the overall PROVE population, the prevalence of previous DVT, PE and superficial leg thrombosis were similar in the Indian population. However, very few enrolled patients (5%) had received prophylaxis before their event. This finding was corroborated by the ENDORSE study as well, which found that only 17.4% of the Indian patients received prophylaxis recommended by the American College of Chest Physicians (ACCP) compared to 50.2% of the global population.[6]

Estimates about the healthcare costs related to VTE ranges from \$7,594 to \$16,644. Thus, the annual estimated costs come to around \$2 billion to \$10 billion annually.[7]

Thus, VTE is a cause of significant morbidity and mortality, and consequently, increased costs and resource utilization. It makes clinical and financial sense to prevent and detect thrombotic events at an early stage before they progress to more disastrous consequences.

A certain diagnosis of DVT can only be made on the basis of imaging. Contrast venography is considered the gold standard for the diagnosis of lower extremity DVT.[8] The diagnosis is based on the presence of a constant intraluminal filling defect on at least two projections. Treatment with anticoagulants could be safely withheld in patients with a technically normal venogram, and only 1.3% patients developed symptomatic DVT within 3 months.[9] The drawbacks of venography include it being an invasive and uncomfortable procedure, not available everywhere, restricted usability in patients having kidney insufficiency or those allergic to contrast dye. The technical difficulties which may be encountered include difficult cannulation (5%) and difficulty in visualizing the venous segments, which might occur in up to 20% cases.

Even with an adequate venogram, interpretation can be difficult at times with considerable interobserver and intraobserver variation.[10] Taking into consideration all the aforementioned drawbacks, it is easy to understand why venography is seldom used in clinical practice.

Venography has been replaced by newer imaging modalities, such as compression ultrasonography (CUS) and computed tomography pulmonary angiography (CTPA). Both of these techniques have been implemented as part of diagnostic algorithms. We will be concentrating on the merits and demerits of the ultrasound techniques.

ULTRASOUND

Ultrasound imaging, as has been the trend in medicine in general, has become a quick and easy way to diagnose and manage DVT, both on the floors and the critical care units. Venous ultrasound is the most extensively used study for the diagnosis of DVT.[11] It has largely replaced technically difficult procedures, such as venography.

Two techniques have gained importance in the diagnosis of DVT, namely CUS and duplex ultrasonography.

Duplex ultrasonography is carried out by specialized technicians or radiologists and is a comprehensive examination. It involves a variety of maneuvers as well as color flow and Doppler techniques. Color Doppler can be used to visualize proximal veins of the pelvis and abdomen where CUS is not feasible.

Compression ultrasound is, simply put, compressing the vessel while visualizing it using the ultrasound probe. If the vein can be easily compressed with gentle pressure, it implies absence of thrombus in the lumen.

As has been mentioned, duplex USG is a specialized technique and is best left to the radiologist. CUS is the technique which can be easily adopted by intensivists, anesthesiologists and physicians of other shades with minimal fuss. CUS has a gentle learning curve and when applied with known caveats in mind, is a sensitive modality for diagnosing DVT.

The diagnosis of DVT based purely on nonspecific clinical criteria is difficult. CUS, in combination with pretest probability and D-dimer has been extensively investigated and widely used in the diagnosis of DVT. ACCP recommends utilizing 2-point CUS, along with pretest probability scores and D-dimer levels for the management of DVT.[12]

Pretest probability scores based on well-defined clinical characteristics and risk factors, such as Wells score;[13] categorize patients into groups with high, moderate or low pretest probability of having DVT (Tables 1 and 2). A modification of the Wells score, groups patients as being likely or unlikely to have DVT, although this score has not been validated in large patient populations.

D-dimer is a fibrin degradation product, typically elevated in acute DVT. D-dimer is a sensitive albeit nonspecific marker for venous thromboembolic events. In essence, a positive D-dimer test does not confirm the diagnosis of VTE, but a negative test essentially rules out the diagnosis.

A detailed description of pretest probability assessment and D-dimer is outside the purview of this chapter. Therefore, a few references for the curious are provided.[14-18]

Deep Vein Thrombosis of the Lower Limbs

The lower limbs can be examined for DVT using two variations of CUS, two-point CUS and whole-leg CUS.

Two-point or proximal leg CUS entails examination of the proximal veins in the groin and the popliteal fossa. The advantages of this technique are its ease of use, rapidity and reproducibility. The major drawback is the need to repeat the examination after one week, if the first examination is negative, which inconveniences the patient. This is done to rule out

Table 1: Wells' score for DVT.[13]

Points	Clinical findings
1	Paralysis, paresis or recent orthopedic casting of lower extremity
1	Recently bedridden (more than 3 days) or major surgery within past 4 weeks
1	Localized tenderness in deep vein system
1	Swelling of entire leg
1	Calf swelling 3 cm greater than other leg (measured 10 cm below the tibial tuberosity)
1	Pitting edema greater in the symptomatic leg
1	Collateral nonvaricose superficial veins
1	Active cancer or cancer treated within 6 months
−2	Alternative diagnosis more likely than DVT (Baker's cyst, cellulitis, muscle damage, superficial venous thrombosis, post phlebitic syndrome, inguinal lymphadenopathy, external venous compression)

Table 2: DVT risk score interpretation.

3–8 Points	High probability
1–2 Points	Moderate probability
−2–0 Points	Low probability

extension of a previously unvisualized distal thrombus proximally, even though the reported detection rate of venous thrombosis after an initial negative examination is less than 2%.[19]

Whole-leg CUS is a more comprehensive examination of the deep venous system of the lower limbs. Multiple studies demonstrate that when the examination is normal, anticoagulation can be safely withheld.[20-23] The main advantage of this technique over proximal leg CUS is that there is no need of a follow-up examination. On the flip side, it is a time consuming process, needs technical expertise[24] and more often than not, is not available after hours in many set-ups. Understandably, when it comes to diagnosing DVT in the intensive care unit and perioperative setting, anesthesiologists and intensivists prefer two-point or proximal CUS.

Gibson et al.[25] in their study comparing the sensitivity and specificity of 2-point CUS and whole-leg CUS, concluded that both diagnostic strategies were comparable and efficient. Use of whole-leg CUS, as has been mentioned before, is technically more demanding and requires more time. Another caveat associated with whole-leg CUS is that, given the uncertainty associated with the treatment of isolated calf vein thrombus, it puts the physician in a diagnostic dilemma.

The American College of Chest Physicians (ACCP) recommends use of proximal CUS after moderately or highly sensitive D-dimer testing for suspected first occurrence of DVT of the lower limbs.[12]

Thus, proximal CUS examination is an easy-to-perform modality which is sensitive and specific for picking up DVT of the lower limbs.

Recurrent Deep Vein Thrombosis of the Lower Limbs

Compression ultrasonography has comparable sensitivity and specificity for the diagnosis of recurrent contralateral DVT. But, a suspected diagnosis of recurrent ipsilateral DVT often puts clinicians in a quandary, and for good reason. Recurrent pain is a common complaint in patient with acute DVT, and it may be because of new thrombus, post-thrombotic syndrome or nonthrombotic problems. Incorrectly ruling out recurrent disease puts the patient at risk of pulmonary embolism. While labeling a patient as having recurrent disease, when they have none, commits them to prolonged anticoagulation.

Studies demonstrate presence of incompressibility on CUS in 80% patients at 3 months, and 50% patients at 1 year, after an episode of proximal DVT.[26,27] Although, finding of new incompressibility in comparison to older examinations confirms the diagnosis of recurrent DVT, it is reported in only 10-20% cases of recurrent thrombosis.[26] Some authors suggest an increase in the venous diameter by 4 mm during compression compared with previous result as a better threshold for the diagnosis of recurrence.[28] But this too suffers from interobserver variability, and thus may result in false positives.

Consensus is lacking on the utility of CUS alone in the diagnosis of recurrent DVT. The ACCP recommends initial evaluation with proximal CUS or highly sensitive D-dimer in cases of suspected recurrent DVT.[12]

Deep Vein Thrombosis in Pregnancy

Proximal CUS is also recommended as the initial modality in the diagnosis of DVT in pregnant patients. In patients suspected of having isolated iliac vein thrombosis with no evidence of DVT on proximal CUS, Doppler ultrasound of the iliac vein is better than serial proximal CUS.

Upper Extremity: Deep Vein Thrombosis

Thrombosis of the veins of the upper extremity is uncommon, but when it does happen, it is usually secondary to presence of central venous catheters, pacemaker wires or malignancy. The anatomy of the upper limbs makes the usage of ultrasound challenging as the ability to compress vessels is impeded by the presence of osseous structures.

Deep vein thrombosis in the upper limbs is diagnosed by the presence of noncompressibility of the vessels (when possible) or by using color Doppler ultrasound and inability to visualize flow in the venous system. The sensitivity and specificity of color Doppler in combination with CUS, or color Doppler alone was similar to CUS.[29]

Pulmonary Embolism

A scan of the lungs can provide additional knowledge in patients suspected of having a PTE. Subpleural infarcts may be visualized on longitudinal scanning which are pleural based, echo-poor, triangular or rounded consolidations. Of course, this would be additional to a transthoracic echocardiogram which may demonstrate right heart abnormalities.

Performance of CUS of the lower limbs is not expected to aid in the diagnosis, as more often than not, residual DVT is absent in patients of acute pulmonary embolism.[30]

HOW I SCAN THE LEG?

For performing a proximal or 2-point CUS, the venous system of the lower limb needs to be assessed at two sites, the common femoral vein in the proximal thigh and the popliteal vein in the popliteal fossa (Fig. 1).

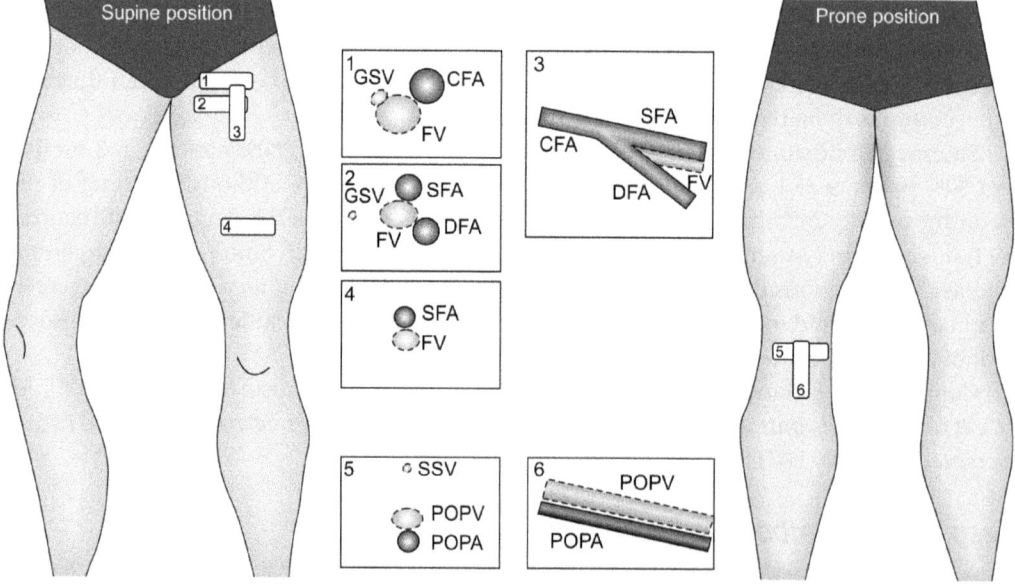

Fig. 1: Transducer positions on the leg.
Abbreviations: GSV: Great saphenous vein; CFA: Common femoral artery; FV: Femoral vein; SFA: Superficial femoral artery; DFA: Deep femoral artery; SSV: Superficial saphenous vein; POPV: Popliteal vein; POPA: Popliteal artery.

A linear array probe of frequency 6–10 MHz is often ideal. For obese patients, we may need to use a lower frequency probe. Sometimes, using an abdominal probe might be necessary.

I prefer to place patients supine with the bed in reverse Trendelenburg position, and a pillow below the knee. Depending on the particular patient, mild external rotation may also be added.

I start scanning from the inguinal ligament with the probe transverse to the leg. The common femoral vein is located medial to the femoral artery in this location (Fig. 2), but quickly becomes posterior to the artery within 2–3 cm. It is prudent to confirm that it is indeed the vein by color or Doppler (if available); or by the physical characteristics (vein is larger than the artery, more ovoid, compressible).

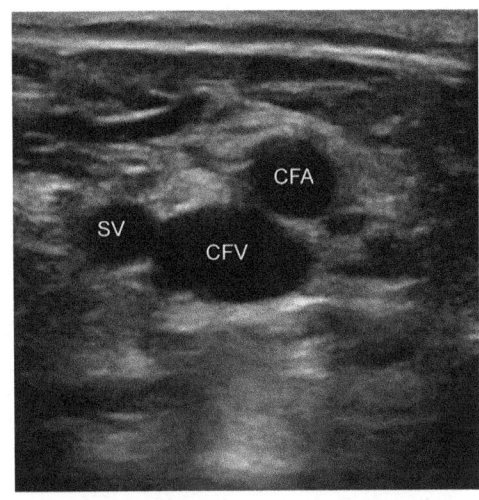

Fig. 2: Vessel position distal to inguinal ligament.

Once identity is confirmed, I start scanning distally, while compressing at 1 cm intervals till about 2 cm beyond the confluence of the superficial and deep femoral veins (Fig. 3). This is important because branch points are especially susceptible to thrombus formation. Changes may need to be made to the gain to acquire acceptable images.

For scanning the popliteal vein, I prefer to make the patient prone, if possible, with a pillow supporting the ankle, so as to flex the knee to about 10–15° and place the bed in reverse Trendelenburg. If it is difficult to place the patient in the prone position, I place the patient in a semi-sitting position with the knee is flexed to around 30°.

The popliteal vein lies posterior/superficial to the popliteal artery in the popliteal fossa (Fig. 4). After confirming that it indeed is a vein, I proceed to scan the entire length of the popliteal vein in the popliteal fossa, compressing in 1 cm intervals.

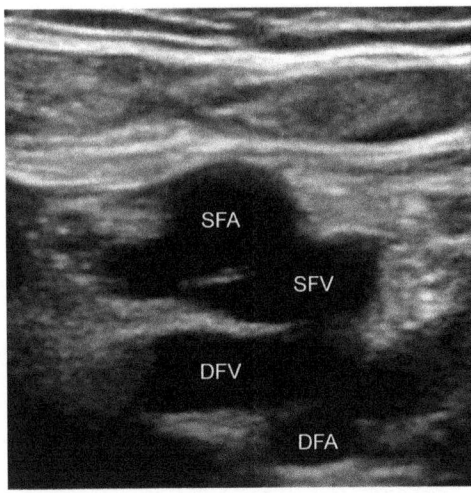

Fig. 3: Confluence of superficial and deep femoral veins.

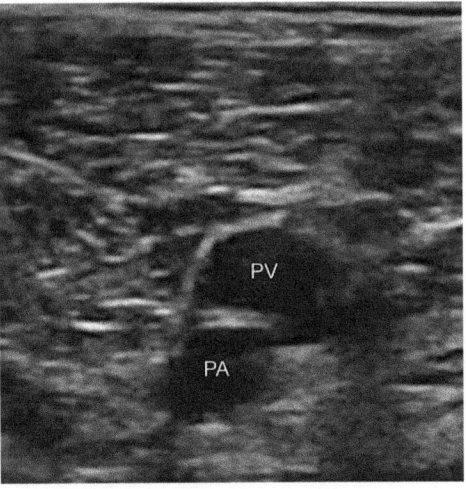

Fig. 4: Vein position in popliteal fossa.

Noncompressibility is a sign of presence of thrombus in the vein. It is not necessary to visualize the thrombus as some clots may be isoechoic to blood, and thus difficult to image on ultrasound.

REFERENCES

1. Heit JA, O'Fallon WM, Petterson TM, et al. Relative impact of risk factors for deep vein thrombosis and pulmonary embolism: a population-based study. Arch Intern Med. 2002;162(11):1245-8.
2. Beckman MG, Hooper WC, Critchley SE, et al. Venous thromboembolism: a public health concern. Am J Prev Med. 2010;38(4 Suppl):S495-501.
3. Deitelzweig SB, Johnson BH, Lin J, et al. Prevalence of clinical venous thromboembolism in the USA: current trends and future projections. Am J Hematol. 2011;86(2):217-20.
4. Cohen AT, Agnelli G, Anderson FA, et al. Venous thromboembolism (VTE) in Europe. The number of VTE events and associated morbidity and mortality. Thromb Haemost. 2007;98(4):756-64.
5. Pinjala RK, Agarwal MB, Turpie AGG. A Characterisation of patients with symptomatic DVT in India; for PROVE investigators. J Thromb Haemost. 2005;3:1043.
6. Pinjala R. Venous thromboembolism risk and prophylaxis in the acute hospital care setting (ENDORSE), a multinational cross-sectional study: results from the Indian subset data. Indian J Med Res. 2012;136(1):60-7.
7. Spyropoulos AC, Lin J. Direct medical costs of venous thromboembolism and subsequent hospital readmission rates: an administrative claims analysis from 30 managed care organizations. J Manag Care Pharm. 2007;13(6):475-86.
8. de Valois JC, van Schaik CC, Verzijlbergen F, et al. Contrast venography: from gold standard to 'golden backup' in clinically suspected deep vein thrombosis. Eur J Radiol. 1990;11(2):131-7.
9. Hull R, Hirsh J, Sackett DL, et al. Clinical validity of a negative venogram in patients with clinically suspected venous thrombosis. Circulation. 1981;64(3):622-5.
10. Lensing AW, Buller HR, Prandoni P, et al. Contrast venography, the gold standard for the diagnosis of deep-vein thrombosis: improvement in observer agreement. Thromb Haemost. 1992;67(1):8-12.
11. Kearon C, Julian JA, Newman TE, et al. Noninvasive diagnosis of deep venous thrombosis. McMaster Diagnostic Imaging Practice Guidelines Initiative. Ann Intern Med. 1998;128(8):663-77.
12. Bates SM, Jaeschke R, Stevens SM, et al. Diagnosis of DVT: Antithrombotic Therapy and Prevention of Thrombosis, 9th ed: American College of Chest Physicians Evidence-Based Clinical Practice Guidelines. Chest. 2012;141(2 Suppl):e351S-418S.
13. Wells PS, Hirsh J, Anderson DR, et al. Accuracy of clinical assessment of deep-vein thrombosis. Lancet. 1995;345(8961):1326-30.
14. Wells PS, Owen C, Doucette S, et al. Does this patient have deep vein thrombosis? JAMA. 2006;295(2):199-207.
15. Wells PS, Anderson DR, Bormanis J, et al. Value of assessment of pretest probability of deep-vein thrombosis in clinical management. Lancet. 1997;350(9094):1795-8.
16. Wells PS, Anderson DR, Rodger M, et al. Evaluation of D-dimer in the diagnosis of suspected deep-vein thrombosis. N Engl J Med. 2003;349(13):1227-35.
17. Constans J, Nelzy ML, Salmi LR, et al. Clinical prediction of lower limb deep vein thrombosis in symptomatic hospitalized patients. Thromb Haemost. 2001;86(4):985-90.
18. Di Nisio M, Squizzato A, Rutjes AW, et al. Diagnostic accuracy of D-dimer test for exclusion of venous thromboembolism: a systematic review. J Thromb Haemost. 2007;5(2):296-304.
19. Michiels JJ, Gadisseur A, Van Der Planken M, et al. A critical appraisal of non-invasive diagnosis and exclusion of deep vein thrombosis and pulmonary embolism in outpatients with suspected deep vein thrombosis or pulmonary embolism: how many tests do we need? Int Angiol. 2005;24(1):27-39.
20. Cornuz J, Pearson SD, Polak JF. Deep venous thrombosis: complete lower extremity venous US evaluation in patients without known risk factors—outcome study. Radiology. 1999;211(3):637-41.

21. Elias A, Mallard L, Elias M, et al. A single complete ultrasound investigation of the venous network for the diagnostic management of patients with a clinically suspected first episode of deep venous thrombosis of the lower limbs. Thromb Haemost. 2003;89(2):221-7.
22. Schellong SM, Schwarz T, Halbritter K, et al. Complete compression ultrasonography of the leg veins as a single test for the diagnosis of deep vein thrombosis. Thromb Haemost. 2003;89(2): 228-34.
23. Stevens SM, Elliott CG, Chan KJ, et al. Withholding anticoagulation after a negative result on duplex ultrasonography for suspected symptomatic deep venous thrombosis. Ann Intern Med. 2004;140(12):985-91.
24. Schellong SM. Distal DVT: worth diagnosing? Yes. J Thromb Haemost. 2007;5 Suppl 1:51-4.
25. Gibson NS, Schellong SM, Kheir DY, et al. Safety and sensitivity of two ultrasound strategies in patients with clinically suspected deep venous thrombosis: a prospective management study. J Thromb Haemost. 2009;7(12):2035-41.
26. Prandoni P, Cogo A, Bernardi E, et al. A simple ultrasound approach for detection of recurrent proximal-vein thrombosis. Circulation. 1993;88(4 Pt 1):1730-5.
27. Heijboer H, Jongbloets LM, Buller HR, et al. Clinical utility of real-time compression ultrasonography for diagnostic management of patients with recurrent venous thrombosis. Acta Radiol. 1992;33(4):297-300.
28. Prandoni P, Lensing AW, Bernardi E, et al. The diagnostic value of compression ultrasonography in patients with suspected recurrent deep vein thrombosis. Thromb Haemost. 2002;88(3):402-6.
29. Baxter GM, Kincaid W, Jeffrey RF, et al. Comparison of colour Doppler ultrasound with venography in the diagnosis of axillary and subclavian vein thrombosis. Br J Radiol. 1991;64(765):777-81.
30. van Langevelde K, Sramek A, Vincken PW, et al. Finding the origin of pulmonary emboli with a total-body magnetic resonance direct thrombus imaging technique. Haematologica. 2013;98(2):309-15.

Chapter
15D
Clinical Case Scenarios

Rahul K Anand, Kapil Dev Soni, Ajay Singh, Rashmi Bhatt

CASE 1

A 28-year-old female was admitted to ICU with complicated intra-abdominal infection and suspected catheter related blood stream infection. The central venous catheter in right internal jugular vein was removed and left subclavian vein cannulated with triple lumen catheter by anatomical landmark technique, all three ports were checked for blood aspiration. A requisition for chest X-ray for the confirmation of central venous catheter (CVC) position was sent. One hour after the catheter insertion, patient developed tachycardia and hypotension without any change in airway pressures or any respiratory symptoms, which was managed with a fluid bolus and vasopressors. Her clinical condition deteriorated and she developed bradycardia followed by asystole. Cardiopulmonary cerebral resuscitation (CPCR) was immediately instituted as per standard protocol. Chest auscultation confirmed bilateral air entry. During cardiopulmonary cerebral resuscitation lung and heart ultrasonography was done (Figs. 1 to 3). Figures 1A and B show the scans obtained.

Discussion

Patient underwent CVC insertion which was followed by hypotension and tachycardia. Differential diagnosis includes pneumothorax, hemothorax and pericardial tamponade. Fluid bolus and vasopressors were started. There was no increase in airway pressure and bilateral air entry was also present. Lung scan was found to be normal with sliding sign and seashore sign (Figs. 1A and B). These findings rule out pneumothorax effectively. Bilateral cardiophrenic angles are free which excludes hypotension secondary to hemothorax. Focused cardiac ultrasound is suggestive of diastolic collapse of right sided chambers (Fig. 1C), which is a new finding and suggestive of pericardial tamponade secondary to CVC insertion. It was drained by inserting a USG-guided single lumen catheter.

The incidence of pericardial tamponade secondary to CVC insertion ranges from 0.0001% to 1.4% but carries a high mortality of 65% to 100% in adults. The signs and symptoms are nonspecific, the classical Beck's triad (hypotension, muffled heart sound and jugular engorgement) is not present in 29% cases and may manifest within few minutes after CVC insertion or up to several months after placement. It is more common with left-sided CVC insertion because of tortuous course of the vessel. There are different mechanisms involved in post CVC cardiac tamponade; direct trauma at the time of insertion, catheter migration and mechanical and chemical erosion. It may be prevented by using catheter with shortest length possible,

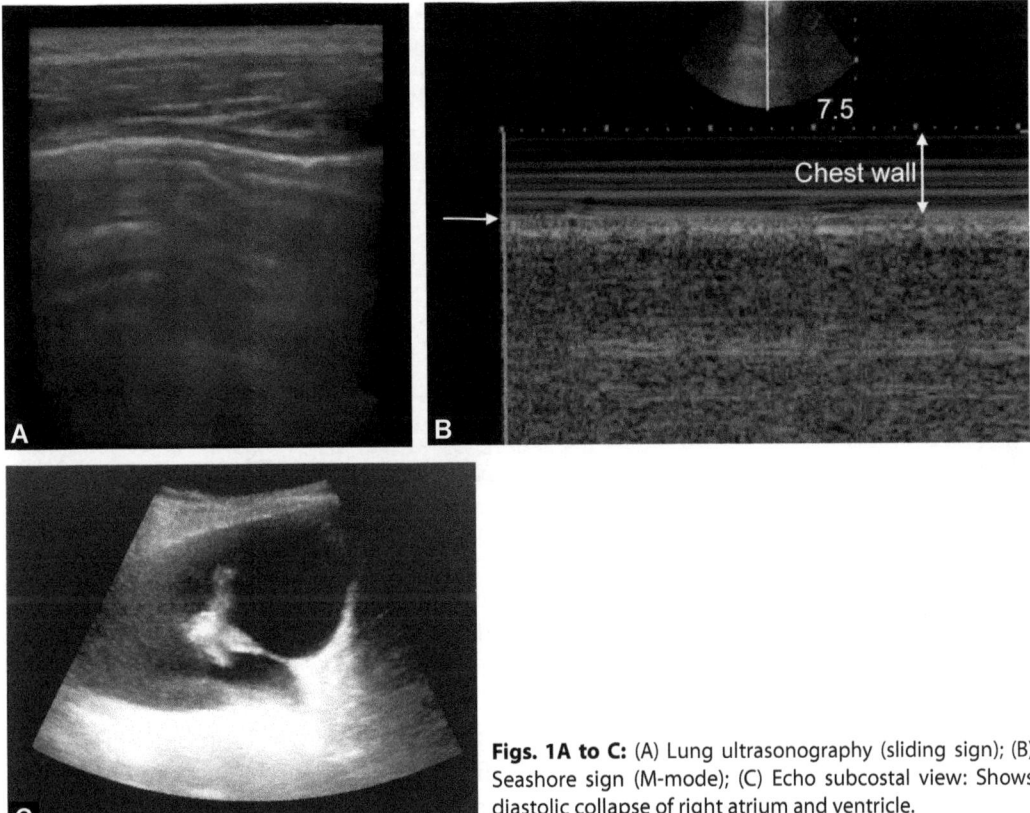

Figs. 1A to C: (A) Lung ultrasonography (sliding sign); (B) Seashore sign (M-mode); (C) Echo subcostal view: Shows diastolic collapse of right atrium and ventricle.

use of ultrasound guidance for catheter insertion and confirmation of tip placement by chest X-ray, fluoroscopy and endocavitary electrocardiography. Successful management depends on early diagnosis and prompt treatment, which can be hastened by use of bedside ultrasound. It should be suspected in all cases of post CVC insertion presenting with clinical deterioration, early clinical suspicion is key to success.

CASE 2

A 25-year-old male patient admitted with severe community acquired pneumonia. Patient's trachea intubated due to progressive respiratory distress. A chest X-ray was ordered and bedside lung ultrasound performed.

Discussion

Ultrasonography lung shows evidence of thickened pleura, consolidation, confluent B lines and air and fluid bronchogram which is suggestive of pneumonia as the primary pathology. Hypoechoic shadows with hematocrit sign may also be present which appear on static imaging as layered as cells collect in a dependent fashion by gravity, seen in highly cellular effusions.

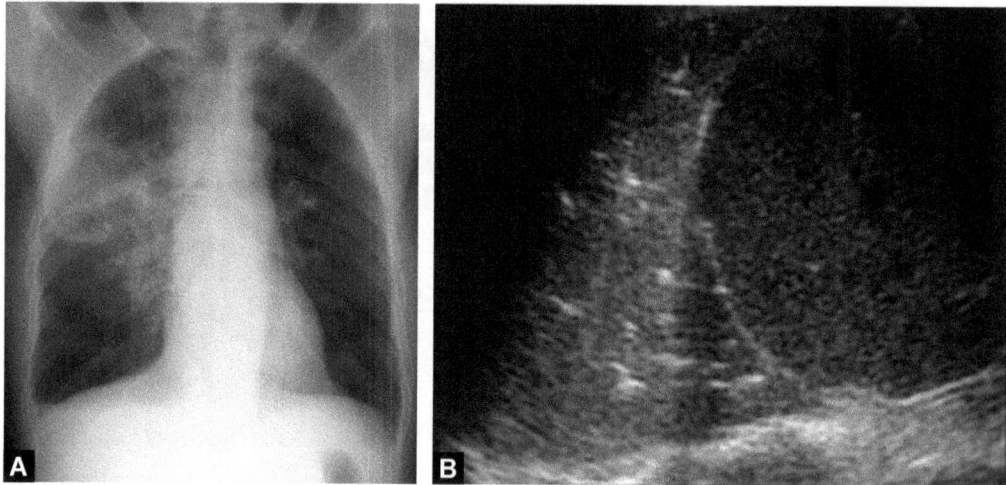

Figs. 2A and B: (A) Chest X-ray; (B) USG showing consolidation as air bronchogram.

Later intercostal chest drainage tube (ICD) was inserted and 750 mL of thick purulent fluid drained. Here in this case, ultrasound differentiated between consolidation with pleural effusion from consolidation alone. This differentiation is of help to avoid unnecessary ICD insertion.

The diagnosis of pneumonia has conventionally depended on chest X-ray or chest CT. Both the modalities are time taking and it may be difficult to obtain good images in critically ill patients. Besides, it is difficult to distinguish between pneumonia and pulmonary embolism, acute respiratory distress syndrome, and pulmonary fibrosis on X-ray only. Over the last several years, ultrasound has been shown to be highly effective in evaluating a range of pathologic pulmonary conditions, offering a quick and easily available diagnostic tool.

USG can be used to diagnose segmental consolidation and it can be followed up with sequential scans to monitor the course of disease. Atelectasis or lung collapse has a similar appearance to consolidation. A number of associated features can potentially distinguish pneumonic consolidation from collapse, yet it may not always be reliable. USG is particularly sensitive for identifying even very small effusions and can also be used to characterize them. False positive cases with lung ultrasound are a possibility due to the presence of small subpleural consolidation (<1 cm) not detectable with a chest X-ray. The most common pathogen causing community acquired pneumonia in Indian population is *Streptococcus pneumoniae* followed by *Klebsiella pneumoniae* and *Staphylococcus aureus*.

CASE 3

An 80-year-old female patient was admitted with community acquired pneumonia and shock. Fluid resuscitation was started along followed by vasopressors. Besides an elevated leucocyte count her other investigations were unremarkable, including the coagulation profile. USG-guided left sided internal jugular vein CVC was inserted. 5 minutes after the procedure, she collapsed. CPCR was started and lung ultrasonography performed.

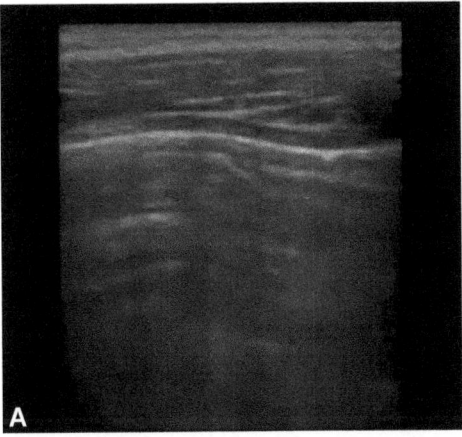 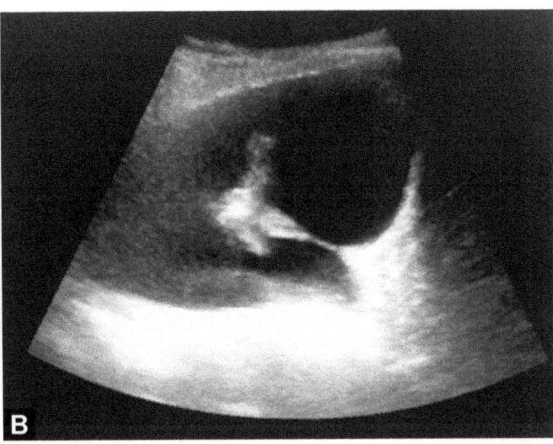

Figs. 3A and B: (A) Normal looking pleura with seashore sign ruling out pneumothorax; (B) Right costophrenic angle (pleural effusion). Hematocrit sign visible.

Discussion

Cardiopulmonary cerebral resuscitation was started and chest X-ray was immediately ordered. Lung ultrasound was done to rule out pneumothorax or hemothorax. Normal lung sliding sign was seen on the left side and a thin rim of fluid was noticed in anterior side of right chest. Right side lateral side was scanned which was suggestive of massive pleural effusion with hematocrit sign. In an emergency, it is difficult to get a diagnosis with chest X-ray as it is time taking whereas USG is a handy instrument available on the bedside which can be used for early diagnosis and treatment.

Patient developed hemothorax post CVC insertion which is a rare, life-threatening complication. Incidence of developing hemothorax after CVC insertion is 0.1–1%. Perforation of the superior vena cava by a central venous catheter proximal to the junction of the superior vena cava and right atrium may result in hydrothorax and hydromediastinum. Perforation beyond this point leads to cardiac tamponade. A catheter from which blood cannot be aspirated after placement in internal jugular or subclavian vein, carries a risk of hemothorax and should not be removed until suitable imaging is obtained and a plan to control any bleeding is in place. Pressure monitoring transducer tracing and pressure monitoring may not be able to diagnose a CVC migrating into the pleural cavity. Chest tube should be placed to drain the blood and fluid for definitive management.

Indications of thoracotomy include immediate drainage of 1000–1500 mL of blood, termed as massive hemorrhage, continued bleeding as 150–200 mL/hr for 2–4 hours, repeated blood transfusion is required to maintain hemodynamic stability. Activation of massive blood transfusion protocol (MTP) is done by clinician in case of massive hemorrhage.

CASE 4

A 30-year-old previously healthy female patient, presents with fever, dyspnea after 2 days of lower segment cesarean section. Her hemoglobin is 10 g%, total lymphocyte count (TLC)

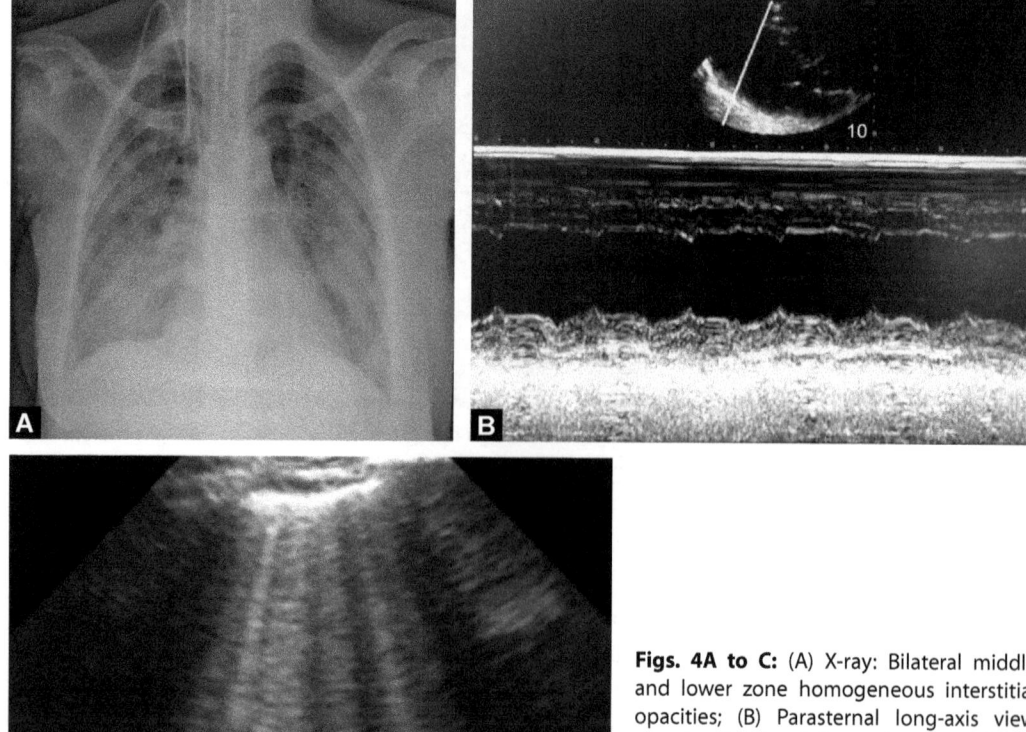

Figs. 4A to C: (A) X-ray: Bilateral middle and lower zone homogeneous interstitial opacities; (B) Parasternal long-axis view showing fractional shortening; (C) Diffuse B lines suggestive of pulmonary edema.

21,000/dL and platelet 102,000/dL. Kidney function and liver functions are within normal limits. She is conscious but anxious. Trachea intubated due to worsening respiratory distress. Surveillance cultures sent and antibiotics started. Differential diagnoses include postoperative pneumonia, pulmonary embolism or peripartum cardiomyopathy.

Discussion

Respiratory distress two days after lower segment cesarean section may be attributed to postoperative pneumonia, pulmonary embolism or PPCM. On the basis of chest X-ray (Fig. 4A)alone, it is difficult to differentiate among these conditions. In lung ultrasound, pleura is smooth, with diffuse B lines in all lung fields. Smooth pleura rules out any infective lung disease like pneumonia and B lines with lung sliding rules out any pneumothorax. These diffuse B lines suggest that patient is in pulmonary edema (Fig. 4C). Inferior vena cava is full with no respiratory variations. In parasternal window fractional shortening is found to be significantly less, with global hypokinesia (Fig. 4B). In apical four chamber view, right ventricle size is less than 0.6 of left ventricle. In deep vein thrombosis (DVT) scan femoral vessels are compressible. These findings with clinical presentation strongly point towards the diagnosis of PPCM.

Peripartum cardiomyopathy (PPCM, also called pregnancy-associated cardiomyopathy) is a rare cause of heart failure (HF) that affects women late in pregnancy or in the early puerperium. The 2010 ESC Working Group defined PPCM as an idiopathic cardiomyopathy with the following characteristics:
- Development of heart failure (HF) toward the end of pregnancy or in the months following delivery
- Absence of another identifiable cause for the HF
- Left ventricular (LV) systolic dysfunction with an LV ejection fraction (LVEF) nearly always less than 45%. The left ventricle may or may not be dilated.

Despite many attempts to uncover a distinct etiology of PPCM, the cause remains unknown and may be multifactorial. While several potential factors have been evaluated and may contribute, experimental research suggests that these multiple factors result in a common final pathway with enhanced oxidative stress, cleavage of prolactin to an angiostatic N-terminal 16 kDA prolactin fragment, and impaired vascular endothelial growth factor (VEGF) signaling because of upregulated soluble fms-like tyrosine kinase (sFLT1).

Risk Factors

Although the etiology of PPCM remains unclear, the following are among the factors associated with increased risk of PPCM:
- Age greater than 30 years
- African descent
- A history of pre-eclampsia, eclampsia, or postpartum hypertension
- Maternal cocaine abuse
- Long-term (>4 weeks) oral tocolytic therapy with beta adrenergic agonists, such as terbutaline.

Its treatment includes medical therapy which is like any case of heart failure due to systolic dysfunction with special consideration during pregnancy. Objective of treatment is to prevent extra fluid from collecting in the lungs and to help the heart to recover as much as possible. The following are used in treatment of PPCM that are safe in breastfeeding women:
- Angiotensin converting enzyme inhibitors: improves cardiac function
- Beta blockers: decrease cardiac oxygen demand and improves recovery
- Diuretics: reduces fluid retention
- Digitalis: improves cardiac contractility
- Anticoagulants: decrease the incidence of DVT.

CASE 5

A 52-year-old male patient shifted to post anesthesia care unit after percutaneous nephrolithotomy (PCNL) for right sided staghorn calculus under general anesthesia. Ten minutes after

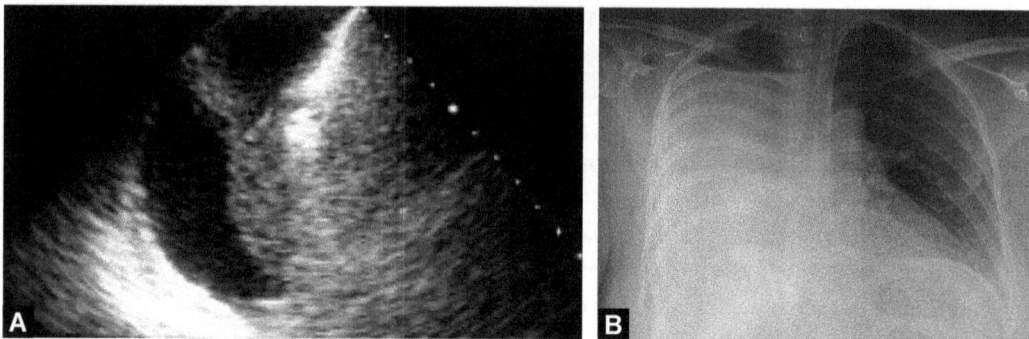

Figs. 5A and B: (A) Pleural ultrasound (right chest lateral area); (B) Chest X-ray showing right massive hemothorax.

shifting to recovery room, he became tachypnoeic and anxious. Room air saturation is 88%, heart rate 128/minute and blood pressure 102/76 mm Hg. Oxygen therapy started by facemask. On auscultation air entry is slightly diminished on the right side with dull percussion note on lateral side of chest. Chest X-ray ordered immediately and bedside lung ultrasound was performed.

Discussion

Patient complained of shortness of breath. Tachypnea, desaturation and tachycardia are present. Given the clinical history of PCNL, the patient has a likelihood of pulmonary complications like pneumothorax, pleural effusion, hemothorax, etc. the bedside lung USG depicts a normal lung sliding and presence of seashore sign which rules out pneumothorax. The right lateral lung scan is suggestive of massive fluid collection which could be pleural effusion or hemothorax (Fig. 5A). Meantime chest X-ray (Fig. 5B) was done which showed right sided white out suggestive of massive hemothorax. Intercostal chest drainage tube inserted followed by clinical improvement in the condition of patient.

The main postoperative complication following PCNL is pulmonary, affecting about 8% of the patients. A high index of suspicion and the use of point of care USG can help early detection as well-management.

CASE 6

A 30-year-old male was brought to the emergency following a road traffic injury. On arrival his airway was patent, he was breathing at the rate of 35 breaths per minute with bilateral equal entry in both the lung fields. His heart rate was 140 beats per min with a blood pressure of 80/55 mm Hg. He was conscious and moving all limbs. Immediately on primary survey, in view of hemodynamic instability, he was intubated and put on ventilator support. Two large bore iv lines were secured and one litre bolus of ringer lactate was initiated after drawing blood for investigations. EFAST was performed to identify the source of bleeding as well as diagnose other injuries due to trauma, while arrangements were made for imaging work up.

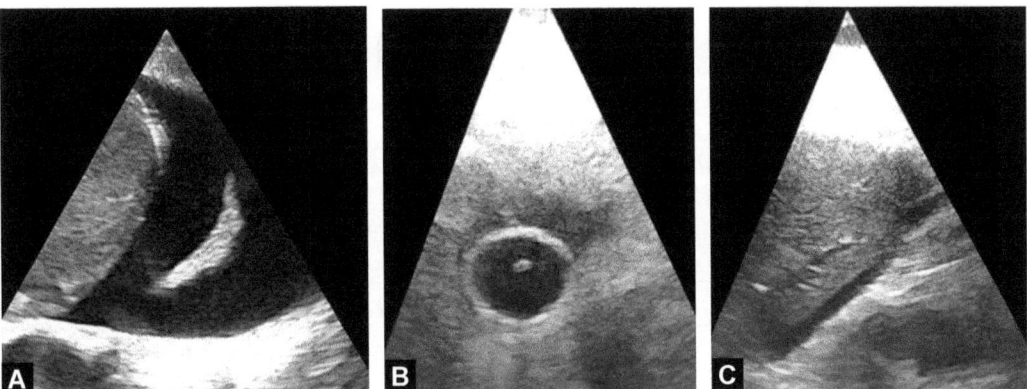

Figs. 6A to C: (A) Hepatorenal space, (B) Pelvic cavity, (C) Thoracic cavity.

Discussion

The EFAST was positive and fluid were present in hepatorenal space, pelvic cavity and thoracic cavity suggestive of hemoperitoneum and hemothorax (Figs. 6A to C).

The management of patient depends upon whether the patient is in shock/hemodynamically unstable. The most common cause of shock in these patients is hemorrhagic shock unless proved otherwise. A patient may bleed in five regions: externally, thoracic cavity, abdominal cavity, pelvis and around long bones. EFAST has become a standard examination in primary survey of trauma patients. If the patient is EFAST positive for hemoperitoneum and non-responder to fluid bolus then he is taken for exploratory laparotomy immediately. If EFAST is positive for hemoperitoneum and patient is hemodynamically stable then further examination is carried out, such as CECT thorax and abdomen, to locate the source of bleeding. For hemothorax, most are managed with insertion of intercostal drain and monitoring. The thoracotomy decision depends upon the amount and rate of bleed from the thorax.

In the above case after securing airway and wide bore intravenous, the patient responded to fluid bolus and became hemodynamically stable. 500 mL of hemorrhagic fluid was drained after insertion of Intercostal drain. CECT was done in secondary survey which revealed grade 3 liver laceration. The patient was angioembolized for liver laceration and managed non operatively with closed monitoring in ICU. The rest of the clinical course was uneventful and patient was discharged after five days.

Chapter 15E

Infection Control and Legal Issues in ICU Sonography

Puneet Khanna, Rajkumar Subramanian

INTRODUCTION

Ultrasonography (USG) is a noninvasive and simple bedside diagnostic modality, which helps in identifying infectious foci in a critically ill-patient. Optimal use of ultrasound at bedside can decrease the need to transfer the patient to other imaging and interventional suites, which has its own implications. Among the numerous applications of USG, its ability to visualize and perform an interventional procedure at the bedside, is the most appealing feature. Decontamination or disinfection of the ultrasound machine and the probe is an essential practice in intensive care unit (ICU) setting, to prevent the transmission of flora between patients. A proper cleaning protocol for the machine and probe has to be devised and followed in every ICU setting.

ORGANISMS OF CONCERN

Nosocomial infections are the major causes in hospital morbidity and mortality worldwide. According to centers for disease control and prevention (CDC), healthcare associated infections amount to 1.7 million infections and approximately 1 lakh deaths in the United States every year.[1] The incidence of nosocomial infections are at least three times higher in adult and pediatric ICU.[2] These infections, especially the multidrug resistant ones, are of major concern to the intensivists and all healthcare specialists involved in patient care. For the clinicians dealing with critically ill-patients, infections caused by multidrug-resistant (MDR) bacteria are a worrisome healthcare problem. Infections in health care are acquired from infected air, surface and water. Thus, contaminated surfaces of inanimate surfaces including USG machines are potential sources of cross-infection in critical care.

As the ultrasound machine is used for scanning many patients within a short span of time, every part of the ultrasound (US) machine including the probes, keyboard, cord, control settings, gel and gel bottles may act as vector for infection.

Though ultrasound probes are disinfected or wrapped with sterile barriers during and after every use, it is unlikely that the entire device is disinfected or protected before every scan. *Acinetobacter, Clostridium difficile* spores, *Escherichia coli*, enterococci and methicillin-resistant *Staphylococcus aureus* persist on inanimate surfaces for more than 4–5 months (Table 1 and Box 1).[3,4]

Table 1: Organism and maximum persistence period.

Organism	Maximum persistence reported
Acinetobacter	5 months
Candida albicans	4 months
Candida parapsilosis	2 weeks
Clostridium difficile	5 months
Corynebacterium diphtheriae	6 months
Escherichia coli	16 months
Enterococci	4 months
Klebsiella	>30 months
Mycobacterium tuberculosis	4 months
Pseudomonas aeruginosa	16 months
Salmonella typhi	4 weeks
Salmonella typhimurium	4 years
Methicillin-resistant Staphylococcus aureus	7 months
Streptococcus pneumoniae	20 days
Streptococcus pyogenes	6 months
Torulopsis glabrata	5 months

Box 1: Organisms reported to be transmitted by ultrasound equipment.[4]

Methicillin-sensitive *Staphylococcus aureus*
Methicillin-resistant *Staphylococcus aureus*[5]
Coagulase-negative staphylococci (CoNS)
Pseudomonas aeruginosa
Corynebacterium species
Acinetobacter
Bacillus
Extended spectrum beta-lactamase *Klebsiella pneumoniae*–from gel
Multidrug-resistant *Pseudomonas*—Transesophageal probe (TEE)

The scope of USG in intensive care is widening. Day by day more and more invasive procedures like central venous and arterial access are done under ultrasound guidance, so the problem of contamination by ultrasound probes has to be taken seriously.
Basic measures like the following help in decreasing contamination and cross-infection:[6]
- Hand hygiene before and after examination
- Use of personal protective equipment including caps, masks and gloves
- Use one hand for handling probe and one for keyboard control
- Avoid switching of hands
- Avoid bending over the patient during the examination
- Cleaning the probe before and after examination

Table 2: Spaulding classification.

Spaulding classification	Critical	Semicritical	Noncritical
Probe to patient	Device enters • Sterile tissue • Vascular system	Device contacts • Mucous membrane • Nonintact skin	Device contacts • Intact skin
Examples	Probes contacting • Vascular system • Puncture of skin	• Transvaginal • Transrectal • Transesophageal echocardiography	• Surface examination only
Disinfection or sterilization	Sterilization	High-level disinfection (chemical disinfectants)	Low- or mid-level disinfection
Disinfecting or sterilizing process	• Autoclave, if possible • Cold-soak sterilants	• Ortho-phthalaldehyde (OPA – 0.55%) • Glutaraldehyde (≥ 2%) • Hydrogen peroxide (7.5%) • Peracetic acid (0.2%)	• Quaternary-ammonium compounds • Germicidal wipes
Other	• Sterile probe cover • Sterile patient drape • Sterile gel • Sterile needle guide • Sterile system drape	• Sterile probe cover • Sterile endocavity guide • Sterile gel packet • Temperature-monitored high-level disinfectant • Fume-free soaking system	• Disinfection wipes • Nonsterile probe covers • Single-use gel packet

- Use of disposable waterproof transparent keyboard cover
- Sterilization of gel or use of single use gel sachet.

In addition Earle H Spaulding created a simple approach for sterilization and disinfection of medical equipment. Because of its simplicity, it is being adapted for almost all equipment including ultrasound. Adapted from Spaulding classification (Table 2).[7]

LEGAL ISSUES IN ICU SONOGRAPHY

Pre-Natal Diagnostic Techniques (PNDT) Act, 1994 was enacted to prevent prenatal sex determination and thereby female feticide.[8] The act also provides restrictions for holding a ultrasound machine and performance of ultrasound by a health care provider. Few pertinent points are stated below:

1. USG machine are to be sold to qualified medical practitioners.
2. Ultrasound sellers are supposed to maintain details of their clients.
3. All ultrasound machine should be registered, if multiple machines are available in a center then all are to be individually registered.
4. The PNDT registration certificate has to be displayed at the place where ultrasound machine is used.
5. Radiologist or intensivist who use the ultrasound machine should have their names in a separate sheet along with the machine registration certificate—to be displayed whenever required.

6. Other doctors may use the machine after notifying the authority informing name of the doctor, personal details and registration number.
7. Board displaying "Sex determination is not done in this center" should be placed at main place.
8. Registration certificate is to be renewed within 30 days.
9. Form F should be duly filled by the medical personnel who perform scan on a pregnant patient.
10. Form F should be submitted to the appropriate authority before 5th of every month and these forms are to be preserved for 3 years.
11. State inspection and monitoring committee pays regular visits and has the authority to seize the equipment, if the set-up does not comply with PNDT Act.

REFERENCES

1. Klevens RM, Edwards JR, Richards CL, et al. Estimating Health care-associated infections and deaths in US Hospitals, 2002. Public Health Rep. 2007;122(2):160-6.
2. Weinstein RA. (1998). Nosocomial Infection Update. [online]. Emerging Infectious Disease Journal-CDC. Available from https://wwwnc.cdc.gov/eid/article/4/3/98-0320_article. [Accessed November 2017].
3. Kramer A, Schwebke I, Kampf G. How long do nosocomial pathogens persist on inanimate surfaces? A systematic review. BMC Infect Dis. 2006;6:130.
4. Odonkor ST, Sackey T, Mahami T. (2015). Evidence of Cross Contamination of Ultrasound Equipment: A call for Infection Prevention Strategy in the Use of Diagnostic Tools. [online]. IJCMAS Website. Available from https://www.ijcmas.com/vol-4-5/Stephen%20T.%20Odonkor,%20 et%20al.pdf. [Accessed November 2017].
5. Shokoohi H, Armstrong P, Tansek R. Emergency department ultrasound probe infection control: challenges and solutions. Open Access Emerg Med OAEM. 2015;7:1-9.
6. GE Healthcare. (2015). Transducer Cleaning and Disinfection Guidelines [online]. Available from: http://www3.gehealthcare.com/static/ge-transducers/GEHC-Guidelines-Transducer_Cleaning_Disinfection_Guidelines.pdf [Accessed November 2017].
7. Infection Control Brochure. [online]. Available from: http://www.chsinterventional.com/DATA/TEXTEDOC/Infection-Control-Brochure-FINAL--2-.pdf. [Accessed Feb 2017].
8. Style and Technique. [online]. Available from: http://www.pndt.gov.in/writereaddata/mainlinkFile/File50.pdf. [Accessed Feb 2017].

Index

Page numbers followed by *f* refer to figure, *fc* refer to flowchart, and *t* refer to table.

A

Abdomen, regions of 184*f*
Abdominal transverse 155*f*
Abdominal ultrasound 25
Abscess, spread of 193
Acalculous cholecystitis 191*f*
Accessory inspiratory muscles 241
Acinetobacter 276, 277
Acoustic enhancement, posterior 47
Acoustic impedance 35, 35*t*
Acoustic shadow 72
    posterior 47
Acute hypotension, causes of 154
Acute respiratory
    distress syndrome, prognosis
        of 111
    failure
        causes of 15*fc*
        differential diagnosis of 3
Adenomyomatosis 49*f*
Advancing needle 173*f*
Air 35
    bronchogram 103*f*, 270*f*
    mucosal
        interface 72, 74
        surface 84*f*
Airway
    anatomy 68
    imaging, challenges of 68
    management 78, 158
        during resuscitation 158
    pathology 78
    prediction of difficult 74
    ultrasonography of 13
    ultrasound 68
        imaging of 70
American College of Chest
    Physicians 261
American Society of
    Anesthesiologists 57
Analgesia 169
Anechoic stripe 145
Aneurysm 182
    formation 188
Antibiotic 169
    prophylaxis 188
Aorta
    abdominal 147, 223, 223*f*, 224*f*
    ascending 118*f*
    discending 19
Aortic aneurysm, abdominal 154
Aortic blood velocity 142
Aortic dissection, suspected 115
Aortic opening 239
Aortic root 118*f*
Aortic stenosis 121
Aortic valve 118, 118*f*, 119*f*, 121, 126*f*
    morphology of 118
    open 125
Apical approach 182
Apical four-chamber view 120
Apical pneumothorax 98*f*
Arcuate ligament 238
    lateral 238
    medial 238
Arm, muscle of 253
Arterial cannulation 65
Arterial vasculature 147
Artery 60
Artifact
    attenuation-related 47
    mirror-image 48
    propagation-related 47
    resolution-related 47
    ring-down 48
    ultrasound beam-related 47, 50
Ascites, ultrasonographic
    appearance of 185*f*
Asthma 112
Atelectasis 173, 239
Atrial collapse, right 133
Atrial contraction 127
Atrial fibrillation 163
Atrial septal defect 129
Atrial thrombus, left 118
Axillary vein puncture 63*f*
Azimuthal resolution 44

B

Bacillus 277
Bacteremia 193
Barcode sign 98*f*, 157*f*
Bat sign 95*f*, 230
Beck's triad 268
Bedside lung ultrasound in emergency
    3, 14, 15*f*, 108, 229
Black hypoechoic stripe 145
Bladder 220*f*
Blood muscle 35
Body mass index 87, 252
Bone 35
    muscle 35
Bowel injury and infections 187
Brachial artery 53*f*, 55*f*
Brain
    death 196
    midline shift of 199
Branch pulmonary arteries 119*f*
Breast, ultrasound image of 45*f*
Breathing pattern, deep 248*f*
Breathing, apposition during
    normal 245*f*

C

Candida albicans 277
Candida parapsilosis 277
Cannulation during resuscitation
    158
Cardiac abnormalities 136
Cardiac arrest 150, 158
    causes of 151
    etiology for 150, 158
    situation 150
        application in 4
    ultrasound exam 159
    victim, ultrasound in 159
Cardiac chambers 234
Cardiac compression 158, 159
Cardiac critical care ultrasound 111
Cardiac evaluation, ultrasonography
    in 115
Cardiac failure 173, 183
Cardiac output, measurement of
    21, 122
Cardiac preload 140
Cardiac pump failure 229
Cardiac standstill 158
Cardiac tamponade 151, 152, 152*f*,
    156, 161, 178
Cardiogenic pulmonary edema 112
Cardiopulmonary
    arrest 160*fc*
    cerebral resuscitation 268, 271

resuscitation 76, 161, 235
    ultrasound in 150
Cardiothoracic surgery 240
Carotid artery 59
    common 13f, 61
    internal 74
Carotid jugular vessels 70
Catheter
    in situ 63f, 172f
    management 193
    selection 168
    size, correct 194
Cellular debris, sedimentation of 101f
Cellular effusions 269
Central nervous system 196
Central tendon 239
Central venous
    access, history of 57
    catheter 268
        placement of 173
        pressure 138-140
Cerebral artery 201f
    anterior 202
Cerebral blood flow 196, 199
    velocity 196
Cerebral circulatory arrest,
    diagnosis of 202
Cerebral ischemia, delayed 200
Cerebral spinal fluid 197
Cerebrocirculatory arrest 196
Chamber pressures, measurement of 22
Chest
    conditions 229
    drainage tube 270
    radiography 241
    ultrasound 110
    wall 99f
        anterior 235
    X-ray 270f
        diagnosis with 271
Cholecystitis, acute 190
Classic Beck's triad 178
Classic Seldinger technique 170
Clostridium difficile 277
    spores 276
Coagulase-negative staphylococci 277
Collapsed bladder 222f
Color Doppler flow imaging 52
Comet tail
    artifact 72, 48
    image 14
Contrast-enhanced computed tomography 20

Coronary artery
    disease 129
    puncture 182
Coronary cusp, right 19
Coronary embolus 151
Coronary thrombosis 151
Corticomedullary differentiation, loss of 222f
Corynebacterium diphtheriae 277
Corynebacterium species 277
Costophrenic angle, right 271f
Cricoid cartilage 69f, 70, 72, 72f
Cricothyroid membrane 69f, 70, 72, 77, 88
Cricothyrotomy
    role of ultrasound in 88
    technique 87
    ultrasound-guided 82, 87
Critical care
    basic views in 19
    echocardiography 31
        basic 31
        training in advanced 32
    medicine, applications in 10
    ultrasound 31
    ward, echocardiography in 19
Critically ill
    hypotensive patients 138
    patient 78
    echocardiography in 115
Curtain sign 100
Curvilinear low frequency probe 242
Curvilinear transducers 37

D

Deep femoral veins, confluence of 265f
Deep vein thrombosis 4, 23, 24f, 233, 260, 262t, 264, 272
    diagnosis of 261
    in pregnancy 263
    of lower limbs 262
    ultrasonography in 260
Diagnostic peritoneal lavage 208
Diaphragm 238, 255, 258
    anatomy of 238f
    applied anatomy of 238
    change in thickness of 245
    dysfunction 240
    in intensive care unit,
        echogenicity of 248
    peripheral part of 239
    thickening fraction, role of 258
    thickness of 243, 245

    ultrasonography for 255
    ultrasound of 238
    with intercostal approach 245f
Diaphragmatic palsy
    bilateral 241
    unilateral 240
Diaphragmatic slip 239
Diaphragmatic velocity 248
Diastasis 127
Diastolic collapse 153f, 269f
    of right ventricle 133
Diastolic dysfunction 5
    characterization of 127
Diastolic function 122
    of left ventricle, assessment of 127
Diffuse reflection 36
Distensibility index 143
Doppler imaging 51
    modes of 51
Doppler phenomenon 34
Drainage technique 170
Drug over dosages 151
Duplex scanning 52
Dynamic air bronchograms 110
Dynamic parameters 138
Dynamic range compensation 38
Dysfunction, causes of 129
Dysrhythmias 182

E

Echocardiography 115
    in intensive care unit
        advantages of 116, 116t
        disadvantages of 116, 116t
Echogenic muscle, less 256f
Ejection fraction, normal 120
Electrolyte 151
Electromyography 241, 252
Elevation resolution 44
Emergency medicine residents 209
Emphysema 103f
Empyema 229, 230
End diastolic velocity 200
Endocavitary probe 64f
Endothelial growth factor 273
Endotracheal intubation 158
Endotracheal tube 75
    placement, confirmation of 76
End-systolic volume 125f
Enterococci 277
Epiglottis 70, 71, 73
Escherichia coli 276, 277
Esophageal mesentery 242

Esophageal opening 239
Esophageal transducer 241
Esophagus 70, 74, 239
European Society of Intensive Care Medicine 31
Euvolemia 148
Evaluating diaphragm, techniques for 241, 241*t*
Extravascular lung water 18, 22, 104, 111, 139
Exudative pleural effusion 102*f*
Eyeball 198*f*
    junction of 198*f*

F

Fat 35
    muscle 35
        interface, superficial 254*f*, 255*f*
Female feticide 278
Femoral artery 24*f*
    common 264
    deep 264
    superficial 264
Femoral vein 264
    cannulation 64
    normal sonography of 24*f*
Fibroadenoma 45*f*
Fibrous pericardium 239
Fistulous communications 193
Flapping lung 100
Floor of mouth 71
Floor of tongue, muscle of 70
Fluid bolus 268
Fluid collection 145*f*
    abnormal 187
Fluid optimization 139*f*
Fluid replacement 154
Fluid response 132
Fluid responsiveness 138
    prediction of 139*f*, 143, 144
Fluoroscopy 241
Foley's catheter balloon 222*f*
Fractal sign 231*f*

G

Gallbladder 48*f*, 49*f*, 190, 225
Gallstone 48*f*
    sludge 190
Gastric
    content, evaluation of 78
    transducer 241
Gastroesophageal reflux 240

Glaucoma 199
Global ejection fraction 139
Global end-diastolic volume 140
    index 140
Gray myocardium 144
Great saphenous vein 264
Guidewire in situ 171*f*
Guiding therapeutic procedures 226

H

Heart
    evaluation of 5
    failure 273
Heart-lung interaction, physiological basis of 141
Hematocrit sign 100, 101*f*, 271
    visible 271*f*
Hematoma formation 188, 193
Hemidiaphragm, right 239
Hemodynamic edema 234
Hemodynamic instability 187
Hemodynamic monitoring 139*f*, 186
Hemoglobin 139
Hemoperitoneum 275
Hemorrhage 186, 193
Hemothorax 178, 182, 229, 268, 271, 274, 275
Hepatic abscess 189, 190*f*
Hepatic cirrhosis 183
Hepatorenal space, scanning 145
Herpes zoster 240
Hiccups 240
Hockey stick probe 62, 64*f*
Homogeneity 111
Homogenous interstitial opacities 272*f*
Hydronephrosis, severe 192*f*
Hydrothorax 271
Hyoid bone 70, 71
Hyperdynamic left ventricle 140
Hyperechoic fibroadipose septae 242
Hyperechoic pericardium 144
Hyperkinetic left ventricular 154
Hyperthyroidism with exophthalmos 199
Hypertrophic crus 239
Hypervolemia 138
Hypoalbuminemia, severe 183
Hypoechoic muscle fibers 242
Hypoglycemia 151
Hypoperfusion 229
Hypopharynx 70

Hypotension 144, 178, 268
    causes of 4*f*
    rapid ultrasound for 154*f*
Hypotensive patient 22
    approach to 24*fc*
Hypothermia 151, 240
Hypovolemia 131, 138, 140, 151, 152*f*, 154, 229
    assessment of 22
Hypovolemic left ventricle 120
Hypovolemic shock 235
Hypovolemic ventricle
    in diastole 132*f*
    in systole-cavity 132*f*
Hypoxemia
    acute 229
    causes of 114, 234
    etiologies of 113*fc*
    severe 230
    severity of 112
Hypoxia 151

I

Infection 178, 240
    control 276
Infective pathologies 173
Inferior vena cava 3, 4, 10, 23*f*, 24, 32, 41*f*, 122, 133*f*, 142, 147*f*, 155, 155*f*, 208, 223, 225, 225*f*
    calculation of 148*f*
    diameter 142
    view, subcostal 122
Inflammatory disorders 240
Inguinal ligament 265*f*
Inhaled nitric oxide, dose of 113
Injury severity score 209
Intensive care unit 1, 32, 117, 216, 229, 252
    abdominal ultrasonography in 216
    fever in 5
    sonography, legal issues in 278
Interatrial septum 120*f*, 121*f*
Intercostal chest drainage 274
Intercostal nerves 239
Interstitial edema 229, 232, 234
    indicate 232
Interstitial pneumonitis 102
Interstitial syndrome 102
Interventricular septum 118*f*, 120*f*, 121*f*
Intra-abdominal collections, percutaneous drainage of 187

Intraabdominal pressure 240
Intracranial pressure 3, 5, 10, 196
Intracranial vasospasm 200
Intravascular volume, assessment
  of 148
Isovolumetric relaxation 127

J

Jellyfish sign 100
Jugular engorgement 268
Jugular vein
  cannulation, external 64
  external 64
  internal 11, 13*f*, 59, 61
  ultrasound assessment of 13*f*
Jugular vessels, internal 70

K

Kidney 35
  hydronephrosis of 222*f*
  right 121*f*
  ultrasound of 221
Kissing papillary muscles 132*f*
Kissing ventricles 140
Kissing walls 131
Klebsiella 277
  pneumoniae 270

L

Lead zirconate titanate 35
Left ventricle
  end of systole 123
  end-diastolic
    area 140
    diameter 123
    volume 138, 140
  outflow tract 121
Left ventricular
  end-diastolic 154
    area 163
  end-systolic area 163
  outflow tract 3, 19, 20
Leg
  raising, passive 143
  veins 143
Limb muscles 253
Liver 35
  muscle 35
Local skin site infection 193
Lower limb 263
  venous system of 264

Lumbar vertebrae 238
Lung 32, 159
  collapse 230
  comet 18
  consolidations 230
  hepatization of 103*f*
  interfaces 229
  morphology 110
  normal 14, 94, 146
  profiles 233
  re-expansion of 177
  right 99*f*
  sliding 16*f*, 230
  sonography of 4, 18*f*, 234
  surface, anterior 233
  ultrasonography 146*f*, 230, 269*f*
  ultrasound 93, 94, 99, 111, 234
    equipment for 93
    in intensive care unit 105
    limitations of 108
    physics of 93
Lyme disease 240

M

Malecot catheter 168, 169*f*
Massive pulmonary embolism 229
McConnell's sign 133
Mean arterial pressure 139, 141
Mechanical ventilation 107
Mediastinal tumors 240
Metabolic imbalances 151
Methicillin-resistant *Staphylococcus*
  aureus 276, 277
Mitral valve 118, 118*f*-121*f*, 124
  leaflet
    anterior 19
    posterior 19
Mixed echogenic appearance 242
Morrison's pouch 26, 218
Morrison's space 209, 210
Motor neuron disease 240
Mucosal interface 69*f*
Muffled heart sound 268
Multidrug-resistant
  bacteria 276
  pseudomonas 277
Multiorgan failure 187
Multiple organ failure 187
Muscle 35, 238
  assessment of 257
  dysfunction 252
    ultrasonography of 252
  echogenicity 248
  function 253

  assessment of 252, 253, 257
    in intensive care unit,
      assessment of 258
  surrogate of 253
  ultrasonography of 253
  mass 259
    decreased 252
    primary 240
  thickness 253, 254*f*-256*f*
Muscular body wall 242
Muscular echotexture 242
Mycobacterium tuberculosis 277
Myocardial disease 115
Myocardial infarct, acute 163
Myocardial infarction 229

N

Nasogastric tube 226
  placement 226, 227
    confirmation of 78
Neck
  surgery 240
  ultrasonographic anatomy of 83
Needle selection 168
Needle tip 59
Needle tracking 171
Neoplasms 173
Nerve blocks related to airway 78
Neurocritical care, ultrasonography
  in 196
Neuromuscular junction, diseases
  of 240
Neurovascular bundle 176
  injury to 176*f*
New York Heart Association 18
Non-coronary cusp 19
Nonstarling pulmonary edema 111
Nontraumatic cardiac arrest, causes
  of 159
Normokinetic apex 133
Nosocomial infection 252, 276

O

Obstructive pulmonary disease,
  chronic 15, 112, 234
Optic nerve 11*f*, 198*f*
  atrophy 199
  junction of 198*f*
  sheath 5
    diameter 10, 196
    diameter, measurement of
      196
    method of 197

Index

Optic neuritis 199
Orbital hematoma 199
Orbital trauma 199
Oxygen
    consumption 139
    delivery 139

P

Palate 71, 72
Papillary muscle 118*f*, 120*f*
    level of 154*f*
Paracentesis 183
    complications 186
    contraindications 184
    indications 184
    needle insertion for 185*f*, 186*f*
    postprocedure care 186
    technique 184
Paralysis 239
Parapneumonic effusion 102*f*, 173
Parasternal sternal short-axis view 119*f*
Pearl technique, string of 89
Pelvic
    abscess 189
    cavity 275*f*
Percutaneous cholecystostomy 189, 191*f*
Percutaneous dilatational
    technique 82
    tracheostomy 77, 82
    ultrasound-guided 82
Percutaneous nephrolithotomy 273
Percutaneous nephrostomy 189, 192
Percutaneous transhepatic biliary drainage 48
Pericardial effusion 21, 118, 121, 133, 152, 153, 153*f*, 162, 179, 236
Pericardial fluid 32
Pericardial space, scanning 144
Pericardial tamponade 32, 134*f*, 229, 236, 268
    incidence of 268
Pericardiocentesis 178, 181*f*
    clinical pearls 183
    complications 182
    contraindications 179
    indications 179
    kit 181
    postprocedure care 182
    technique 179
Pericholecystic fluid 190
Perihepatic fluid 100*f*

Perimysium 242
Periosteum 255*f*
Peripartum cardiomyopathy 272
Peripheral muscle 253, 258
    ultrasonography for 253
Peripheral venous cannulation 65
Phrenic nerve 240
    palsy 239
Pigtail catheter 168
    in situ 190*f*
Pigtail shaped distal loop 169*f*
Pigtail thoracocentesis,
    representation of 177*f*
Plankton sign 100
Pleth variability index 141
Pleura 70
    thickened 269
    ultrasound of 93
Pleural cavity 208
Pleural effusion 17, 99, 111, 175*f*, 229, 230, 231*f*, 271*f*, 274
    in right lung 145*f*
    large right-sided 46*f*
Pleural fluid 99*f*
    producing hematocrit sign 101*f*
Pleural space, right 145
Pleural syndrome 108, 111, 112
Pleural ultrasound 14*f*
Pleuroperitoneal folds 242
Pneumonia 111, 112, 229, 269
    diagnosis of 270
    postoperative 272
Pneumoperitoneum, detection of 226
Pneumothorax 15, 98, 112, 146, 156-158, 178, 182, 208, 229, 230, 235, 268, 271, 271*f*, 274
    diagnosis of 79
    presence of 233
Point-of-care ultrasound 1, 2
    characteristics of 1
    history of 1
Polyvinylidene difluoride 37
Poor quality of life 252
Popliteal artery 264, 265
Popliteal fossa 265
    vein position in 265*f*
Popliteal vein 264, 265
Portal venous flow 53*f*
Positive end-expiratory pressure 107
Positive pressure ventilation 140
Posterolateral alveolar syndrome 108, 111, 112
Postextubation stridor, prediction of 78

Postsurgical fluid collections 187
Pre-natal Diagnostic Techniques Act 278
Pretest probability scores 262
Pre-tracheal tissue 84*f*
Probes, types of 242*t*
Propagation velocity 34
Pseudomonas aeruginosa 277
Pulmonary
    artery 119*f*, 133
        bifurcation of 119*f*
    catheters 148
    hypertension 129
    occlusion pressure 138, 140
    pressure 134
    systolic pressure 136*f*
    wedge pressure 140
    blood volume 140
    edema 18, 18*f*, 102, 229, 232, 234, 272, 272*f*
    re-expansion 178
    embolism 20, 112, 118, 260, 264, 272
        acute 20, 133
    embolus 151, 152*f*, 156
    fibrosis 102, 111, 270
    function test 241
    hypertension 134, 135
    masses 111
    regurgitation 129
    thromboembolism 260
    valve 118, 119*f*
    vascular dysfunction 112, 139
    vein 120*f*, 121
        normal 130*f*
        pulsed-wave doppler 127
Pulse
    Doppler 52
    pressure variation 141
    repetition frequency 43
Pulsed-wave doppler 127
Pulseless electrical activity 151, 158
Pupillary light reflex 196, 203
Pupillary reaction 10
Pupillary size 12*f*

Q

Quad sign 231
Quinones method 124

R

Rectovesicular space, scanning 145
Reflection 36

Refraction 36
Renal failure 183
Renal pelvis, dilated 222*f*
Respiratory distress 272
 syndrome, acute 17, 106, 110, 111, 234
Respiratory failure
 acute 108
 causes of 111, 234
Resuscitation, causes during 152
Reverberation artifact 96*f*, 97*f*, 227*f*
Rib cage 241
Right atrium, draining into 155*f*
Right ventricle
 dilatation of 129
 dilatation, causes of 129
 end-diastolic volume 138, 140
 in pericardial tamponade 153*f*
 thrombus in 135*f*, 156*f*
Right ventricular
 dilation 21*f*
 dysfunction 112
 ejection fraction 129, 131
 function, assessment of 129
 size and contractility 32

S

Safety considerations 38
Salmonella typhi 277
Salmonella typhimurium 277
Saphenous vein
 normal sonography of 24*f*
 superficial 264
Sarcopenia 252
Seagull sign 224, 224*f*
Seashore sign 96*f*, 230, 269*f*
Seldinger technique 88, 170, 177, 182, 188, 191
 of catheter insertion 171*f*
Sepsis 193
 severe 136
Sesame protocol 235
Shock 144, 229
 and hypotension, rapid ultrasound for 4, 146, 159
 causes of 163*t*, 229
 evaluation of 4
 rapid ultrasound for 154*f*
 type of 163
Simpson's method 125*f*
 of discs, modified 124
Sinusoid sign 100, 231
Sliding sign 269*f*
Sniff test 143

Society of Interventional Radiology 188
Sonographic air bronchograms 103*f*
Sonography 10
 of trauma, focused assessment 208
Sound, velocity of 35*t*
Spaulding classification 278*t*
Specular reflector 36
Splenorenal recess 218*f*
Splenorenal space, scanning 145
Spontaneously breathing adults 143
Staphylococcus aureus 270
Static parameters 138
Sternocleidomastoid 74
Strap muscles 72-74
Streptococcus pneumoniae 270, 277
Streptococcus pyogenes 277
Stroke volume 139
 increase in 138
 variation 139, 141
Subarachnoid hemorrhage 200
Subclavian vein 10, 11, 63*f*
 cannulation 62
Subcostal approach, posterior 246*f*
Subcostal four-chamber view 121
Subcostal transverse 144
Subdiaphragmatic fluid 100*f*
Subglottic stenosis 89
Sublingual approach 70
Submandibular sagittal view, extended 71*f*
Subphrenic space 211
Subpleural consolidation 106*f*
Subpulmonic effusion 239
Substantial pericardial effusion 234
Subxiphoid cardiac view 220*f*
Sunken eyes 138
Superficial veins, confluence of 265*f*
Superior vena cava 122, 140
 collapsibility index 143
Suprasternal notch 13*f*, 74*f*, 75
Swinging heart 154
Systemic vascular resistance 139
 index 139
Systolic dysfunction 273
Systolic function 122
 of left ventricle, assessment of 123
Systolic pressure variation 141
Systolic spikes 202, 203*f*

T

TACA technique, transverse 88
Tachycardia 274

Tachypnea 274
Tandem trocar technique 170
Target sign 65
Technological advancement in medicine 151
Temporal resolution 44
Tendon 238
Tension pneumothorax 151, 153, 156
Therapeutic procedures, ultrasound-guided 167
Thigh muscle 254
 anterior 255*f*
Thoracic aorta 147
Thoracic cavity 275*f*
Thoracocentesis 173, 176*f*
 ultrasound-guided 173
Thoracotomy, indications of 271
Thromboembolic pulmonary hypertension, chronic 118
Thrombus in main pulmonary artery 135*f*, 157*f*
Thyroid 69
 cartilage 70, 71, 73
 gland 70, 74
 isthmus 75
Tissue
 Doppler imaging 125, 131, 131*f*, 224*f*
 harmonic imaging 44, 45
 like sign 232*f*
Tongue 70-72
Torulopsis glabrata 277
Toxicity 151
Trachea 13*f*, 84*f*
 anterior 85*f*
 localization of 77
Tracheal lumen 84*f*
Tracheal rings 69, 69*f*, 70, 84
Transcranial color Doppler 202
Transcranial Doppler 11, 201*f*
Transcranial sonography 196
Transdiaphragmatic pressure 241
Transducer material 35
Transesophageal echocardiography 32, 113, 148
Transmitral pulse wave Doppler 128*f*
Transthoracic echocardiogram 32
Transthoracic echocardiography 4, 117*f*, 148, 236
 common 117
 probe 116, 117*f*
 subcostal 147*f*, 148*f*
Transudative pleural effusion 49*f*

Transverse submandibular view 72f
Trauma 105
   examination, sonography for 145f, 146
   sonography of 4, 10, 25, 106, 144, 208, 217, 220
   victims 153
Traumatic brain injury 199
Traumatic pericardial effusion 144
Tricuspid annular plane systolic excursion 114, 121, 129, 131
   normal 130f
Tricuspid regurgitation 118, 129
Tricuspid valve 118, 120f, 121f
Trocar technique 170
Typical reflection factors 35t

U

Ultrasonography 10, 82, 196, 216, 241
   limitations of 249
   modes of 39
   recognizing primary 16f
Ultrasound 261
   artifacts 47
   imaging of upper airway, advantage of 68
   instrumentation, basic 37
   of airway, structures visualized in 70
   profiles, accuracy of 112t
   properties of 34
   stethoscope 1
Ultrasound-guided management 151
Upper airway 69f
   ultrasound 74
   interpretation of 70
Upper eyelid 197f
Urethral catheterization 227
Urinary bladder, ultrasound of 221
Uterus 220f

V

Valsalva maneuver 240
Valvular heart diseases 229
Valvular incompetence, severe 32
Vascular injury 178
Vascular probes 93
Vein 60
Velocity time integral 24
Venous disease, chronic 260
Venous thromboembolic events 260
Venous thromboembolism 260
Venous thrombosis, diagnosis of 32
Ventricular function, assessment of 20, 122
Ventricular septal defect 118, 129
Vertebral body 74
Vesicouterine space, anterior 145
Vigileo™ 139f
   monitor, use of 139f
Vocal cords 70, 73, 75

W

Wells' score 262t
White lung 105f
   appearance 105f

X

X-ray absorptiometry, dual energy 253

Z

Zirconate titanate 37
Z-track
   method 186
   technique 186, 186f

EU GSPR Authorised Reprsentative
Logos Europe, 9 rue Nicolas Poussin
1700, La Rochelle, France
Phone: +33 (0) 6 67 93 73 78
E-mail: contact@logoseurope.eu

www.ingramcontent.com/pod-product-compliance
Ingram Content Group UK Ltd.
Pitfield, Milton Keynes, MK11 3LW, UK
UKHW051846210426
5322IPUK00019B/283